Illustrated Encyclopedic Dictionary of Building and Construction Terms

Illustrated Encyclopedic

Dictionary of Building

and Construction Terms

Hugh Brooks

PRENTICE-HALL, INC.   *Englewood Cliffs, N.J.*

Prentice-Hall International, Inc., *London*
Prentice-Hall of Australia, Pty. Ltd., *Sydney*
Prentice-Hall of Canada, Ltd., *Toronto*
Prentice-Hall of India Private Ltd., *New Delhi*
Prentice-Hall of Japan, Inc., *Tokyo*

© 1976 by
Prentice-Hall, Inc.
Englewood Cliffs, N.J.

Twelfth Printing . . . . . May, 1982

Library of Congress Cataloging in Publication Data

Brooks, Hugh,
    Illustrated encyclopedic dictionary of building and
construction terms.

    1. Building--Dictionaries.   I.   Title.
TH9.B76   690'.03   75-4921
ISBN 0-13-451013-5

Printed in the United States of America

to Becky, my wife

to my Aunt Cornelia

who long ago kindled the fire

and to the Reader who under-
stands why James Watt said:

*Of all things, but proverbially
so in mechanics, the supreme
excellence is simplicity*

BENEFITING FROM THIS
ENCYCLOPEDIC DICTIONARY

Test Yourself—Do you know the meaning of these words?

camber	addendum	micron
Hooke's Law	cooler nail	draft curtain
diaphragm	pilaster	class "D" fire
drift	MIG welding	blue top stake
space frame	verge rafter	metal halide lamp
cricket	rift cut veneer	fourwire service
soffit	ledger	feeder
collar beam	mullion	soil stack
chiller	fan coil unit	hydronics
epoxy	post tensioned	thermoplastic
mansard roof	sheet piling	electrostatic painting
expansive soil	section modulus	stinger

These often used words are a part of the language of construction—36 of the 2200 terms to be found in this book—terms basic to the vocabulary of active, informed construction people and terms that come up in conversations, meetings, reports, books, advertisements, and magazine articles.

As with other technologies, fast changes in the building industry have created new words at an enormous rate, the usage of many older words has changed, and many have become obsolete. This book is an up-to-date reference to make communication between us more effective by defining, clearly and concisely, the things we talk about, read about, and write about. Literally hundreds of sources of information—books, pamphlets, magazine articles, and correspondence with persons particularly knowledgeable in specialized areas of the building industry—have been culled and condensed to put the most essential, usable information into one book. Unlike generalized technical and scientific dictionaries, this book is tailored specifically to the needs of those involved with construction. It is today's language by, and for, those using it.

The building industry is huge, its complexity is growing, and each year more people become involved. Building codes become thicker with each new edition, governmental involvement grows, product files expand, and each week literature describing new materials, techniques, and services pour into the mail, creating and emphasizing the need for simplicity and order. This book is comprehensive enough to provide all that is necessary and useful on construction terminology, condensed and simplified to present it to the reader in one volume. Our vocabular framework—the words we use—needs to be pulled together and tied down to make it easier for us to understand each other. The purists may, in many cases, object to oversimplification, but to do otherwise would make the book and its intended usefullness impossible.

The book has been written primarily to fill three needs:

1. To provide a single source of terms in all areas relating to construction including real estate, insurance, mathematics, surveying and engineering.

2. To give an up-to-date reference including the many new terms resulting from our fast changing technology.

3. To present the information in non-technical language making it equally useful to the newcomer to construction and to the non-technically trained whose work requires some construction knowledge.

The result is a ready reference of authoritative information, of broad scope, and clarity. It is designed to save time and the effort of looking through dozens of books for the information needed. It eliminates the disorganization of turned down or paper clipped pages, xerox copies of "save" sheets, and scattered notes. Unfamiliar words heard in a meeting or read in a report can be looked up in seconds. Obscure words and those whose meanings have become common "household knowledge" have been omitted.

This is a book for busy, time-conscious people. Its one volume alphabetical format makes information finding easy and fast. It is planned to work in day-to-day situations—to be at your elbow each time an unfamiliar term occurs or a meaning requires checking. It goes beyond the scope of a dictionary, expanding important terms to give more information. Many terms are given "mini-text" treatment to provide a condensed overview of a subject. Illustrations are used throughout the text to add clarity and understanding or to help the reader "see what it looks like."

FIFTEEN INFORMATION SITUATIONS
WHERE THIS BOOK GIVES ANSWERS

1. Engineering terms which are often baffling to non-engineers are defined in simple, non-technical language—terms such as *Diaphragm, Hooke's Law, Moment of Inertia, Overturning Moment, Section Modulus,* and many others.

2. Basic *Plumbing* terminology is explained. For example, *Drainage and Vent Piping* describes the complete system including the meaning of such terms as *Branch Vent Pipe, Soil Stack, Vent Stack, Waste Pipe.*

3. *Plywood*, one of the most widely used materials, is explained in detail, describing the two basic classifications, how it is specified when ordering, what the grade designations mean, what sizes are available, and other useful data.

4. The many kinds of *Pipe* (steel, plastic, clay, copper, cast iron, etc.) are tabulated and show standard available diameters, lengths, uses, and how they are joined.

5. The concept of *Prestressed Concrete*, is covered, together with diagrams illustrating the principle.

6. *Arc Welding* is explained, and it includes the meaning of *Weld Symbols, TIG Welding*, and *MIG Welding*. Also covered are *Gas Welding and Cutting* and *Resistance Welding*.

7. *Lumber* ordering, specifying, size classification, grading, etc., which can be complex, as well as *Softwoods* and *Hardwoods*, are thoroughly explained.

8. Under *Automatic Fire Sprinklers*, the four basic types of systems are described, plus explaining such terms as *Ordinary Hazard, Calculated System* and *Pendant Head*.

9. *Electric Service* explains how the serving utility company provides electric power, and the combination of voltages and phases available. It includes descriptions of such terms as *Three-Wire Service* and *Four-Wire Service*.

10. *Structural Steel* is specified by ASTM (American Society for Testing and Materials) numbers such as A36, A572, etc. Included is a description of what these numbers mean in terms of properties and characteristics, as well as a general description of structural steel, including definitions of *Mild Steel* and *Alloy Steel*. The coding system for *Structural Steel Shape Designations*, such as W10 x 24, is also included.

11. The basics of *Plastics* are presented, including the most commonly used plastics and their uses and properties, such as *Polyvinyl Chloride, Silicone*, and *Urethane*. (page 273).

12. Twenty-three *Door Types* are described, including *Kalamein Door, Pilot Door, Sectional Door, Labeled Door*, and *Guillotine Door*.

13. The essentials of *Central Air Conditioning* are described, and discussed are *Packaged Air Conditioning, Split System, Single Duct System* and *Double Duct System*; and under *Refrigeration Equipment* is a description of the two basic types of refrigeration.

14. Essential mathematical terms are defined, such as *Parabola, Hypotenuse, Logarithmic* and *Trapezoid*; under *Circle Nomenclature* a diagram shows what is meant by *Arc, Chord, Radian, Segment* and *Sector*, among others.

15. Twenty types of *Concrete Blocks* are described, on pages 115-118, including *Bond Beam Block, Cinder Block, Lintel Block, Scored Block*, and *Split-Face Block*.

The book offers unique features such as:

—An Index by Function. This lists, alphabetically, all the terms in each of 23 subject areas. To get informaiton or check your knowledge in any specific field, simply scan the list and look up the unfamiliar terms.

—Frequently used formulas, charts, and tables including how to find areas and volumes of commonly used shapes, conversion factors to quickly convert feet to meters, B.T.U.'s to kilowatts, liters to gallons, etc., and standard abbreviations used on plans.

—An up-to-date Directory of Construction Related Organizations, listing important organizations by 16 subject areas with current addresses.

—A list of Trade Unions and the various trades they represent.

Although every effort has been made to check, and in most cases double check, each definition, it is a regretable certainty that some will get by to which a knowledgeable reader will object. For this, and for the improvement of possible future editions, comments and suggestions are invited.

HOW THE BOOK IS ORGANIZED

The book is in two sections. First is the Index by Function consisting of 27 pages. Second is the body of the book with alphabetical listing of all the terms thoroughly documented.

The book defines all italicized words and terms, words defined within the body of a more general definition are capitalized.

HOW TO USE THIS BOOK . . .

Begin by thumbing through it to see what's inside; stop at some words you are familiar with to note how they are presented—browse around in the book. Keep it handy—on your desk, in the drafting room, on the car seat, in the field office—ready for information emergencies and for casual spot-reading. Sharpen your memory of known words and better your understanding of other specialized areas. And—keep the book where your secretary can use it; it will answer her questions, save you time, and make her more valuable by increasing her knowledge.

Hugh Brooks

Acknowledgements

A book of this scope could not have been done without a very considerable amount of help. To the expertise of the hundreds of individuals, companies, and organizations who contributed material, reviewed definitions, offered suggestions and in many cases granted permission to use or adapt previously published definitions: my very grateful appreciation.

Special thanks are due Frank Andrews, Orlo Tyler, Hal Thomas, Tom Lake, Bill Gallagher, Gene Pierce, Sanford Paris, and Glenn Gordon for their expertise in their respective fields; and to Jean Warren for very capably helping with the researching, assembling, and typing of the manuscript; and Ken Abang for most of the drawings.

Index by Function

A. Professional Services: Architects, Engineers, Preparation of Plans, Specifications

B. Construction Administration & Management

B. Construction Administration & Management: (cont.)

C. Real Estate, Financing, Insurance, & Bonds

E. Architectural Parts of a Building: (cont.)

F. Structural Parts of a Building (cont.)

G. Engineering & Mathematics

G. Engineering & Mathematics: (cont.)

H. Earthwork & Foundations

H. Earthwork & Foundations: (cont.)

H. Earthwork & Foundations: (cont.)

I. Concrete Work

I. Concrete Work: (cont.)

J. Carpentry, Wood, Millwork, Drywall Work, & Insulation

J. Carpentry, Wood, Millwork, Drywall Work, & Insulation: (cont.)

K. Masonry

K. Masonry: (cont.)

M. Hardware: (cont.)

N. Doors, Windows, Glass & Glazing

N. Doors, Windows, Glass & Glazing: (cont.)

O. Plastering

O. Plastering

P. Roofing, Roof Accessories, & Sheet Metal

Q. Plastics

R. Painting

T. Plumbing: (cont.)

U. Electrical Work & Illumination

W. Fire Protection: (cont.)

Illustrated Encyclopedic
Dictionary of Building
and Construction Terms

A

Abbreviations For commonly used abbreviations see Figure A-1.

Abbreviations see Figure A-1.

Figure A-1

Abbreviations

a.	acre	A.S.M.E.	American Society of Mechanical Engineers
A	Area	A.S.T.M.	American Society for Testing and Materials
A.B.	Anchor bolt		
Abt.	About	Ave.	Avenue
AC	Alternating current	av.	average
		BF	Board feet
A.C.	Asphaltic concrete paving	Bm	Beam
		B.M.	Bench mark
		Bldg.	Building
A.C.I.	American Concrete Institute	B.T.U.	British thermal unit
		B.T.U.H.	British thermal unit per hour
A.I.A.	American Institute of Architects	C	100
		C.E.	Civil Engineer
Al.	Aluminum	Cer.	Ceramic
approx.	approximate	cu. ft.	cubic feet
Arch.	Architect	cfm	cubic feet per minute
A.S.C.E.	American Society of Civil Engineers	cfs	cubic feet per second
		C.I.	Cast iron
A.S.H.R.A.E.	American Society of Heating, Refrigerating and Air Conditioning Engineers	Clg.	Ceiling
		CM	Construction Manager
		C.O.	Cleanout
		Coef.	Coefficient
		Comp.	Composition
		Conc.	Concrete
		Const.	Construction
		cont.	continuous

Contr.	Contractor	horiz.	horizontal
Corr.	Corrugated	h.p.	horsepower
cu.	cubic	hr.	hour
CWT	100 pounds	Hwy.	Highway
cu. yd.	cubic yard	I	Moment of
d	penny		Inertia
Dia.	Diameter	I.C.B.O.	International
Db.	Decibel		Conference of
dbl.	double		Building
DC	Direct current		Officials
Dept.	Department	ID	Inside Diameter
Diag.	Diagonal	in.	inches
Dim.	Dimension	in.-lb.	inch-pound
D.L.	Dead load	Incl.	Include
Dn.	Down	Int.	Interior
Do.	Ditto	I.P.	Iron pipe
Dr.	Drive	K	Kip (1,000 lbs.)
D.S.	Downspout	kwh	kilowatt-hour
Dwg.	Drawing	Lav.	Lavatory
E	Modulus of	lb.	pound
	elasticity	L.O.A.	Length overall
ea.	each	Lin.	Linear
El.	Elevation	L.L.	Live load
Elev.	Elevation	L.S.	Lump sum
Engr.	Engineer	L.S.	Licensed
Eq.	Equal		Surveyor
Exist.	Existing	L & T	Lead and tack
Exp.	Exposed	M	Bending moment
Ext.	Exterior	max.	maximum
Fdn.	Foundation	MBF	1,000 Board feet
Fin.	Finish	Mech.	Mechanical
Fd.	Found	Mezz.	Mezzanine
F.O.B.	Free-on-Board	mfg.	manufacturing
F.S.	Full size	M.H.	Manhole
F.S.	Face of stud	min.	minimum
ft.	feet	misc.	miscellaneous
Ftg.	Footing	M.O.	Masonry opening
ft.-lb.	foot-pound	N.E.C.	National
Ga.	Gage		Electrical
gal.	gallon		Code
Galv.	Galvanized	N.E.M.A.	National
G.I.	Galvanized iron		Electrical
gpm	gallons per		Manufacturers
	minute		Association
Gr.	Grade	neg.	negative
Gyp.	Gypsum	N.I.C.	Not in contract

No.	Number	T & G	Tongue and
N.P.C.	National Plumb-		Groove
	ing Code	Tol.	Tolerance
N.T.S.	Not to scale	Trans.	Transformer
O.C.	On center	TV	Television
OD	Outside diameter	typ.	typical
opp.	opposite	U.B.C.	Uniform Building
oz.	ounce		Code
PBX	Private branch	U.L.	Underwriters
	exchange		Laboratories
pcs.	pieces		
pcf	pounds per	/	per
	cubic foot	"	inches or seconds
Perm.	Permanent	'	feet or minutes
Perp.	Perpendicular	×	by
Pl.	Plate (also ℞)	°	degree
P.L.	Property line	@	at
pos.	positive	#	number
P.P.	Power pole	℞	plate
pr.	pair	₵	center line
psf	pounds per	φ	round
	square foot	□	square
psi	pounds per	⊥	perpendicular
	square inch	//	parallel
Ptn.	Partition	>	greater than
qt.	quart	<	less than
r	Radius of	≅	approximately
	gyration		equal to
ref.	refer (or	∠	angle
	reference)		
req'd.	required		
R.R.	Railroad		
R/W	Right of way		
S	Section Modulus		
Sect.	Section		
sq. ft.	square foot		
sq. in.	square inch		
Sim.	Similar		
Spec.	Specification		
Sprklr.	Sprinkler		
sq.	square		
St.	Street		
Std.	Standard		
Struct.	Structural		
sq. yd.	square yard		
sym.	symmetrical		

Abrasive Any hard, sharp material that can wear away a softer material. Abrasives, for example, form the granular coatings on *Sandpaper* and grinding wheels. Abrasive materials include aluminum oxide, garnet, quartz, silicon carbide, corundum, and diamond; the latter being, in order, the second hardest and the hardest known substances.

Absolute Zero The lowest temperature that anything can be and at which all molecular motion stops. It is 460° below zero on the Fahrenheit scale and -273.1° on the Centigrade scale. It is used as a reference temperature for describing characteristics of various types of mechanical equipment, such as *Compressors*. See also: *Temperature Scales*.

Absorbent A material having a physical or chemical attraction for certain other materials, which changes either chemically or physically while assimilating such materials (as opposed to *Adsorbent*). Examples include: *Calcium Chloride* (a moisture absorbent), and *Lithium Bromide* (as used in *Absorption Refrigeration*).

Absorber In *Air Conditioning*, a device in which an *Absorbent* is used to absorb another substance; for example, the low pressure side of an *Absorption Refrigeration* system in which sprayed *Lithium Bromide* absorbs moisture in vapor form from the *Evaporator*.

Absorption Rate A measure of the amount of moisture absorbed and retained in building materials exposed to water. The term is most applicable to specifications for masonry units.

Absorption Refrigeration See *Refrigeration Equipment*.

Abut The joining of two pieces or surfaces along a common boundary.

Abutment A relatively massive structural element, usually of *Concrete*, at the end of a bridge or *Arch*. Its function is to support or otherwise restrain an end against movement.

A.C.

1. Abbreviation for *Alternating Current*.

2. Abbreviation for *Asphaltic Concrete Paving*.

Accelerated Test A weather exposure test where mechanical means are used to simulate actual conditions, in exaggerated form, to shorten the test period.

Acceleration

1. A mathematical concept describing the rate at which velocity changes; a change of speed. As defined in the equation, force = mass × acceleration, the term is critical in calculating forces exerted on a structure by an earthquake (the acceleration being caused by the ground motion).

2. The standard acceleration of a falling body caused by earth's gravity equals approximately 32.2 ft. per second for each second of elapsed time.

Accelerator A chemical added to a mixture to speed its reaction. Accelerators, for example, are used in *Paint* to speed drying time and in *Concrete* mixes to shorten setting time.

Accelograph An instrument used to measure ground displacement during earthquakes (also *Accelerograph*). They can be installed in tall buildings to provide a record of the structure's response (movement) during an *Earthquake*. A pendulum or other type of seismic trigger is designed to activate a recording device. (See Figure A-2.)

Accents In decoration, elements of color having characteristics widely different from the basic scheme. They are usually bright or brilliant and used sparingly for the purpose of emphasis.

Access Door *See Door Types.*

Access Floor An elevated, finished floor usually for computer rooms such that the space below, which is generally 12 to 18 inches clear height, is used for electric *Cable*, *Conduit*, and *Duct* work. Such floor systems generally consist of 2' square removable panels (hence the term "access" floor) supported on adjustable-height pedestals at each corner. (See Figure A-3.)

Courtesy Teledyne Geotech

FIGURE A-2   Accelerograph

Courtesy Liskey Aluminum, Inc.

FIGURE A-3   Access floor

Accordian Door   See *Door Types*.

Accumulator   In plumbing, a temporary storage chamber for gases or liquids, to smooth out pulsations in a piping system. It is also called a "surge tank" or SURGE DRUM.

Acetone   See *Solvents*.

Acetylene   A gas, almost wholly composed of gaseous carbon, which is explosive in air. When mixed with oxygen it is called oxyacetylene and is used in *Gas Welding and Cutting*.

Acetylene Welding and Cutting   See *Gas Welding and Cutting*.

Acid   A substance considered the chemical opposite of a *Base* or alkali. Generally water-soluble, acids are sour in taste, turn litmus paper red, are often corrosive or explosive in reaction with other substances, and are capable of producing free hydrogen in exchange for a metal or other positive chemical groups to form a salt.

Acoustical Plaster   See *Plaster Finish*.

Acoustical Tile   Manufactured, highly porous mineral, wood fiber, or *Fiberglass* tiles used as a sound absorbing and decorative ceiling finish (see Figure A-4). They are available in many patterns and textures (generally 12″ square and ½″ to ¾″ thick). When applied over a wood or *Gypsum Board* backing, they are attached by clips, staples, or by *Adhesive*. Other types, usually 24″ × 48″ panels, are laid or fitted into metal grid, *Suspended Ceiling* systems.

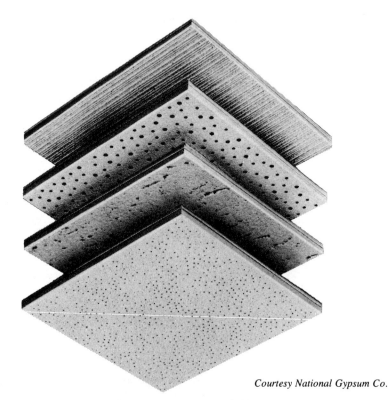

Courtesy National Gypsum Co.

FIGURE A-4   **Acoustical tile**

Acoustic Paint   See *Paint.*

Acoustics   The art and science of sound control.

Acre   A unit of land measurement. One acre equals 43,560 square feet, and there are 640 acres in one square mile. An area of land 208.71 feet on each side (square) is equal to one acre.

Acre-Foot   A unit of measurement of stored water, as in a reservoir or behind a dam. It is the equivalent of 43,560 cubic feet—a one foot depth of water over an area of one acre.

Acrylic   A type of *Thermosetting Plastic* used in sheet form for skylights, light fixtures, facings of internally-lighted signs, and similar uses requiring high light transmission, and weather and shatter resistance.

Acrylic Paint   See *Paint.*

Activated Alumina   A moisture *Adsorbent.* A form of aluminum oxide used in *Dehumidifiers* as a drying agent.

Activated Carbon   In *Air Conditioning,* a specially processed form of carbon used as an *Adsorbent* for odorous or harmful materials in a gaseous or vaporized state—materials that cannot be removed by particular filters. It is also called ACTIVATED CHARCOAL.

Activated Charcoal   See *Activated Carbon.*

Active Door   In a pair of doors, the door that opens first and the one to which the lock is applied.

Active Earth Pressure   See *Earth Pressure.*

Act of God   In insurance terminology, a destructive occurrence which cannot be prevented or foreseen, such as an *Earthquake* or flood.

Actuated Tools   See *Power Actuated Tools, Power Activated Tools.*

Addendum   An added part, or change, in the *Drawings, Specifications,* or other *Contract Documents* to clarify, correct, add to, or change some part or parts of the original contract documents. It is issued after the original documents have been circulated to bidders, but before *Bids* have been received. It differs from a *Change Order* in that the latter is issued to effect changes during construction, after award of a contract.

Addition   A term used in *Building Codes* to mean new construction which adds to the physical size or floor area of an existing building or structure, as opposed to an *Alteration.*

Adhesive   A substance capable of holding materials together by surface attachment; a general term that includes *Cements, Glues,* etc.

Adiabatic   An adjective describing a process wherein no heat is added to, or taken from, the system undergoing the process; an example is *Evaporative Cooling.*

Adjustable Triangle   A drafting tool similar to a *Triangle,* but having one side adjustable with a calibrated scale so a line of any degree (slope) can be drawn. For illustration see *Drafting Tools.*

Admixture   A substance added to a *cement* base mixture (such as *Concrete* or *Plaster*) to cause changes in its properties, such as improved workability, faster setting time, increased strength, greater watertightness, etc. Examples are *Accelerators, Air Entraining Agents, Plasticizers, Waterproofing* agents, water reducing agents, coloring agents, etc.

Adobe   A general term applied to various *Clay* soils of medium to high *Plasticity* which occur throughout the country but particularly in the south and west.

Adobe Bricks See *Bricks.*

Adsorbent A material having an attraction for certain other materials, causing molecules of gasses, liquids, or dissolved solids to adhere to its surfaces, but which does not change either physically or chemically upon adsorbing such materials (as opposed to *Absorbent.*) Examples include *Activated Alumina, Activated Charcoal,* and *Silica Gel.*

Adsorption The surface attachment of one substance by another when they are in contact, such as the adsorption of moisture by *Silica Gel.*

ADT A means for automatic detection of fire or forced entry into a building using devices manufactured by the American District Telegraph Co., hence the abbreviation ADT.

Adulteration The substitution of inferior materials for those represented or specified.

Advertisement for Bids A formal announcement, usually published in a trade magazine or newspaper, soliciting *Bids* for a construction project. Such notice usually gives a brief description of the project, name of the architect or engineer, and the time and date set for the receiving of bids.

A-E Abbreviation for a design firm (Architects and Engineers) offering both architectural and engineering services.

Aeration The exposing of a substance or area to *Air.*

Aerial Surveying In general, any means by which information about the earth's surface is obtained by photographing it from above, as from an airplane, balloon, satellite, etc. Commonly, aerial photographs are used to provide visual

data for complex areas and, by using known distances on the ground, other objects on the photograph can be approximately scaled. *Topographic Surveying* can likewise be accomplished aerially by the use of *Photogrammetry.*

Aerosol An assemblage of small particles, solid or liquid, suspended in *Air.* The diameters of the particles may vary from 100 microns (one micron = one millionth of a meter) down to 0.01 micron or less, for example: dust, fog or smoke. Aerosol spray cans contain a mixture of the material to be sprayed and a liquified gas under pressure. When the valve is opened, the liquified gas vaporizes and discharges as a spray containing the material.

AFL-CIO See *Trade Unions.*

A-Frame In general, any structural support framework in the shape of the letter "A" but usually meaning a building system having sloping side members which act as both walls and roof. Horizontal members near the base act as cross ties and are used to support a floor. It is a popular and economical framing method for vacation cabins. (See Figure A-5).

African Mahogany See *Hardwood.*

Agent A person or firm authorized to act for another person or firm by mutual agreement between them.

Aggregate A collective term for sand, gravel, crushed rock, and similar materials used to make up the bulk of a cementitious mixture such as *Concrete, Plaster,* or *Asphaltic Concrete Paving.* Such materials are bonded into a solid mass by the addition of a paste of *cement* and water, or in the case of asphaltic concrete, by asphalt. The term aggregate is further subdivided by particle size into FINE AGGREGATE (smaller than ¼"), and COARSE AGGREGATE (¼" and

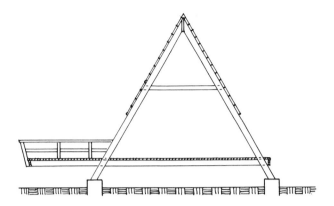

FIGURE A-5   "A" Frame Construction

larger). FINES generally refers to those aggregate particles passing a No. 200 *Sieve (Size)*.

Agreement A term often used interchangeably with *Contract*. It indicates a meeting of minds relative to certain performances and obligations, and may or may not fulfill all the essential elements of an enforceable *Contract*.

Air Air consists of about 78% nitrogen, 21% oxygen, and 1% other gases. In addition, it contains water vapor and particles such as dust and pollen. At sea level air weights about .081 pounds per cubic foot. The total weight at sea level of a "column" of air extending upward to outer space is called *Atmospheric Pressure* and is about 14.7 pounds per square inch (also about 29.92 inches of mercury, barometric pressure) and varies with climatic conditions.

Air, Ambient See *Ambient Temperature*.

Air Balancing In *Air Conditioning*, the adjusting of the flow of conditioned air into various spaces by use of *Dampers*, adjustable *Diffusers*, or other means such that the air distribution maintains proper temperature control and noise conditions.

Air Changes A method of expressing the amount of air supplied to or exhausted from a room or building in terms of the volume of the room or building. For example, to provide two air changes per hour for a room $12' \times 10' \times 10'$ high would require a fan having a capacity of 40 C.F.M. (cubic feet per minute).

$$\left[\frac{12' \times 10' \times 10' \times 2}{60} = 40 \text{ C.F.M.}\right].$$

Air changes are also expressed as the number of minutes required to effect one air change. In the preceding example, it would be 30 minutes

$$\frac{60 \text{ minutes}}{2 \text{ changes}} = 30 \text{ minutes per air change}$$

Air Circulation Movement of air either by natural gravity *Convection* or by a *Fan*, the latter called *Forced Circulation*.

Air Cleaner A device used within, or in conjuction with, an air moving system to remove air-borne impurities such as dust, gases, vapors, fumes, and smoke. Types of air cleaners include *Air Filters*,

Activated Carbon (for deodorizing), *Air Washers* (also called "scrubbers"), and *Electrostatic Precipitators*.

Air Compressor A machine to compress air, usually for the purpose of operating *Pneumatic* tools. Most commonly they consist of an electric or gasoline motor driving a reciprocating piston in which air is compressed and delivered under pressure to a storage tank from where it is then distributed through a hose. See also *Compressor* (as used in *Refrigeration Equipment*).

Air Conditioning The process of treating air so as to simultaneously control its temperature, *Humidity*, cleanliness, and distribution. (For cleanliness control, odor level, or humidity, see *Air Cleaners, Humidifiers*, and *Dehumidifiers*.) The temperature of the air is changed by blowing the air across *Coils*, through which hot or cold water, or a *Refrigerant*, is flowing, the hot water being from a *Boiler* and the cooling effect either from a

Chiller or from the circulating *Refrigerant* as it vaporizes in the coils. Heated air, as part of an air conditioning system, can also be provided by blowing the air across electrically heated wires, or by a *Furnace*. The two basic types of air conditioning are *Room Air Conditioning* which is a through-the-wall type having little or no *Duct*work, where the heat withdrawn from the room is dissipated out the exterior side of the unit, and *Central Air Conditioning*, where the equipment is located remotely from the rooms being served.

Air Cooler A factory-made group of components assembled to reduce the temperature of air as it passes through the equipment. For example, a *Fan Coil Unit* or an *Evaporative Cooler*.

Air Curtain A device to protect a door or window opening against heat loss or passage of insects by providing a high velocity flow of air across the opening. (See Figure A-6).

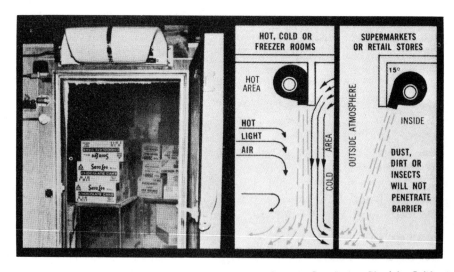

Courtesy Curtainaire, Glendale, California

FIGURE A-6   Air Curtain

FIGURE A-7 *Air Diffuser*

Air Diffuser A device to distribute conditioned air into a room from an outside source, such as from a *Duct*. It is usually located in a ceiling, and has adjustable vanes to control the direction and force of the air discharge. It differs from a *Register* in that the latter is usually placed in a wall or floor. (See Figure A-7).

Air Dried Lumber Lumber dried in normal outside air as contrasted to drying in an artifically heated kiln. See also *Kiln Dried Lumber, Seasoning*.

Air, Dry An *Air Conditioning* term meaning air without any moisture content.

Air Duct See *Duct*.

Air Entraining Agent An *Admixture* for concrete containing a chemical which causes a concrete mix to have microscopic air bubbles throughout, usually for the purpose of improving resistance to freezing and thawing. As the water within the concrete freezes and expands, the resulting pressure can be absorbed by empty air voids thus preventing a build up of damaging pressure. It also improves workability of the mix due to the lubricating effect of the air bubbles.

Air Filter A device placed in an air stream, such as at the entrance to an air conditioning equipment unit, to trap particles such as dust. One of the most common incorporates a pad of spun fiberglass held between perforated metal grids in a fiberboard casing. The fiberglass is treated with a special *Adhesive* so that particles are trapped by impingement and held to the fiberglass. Air filters are basically divided into three types: viscous impingement, dry, and *Electrostatic Precipitators*. Either type can be disposable (per example above), renewable, clean-able, or automatic self-cleaning. See also: *Air Cleaner.*

Air Gap In plumbing, the vertical distance between the flood level rim of a fixture and the lowest opening supplying the fixture. It prevents *Backflow.* If such a distance did not exist, backflow of used water into the supply system could occur.

Air Handling Unit In *Air Condition-ing,* a unit having a *Blower* or *Fan,* cooling and heating *Coils* (which receive chilled or hot water from another source), and *Air Filters,* all in one housing used to distribute hot or cool air to the con-ditioned space. Smaller units, handling under about 3,000 cubic feet per minute, are usually called *Fan Coil Units.*

Air Lock

1. In plumbing, a condition when a gas, such as air or a vapor, is trapped between two liquid surfaces in piping or a container so that the flow of the liquid is impeded or stopped by the trapped gas. This is also called "vapor lock."

2. A vestibule between two rooms, or between a room and outdoors, through which persons must pass and which pro-vides a temperature insulation, or limits air pressure loss, between the two spaces where, without such a vestibule, there would be an excessive pressure or heat loss (or gain) each time the door was opened.

Air, Make-Up In *Air Conditioning,* a supply of air into a space to replace air that has been exhausted (as by an exhaust *Fan*), burned during combustion of a fuel, or otherwise depleted.

Air, Outdoor In *Air Conditioning,* air taken from outdoors as opposed to air which is recirculated.

Air Pollution Contamination of the air by smog, industrial fumes, smoke, or other visible and/or injurious contami-nants.

Air, Return In *Air Conditioning,* air which is withdrawn from a room or space to be reprocessed by an air conditioner and recirculated back into the room or space. Also called "recirculated air."

Air Rights A legal right similar to an *Easement,* whereby the owner of a prop-erty grants to another party the right to construct something above and clear of an existing building or structure. For example, the New York City Pan Am Building is constructed above the Penn Central Railroad tracks.

Air, Saturated Air which contains the maximum amount of moisture it can hold at a specific temperature and pres-sure. This point occurs when dry air and saturated water vapor coexist at the same *Dry Bulb Temperature,* and is equivalent to 100% *Relative Humidity.*

Air Shower An enclosure through which personnel pass before entry into a *Clean Room.* Its purpose is to remove dust and lint from clothing by means of a large volume of turbulent downward air flow within the enclosure. The lint-laden air is filtered and recirculated back through the air shower.

Air Supported Structure A tent-like curved structure held up by air pressure and having an air-tight seal around its base perimeter. A blower provides the air within the structure itself (a positive pres-sure of ¼ lb. per square inch). The mem-brane material is generally of vinyl coated nylon. These structures are used as warehouses, to enclose tennis courts, and other recreational facilities. Access to the enclosure is through intermediate *Air Locks*; they are also available heated and lighted. (See Figure A-8).

Courtesy Birdair Structures, Inc.

FIGURE A-8   **Air Supported Structure**

Air Trap In slow moving or low pressure water piping, the collection of air at high points of the system which could shut off or reduce the normal flow of water.

Air Washer An arrangement of sprays (''water/curtain'') for cleaning and/or *Humidifying* air and which can also act as an *Evaporative Cooler*. See also: *Air Cleaner*.

Aisle Any passage path, for persons, trucks, etc., that is free and clear of objects which would impede the use of such space.

Alabaster A very fine grained variety of *Gypsum* used for ornamental plaster castings.

Alaska Cedar See *Softwood*.

Alclad A process for electrolytically coating various metal products with a layer of *Aluminum* or its alloys to improve corrosion resistance while retaining original strength.

Alcove A room-like area, usually rather small, adjacent to another room but without a dividing wall.

Alder See *Hardwood*.

Alluvium A general term for a soil which at one time was, or presently is, deposited in river beds, flood plains, etc. In addition to soil, the deposits can also include *Gravel, Cobbles,* and *Boulders*.

Alpine Fir See *Softwood*.

Alternating Current See *Electric Current*.

Alteration In *Building Code* terminology, the changing or remodeling of a building without adding to its physical size or floor area, as opposed to an *Addition*.

Alternator

1. An electric generator which produces *Alternating Current*. It differs from a *Generator* in that the latter is usually thought of as producing *Direct Current*.

2. A device used with duplex electric

motor systems (such as for a *Sump Pump* where two motors are used for reliability) to automatically change the starting sequence of the two motors so as to equalize wear.

Altitude of Sun See *Solar Angles*.

Alumina A constituent of certain *Clays* for *Brick* and *Refractory* materials; it is also used as an *Abrasive* material. Chemically, it is aluminum oxide (Al_2O_3).

Aluminum A silvery-gray material characterized by light weight (about one-third that of *Steel*), high corrosion resistance, and electric conductivity (aluminum wire has about 60% the conductivity of copper wire). Its strength is approximately three-fourths that of steel but as with steel, it varies widely depending upon alloy content. Most aluminum used in construction is extruded into a variety of shapes used for door and window frames, trim, railings, etc. Like steel, standard shapes are available in *I-Beams, Channels, Angles, Tees,* and *Tubing*, the latter being round, square, or rectangular. Most architectural aluminum is treated by *Anodizing* to give greater corrosion resistance, surface hardness, or color (generally gold or shades of bronze). Due to the wide range of alloy composition and tempers, a system of numbers and letters is used to specify aluminum. The system uses four digits which, by their combination and meaning, designates aluminum percentage, impurity limits, and alloying elements. The numbers are followed by a letter indicating the temper classification. A standard reference source of properties of aluminum and other technical data, including the specifying index system is the Aluminum Construction Manual published by the Aluminum Association, 750 Third Ave., New York, N.Y. 10017. (See Figure A-9 and A-10.)

Aluminum Foil Thin, paper-like sheets of aluminum (less than .005″ thick) used primarily as insulation. Heat striking the bright surface is almost totally reflected. Such insulation is used in walls and ceilings and usually consists of two layers having an insulative dead air space between them.

Aluminum Paint See *Paint*.

Aluminum Roofing and Siding Aluminum sheets corrugated or otherwise shaped to stiffen the sheets for greater span between supports. Such sheets are usually used as a finish material, without other covering, and are available plain or with various *Baked Enamel* color finishes. Most common thicknesses are .024″ and .032″ (the latter being about 1/32″).

Aluminum Wire Aluminum wire is sometimes used in lieu of *Copper Wire* due to lower cost and ease of installation (since it is lighter in weight) but it has a higher electrical resistance than copper wire. It is available in sizes corresponding to copper wire, but is primarily used in large sizes, such as for *Feeders*.

Ambient Temperature The natural temperature surrounding an object. It is usually in reference to outdoor temperature although it can, under some circumstances, mean an inside temperature.

American Softwood Lumber Standard A product standard established by the U.S. Department of Commerce for voluntary use by the various lumber manufacturers' associations (for example: the Southern Pine Association and the Western Wood Products Association) to serve as a guide by standardizing nomenclature, sizes, descriptions, shipping provisions, grade marking, inspection rules, etc.

American Standard Beams A stan-

Courtesy International Extrusion Corp. Alhambra, Calif. and Farrel Corp., Rochester, N.Y.

FIGURE A-10 Aluminum Extrusion Press

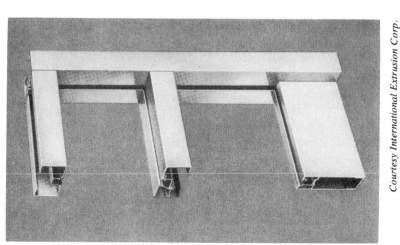

Courtesy International Extrusion Corp. Alhambra, Calif.

FIGURE A-9 Aluminum Door Framing Sections

dard series of rolled steel or extruded aluminum "I" beams characterized by a relatively narrow *Flange,* as opposed to *Wide Flange Beams.* See also: *Structural Steel Shape Designations.*

American Standard Pipe Gage A standard for the dimensions and manufacturing tolerances in the production of piping materials, established by the American National Standard Institute (formerly American Standards Association).

American Standard Screw Gage A standard for the dimensions and manufacturing tolerances in the production of screws and other threaded materials, established by the American National Standards Institute (formerly American Standard Association).

American Standard Wire Gage Standard U.S. system of designating wire sizes for non-ferrous metals. The gage numbers increase with decreasing wire diameters. It is the same as the BROWN AND SHARPE GAGE.

Ammonia In *Air Conditioning,* a gas used as a *Refrigerant,* with water, in the *Absorption Refrigeration* method (*Gas Air Conditioning*).

Ammonia Absorption Machine See *Refrigeration Equipment.*

AMP Short for *Ampere.*

Ampacity In electricity, the current carrying capacity of a *Conductor* expressed in *Amperes.*

Ampere The electrical unit of measurement for the amount of *Current* flowing in a wire. Often abbreviated *Amp.* See also *Ohm's Law.*

Anchorage The connecting or tying of an element of a building to the main structure of the building, usually to prevent lateral movement. For example, the anchorage of a roof to a wall, or *Veneer* to its backing.

Anchor Bolts See *Bolts.*

Anchor Ties In masonry work, any type of fastener used to secure *Masonry* units to a more stable object such as another wall. Its value is as a tension tie and it often is in the form of a manufactured, corrugated sheet metal clip. See also: *Framing Anchors.*

Anemometer An instrument for measuring air velocity.

Anemotherm Velometer An instrument used to measure static pressure, temperature, and air velocity for *Air Balancing.*

Angle Iron See *Angles.*

Angle of Repose The steepest angle measured from the horizontal at which a

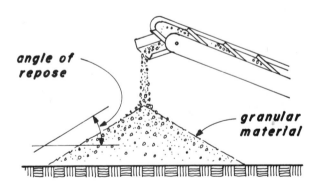

FIGURE A-11   *Angle of Repose*

granular material can be piled without sliding. It varies with the type of material. For dry sand, for example, it is approximately 34°, but can vary with the angularity of the grains. (See Figure A-11).

Angles "L" shaped metal members. They are available in a standardized series of sizes in both steel and aluminum. In steel, they vary from 8″ × 8″ to 1″ × 1″ in equal leg sizes and 9″ × 4″ to 1-¾″ × 1-¼″ in unequal leg sizes with several thickness choices for each size. For illustrations see *Structural Shapes*.

BAR SIZE ANGLES—are those used for lighter framing and are available in sizes down to ½″ × ½″.

Angles, Solar See *Solar Angles*.

Angle Valve See *Valves*.

Angstrom Unit A unit of measurement of the length of light-waves which is equal to one one-hundred millionth of a *Centimeter*.

Angular Measure In surveying, angles are usually indicated by degrees, minutes, and seconds. A full circle is subdivided into 360 degrees. Each degree is further subdivided into 60 minutes, and each minute is divided into 60 seconds. Notationally, an angle of 36 degrees, 41 minutes, and 28 seconds, would be written: 36°-41′-28″. (One second of angular measurement is equivalent to a separation of about ⅛ inch at a distance of one-half mile).

Anhydrous The absence of water in a substance.

Annealed Wire Wire having greater *Ductility* resulting from *Annealing*.

Annealing A heat treating process usually applied to produce softening of metals and alloys. Specific types of annealing generally refer to certain methods of heating and cooling cycles or to par-

ticular properties that are produced by the cycles.

Annual Rings In *Wood* terminology, the alternating of light and dark rings, each cycle of which represents one year of growth. The lighter colored rings represent SPRINGWOOD (also called EARLYWOOD) which is less dense and weaker mechanically than the darker rings called SUMMERWOOD (also LATEWOOD) which is harder and stronger. (See Figure A-12).

Annulus A ringlike shape. In mathematics, the area between two concentric circles lying in the same plane. The area of the annulus equals the difference between the areas of the two circles.

Anodizing An electrolytic oxidation process which forms a protective surface film on the surface of *Aluminum* by anodic *Oxidation*. Oxygen combines with the surface of the metal in a solution to etch the outside film. It can provide varying colors—usually bronze and gold shades—by using different solutions. It is sometimes used to increase surface hardness.

Anticorrosive Paint See *Paint*.

Antique Finish A finish usually applied by brush or cloth to furniture or other woods to give the appearance of age. An antique finish can also be achieved by partial removal (wiping with a cloth) of a second coat of material.

Apartment Nomenclature An apartment is a dwelling unit consisting of two or more units within the same structure. Nomenclature varies with locality, but generally is as follows:

BACHELOR APARTMENT—A unit without a bedroom, having one room used for living and sleeping; a bathroom; and a separate, or alcove, kitchen.

COOPERATIVE—A term indicating a

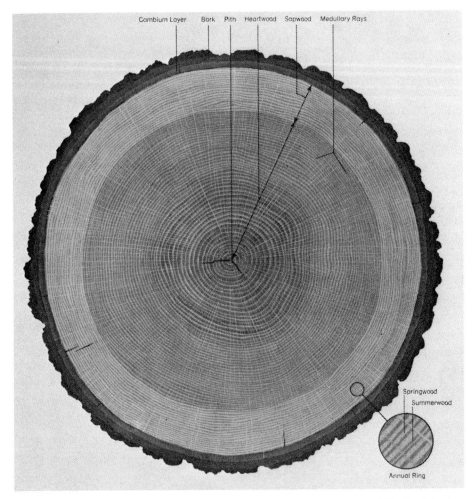

Cambium Layer   Bark   Pith   Heartwood   Sapwood   Medullary Rays

Springwood
Summerwood

Annual Ring

Courtesy Architectural Woodwork Institute

FIGURE A-12   Cross Section of a Log

means of ownership rather than a type of apartment. See separate listing: *Cooperative Ownership*.

CONDOMINIUM—A form of ownership. See description under *Condominium*.

DUPLEX—Two units only, in one building. In some areas a duplex apartment is considered an apartment on two floors.

EFFICIENCY APARTMENT—Same as a *Bachelor* apartment.

FLAT—An apartment on one level; it is a term used principally in the east.

FOURPLEX—Four units all within one building.

GARDEN APARTMENT—Units with an above average emphasis on landscaping, patios, balconies, walks, and greater privacy.

HOUSING—A general term most often used to describe a large scale apartment project which is financed in part or whole by government agencies, generally for lower income groups.

STUDIO APARTMENT—One unit on two floors, usually with the bedrooms on the upper floor. It is sometimes also used to mean a *Bachelor* apartment.

TOWNHOUSE—A group of apartments, usually row-oriented, where each one is like a separate "house" and is usually two story. It generally refers to such units in an urban area: They are often *Condominiums*.

TRIPLEX—Three units all within one building.

WALK-UP APARTMENT—Above the ground floor where access is by stairs rather than an elevator.

Apex The highest point.

Apprentice A person learning a trade; the period before qualifying as a *Journeyman*.

Approved Equal A phrase used on *Plans* and in *Specifications* to indicate that a substitution can be made for a specified brand name or manufacturer.

Apron

1. A wood finish piece under a window *Sill*, in the corner formed at the wall surface.

2. That portion of a driveway forming an access transition between a roadway and a property.

Arbitration The process by which parties to a dispute voluntarily submit their differences to the judgment of one or more impartial arbitrators selected by them to hear the evidence and arguments, with the agreement that the resulting decision will be final and binding upon each side. One procedure for arbitration is that

established by the American Arbitration Association.

Arc (of a Circle) A portion of the *Circumference* of a circle. For illustration see *Circle Nomenclature*.

Arcade A covered passageway, usually extending between two buildings.

Arch A curved structural form whose stability depends primarily upon its being in compression through most or all of its length, although bending also occurs due to unbalanced loading (loading on one side only or wind loading). An arch requires end support to resist outward (horizontal) thrust. Arches, in design, are classified as TWO-HINGED ARCHES or THREE-HINGES ARCHES, depending upon whether they are continuous members from one end to the other, or in two pieces connected at the top.

Architect A specially trained and experienced person who plans, designs, and oversees construction of buildings and related facilities, and who usually leads and coordinates mechanical, electrical, and structural consultants. Use of the title "Architect" requires licensing by the state in which the architect practices. See also: *Architecture*.

Architect's Scale A scale used for drawings which is based on fractions of an inch (such as $\frac{1}{4}'' = 1'-0''$), as opposed to an *Engineer's Scale*, which is based on decimals of an inch.

Architecture Traditionally, the term "architecture" has meant the total art and science of building of all man's habitable environment, including all the design disciplines involved. More recently, however, the term has also come to mean only the visual, aesthetic portion of a total system of building design, which also but separately, includes planning, economic analysis, engineering, etc.

Arcing Electrical current flowing (sparking) across an air gap. Severe arcing can cause considerable damage to equipment and possibly fire. Under controlled conditions arcing is used for lighting, such as for *Mercury Vapor* and *Fluorescent Lamps* and for *Arc Welding*.

Arc Welding Also called "Electric Arc Welding," it is one of the three basic means of welding and the type most frequently used in construction. (See also: *Gas Welding and Cutting* and *Resistance Welding*.) Arc welding is used for *Steel,* titanium, and *Aluminum*. The heat required to melt and fuse the metals is obtained from an electric arc passing between the *Electrode* and the *Base* (or *Parent*) *Metal*. Electrodes are metal rods which are classified as either "consumable" or "non-consumable." In the case of consumable electrodes, metal from the electrode melts and is deposited as a bead along, and within, the joint being welded. The metal thus deposited is called the FILLER METAL and fuses with the base metal. These electrodes are available as wire or rods in diameters of 1/8", 3/16", 1/4" and 5/16", the choice of size depending upon the desired amount of weld material to be deposited in one "pass." The composition of the electrode is approximately the same as the metal being welded, but with variations particularly suited to the material and conditions of the weld. A designation for types of consumable electrodes, established by the American Welding Society, uses an "E" followed by four or five digits. The first two digits (or three in the case of a five digit series) indicates the tensile strength of the deposited metal in thousandths of pounds per square inch. The next to last digit indicates the weld position (overhead, horizontal, flat, etc.) and the last digit indicates information pertaining to

power supply, type of *Slag*, arc, penetration, etc. Consumable electrodes are either bare or coated. Since the composition of molten metals is weakened when there is exposure to oxygen or nitrogen, it is desirable to "shield" the molten area from these elements. One way of accomplishing this is to coat the electrode with *Flux*. As the coating burns, an inert gas from the flux is released which protects the molten weld area from oxygen and nitrogen.

The non-consumable type of electrodes are used only to make the arc, and welding rods, supplying the filler metal, are fed separately into the molten area. Non-consumable electrodes consist of carbon and tungsten.

SUBMERGED ARC WELDING—is a production type of welding whereby the electrode moves through a blanket of granular flux material such that the arc is covered by the flux as it moves, usually automatically, along its weld path. The arc is not visible beneath the flux hence the name "submerged arc."

GAS SHIELDED ARC WELDING—is another means of shielding the molten weld area from oxygen and nitrogen and cosists of a stream of inert gas, usually argon, being discharged from a nozzle over the molten area. There are two basic types of gas shielded arc welding—TIG WELDING and MIG WELDING. (MIG stands for "Metallic-Inert Gas" and TIG stands for "Tungsten-Inert Gas.") MIG welding consists of a consumable electrode in the form of a wire which is fed from the end of a combination nozzle and holder which melts into the molten area caused by the arc. In TIG welding a non-melting tungsten electrode is used. The inert gas is discharged over the molten area as with MIG welding, but the filler metal is fed

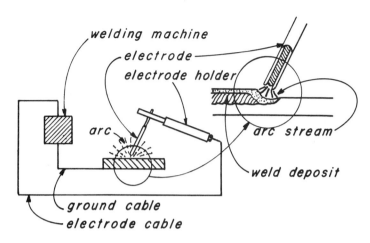

FIGURE A-13   *Arc Welding Circuit*

into the molten area from a separate rod. MIG and TIG welding are used to weld aluminum, magnesium, copper, and titanium. (See Figure A-13.)

Area Drain   A drain (*Catch Basin*) installed to collect surface water.

Areas (of Common Shapes)   For formulas to determine areas of common geometric shapes see Figure A-14.

Areaway   A means of providing light, ventilation, or access to a *Basement* or *Cellar* by an earth-retaining, pit-like recess, out and around a below-ground window or doorway.

Arithmetic Mean   The same as the *Average* of a series of numbers.

Armored   Cable   See   electrical *Conduit*.

Arris   The external edge formed by two surfaces meeting each other.

Artificial Stone   See *Cast Stone*.

Asbestos   A fibrous material obtained from a variety (chrysotile) of the mineral serpentine. The fibers are flexible, moisture resistant, and have very high resis-

tance to heat. They are used in a wide range of building products such as those made from *Asbestos Cement*.

Asbestos   Cement   A mixture of *Asbestos* fibers, *Portland Cement*, and water which can be press-formed into building panels, pipe, and related products having high weather and fire resistance. Transite (trade name of Johns-Manville) is an example.

Asbestos Cement Pipe   See *Pipe*.

Asbestos   Cement   Shingles   See *Shingles*.

Asbestos Sprayed Fireproofing   A sprayed-on fireproofing applied to steel beams and columns. It consists of *Asbestos* fibers, *Rockwool*, and *Cement* with water added at the discharge nozzle. See also: *Plaster Fireproofing*.

As-Built   Drawings   Construction drawings which have been revised during the course of construction to show changes made from the original drawings. Their purpose is to have a permanent record of the actual construction.

Ash   See *Hardwood*.

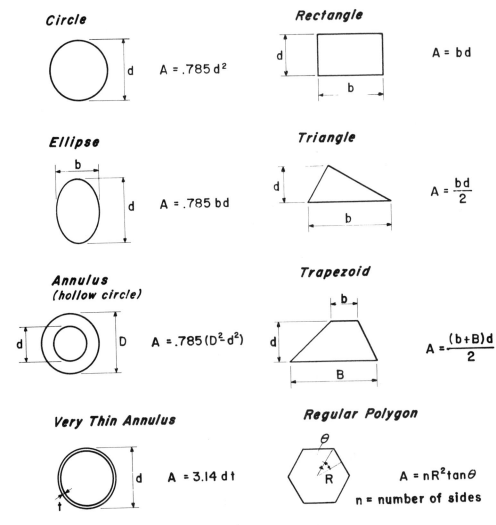

Circle
$A = .785 d^2$

Rectangle
$A = bd$

Ellipse
$A = .785 bd$

Triangle
$A = \dfrac{bd}{2}$

Annulus (hollow circle)
$A = .785 (D^2 - d^2)$

Trapezoid
$A = \dfrac{(b+B)d}{2}$

Very Thin Annulus
$A = 3.14 dt$

Regular Polygon
$A = nR^2 \tan\theta$

n = number of sides

FIGURE A-14   Areas of Common Shapes

Ashlar Masonry Masonry made from stones which have been carefully cut so as to form thin joints as contrasted to irregular *Rubble Masonry*. The term can also apply to any irregular joint pattern in a wall made from manufactured masonry units.

Aspect Ratio In *Air Conditioning*, the ratio of the width of a rectangular *Duct* to its depth.

Asphalt A black insoluble substance, also called *Asphaltum*, which occurs in nature or is obtained from crude petroleum and used as a weather resistant binder ingredient for *Asphaltic Concrete Paving* and a variety of other building products such as *Asphalt Shingles* and *Asphalt Tile Resilient Flooring*.

Asphalt Batch Plant A central drying and mixing plant for preparing

Hollow Ellipse

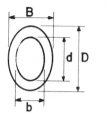

$$A = .785\,(BD-bd)$$

Circular Sector

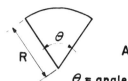

$$A = .0087\,\theta\,R^2$$

θ = angle in degrees

Irregular Shape

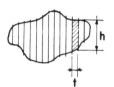

divide into equal
width narrow
strips and add
area (t x h) of
all strips.

To Find Area of any Complex Shape:

Divide into parts and compute areas of each part seperately per examples above and add (or subtract) to obtain total.

Example: Area for illustrated shape = area of rectangle "A" plus area of triangle "B", minus area of quarter circle "C" plus area of rectangle "D"

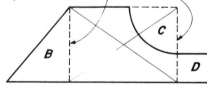

FIGURE A-14   Areas of Common Shapes
(cont.)

Asphaltic Concrete Paving, wherein drying, screening, storing, and mixing units are so arranged as to dry, separate, and recombine proper proportions of various sizes of *Aggregate,* and mix and coat them with the proper grade and proportion of hot liquid *Asphalt* from which the mixture ("batch") is discharged into trucks for transport to the job site. (See Figure A-15.)

Asphaltic Concrete Paving Com-

monly called "asphalt paving" or "black-top," it is a mixture of *Asphalt* and *Aggregate* (*Sand* and *Crushed Rock* or *Gravel*) which is spread while hot onto a prepared surface where it is leveled with a roller to form a finished, weather-resistant surface for streets, driveways and parking areas. Thickness is usually 2 to 4 inches. It can also be mixed and laid cold, termed *Cold Mix*. (See Figure A-16.)

Courtesy Stansteel Corp.

FIGURE A-15   Asphalt Batch Paint

Asphalt Roll Roofing A rag fiber impregnated with *Asphalt* and known also as *Roofing Felt*.

Asphalt Shingles See *Shingles*.

Asphalt Tile See *Resilient Flooring*.

Asphaltum Same as *Asphalt*.

Aspiration An *Air Conditioning* term meaning the movement and mixing of air which takes place when a stream of air (primary air), as from an *Air Diffuser*, discharges into a larger air mass (secondary air), as into a room. It is this circulation that causes dirt particles to collect adjacent to a *Diffuser or Register*.

Assessed Value A value placed upon a property by the county or other taxing authority to be used as a base for taxation

FIGURE A-16 *Asphaltic Concrete Paving*

purposes. It is usually a percentage of the market value as determined by the taxing assessor. For example, in Los Angeles County the assessed value is 25% of the assessor's determined market value.

Assessment A charge or *Lien* on *Real Property* made by a taxing authority to pay for an improvement project (such as sewers or flood control) which benefits the property. These charges are usually paid in fixed installments over a period of years and are in addition to normal real estate taxes. They are sometimes called "Special Assessments."

Astragal On a pair of doors, the lip-like member on the free edge of one door against which the other door closes. It can be either of wood or metal.

Atmospheric Pressure The pressure of the atmosphere, due to its weight, measured by means of a barometer. It is equivalent to 14.7 pounds per square inch

or 29.92 inches of mercury at 32° at sea level. See also: *Air*.

Attenuation A sound reduction process in which the energy of sound is absorbed or diminished in intensity by conversion into motion or heat.

Attic The enclosed space between a roof and the ceiling below.

Attic Fan A *Fan* mounted in the *Attic* of a house to draw outside air through the house to cool it, and exhaust out the warmer air. It can also be a fan in one end of an attic to draw air through the attic only, to cool the house by preventing a heat build-up in the attic.

Auger A screw-like excavation implement which may be "drilled" into the ground for recovering soil samples or making cylindrical holes such as for *Piles*. (See Figure A-17.)

Automatic Closing Door See *Door Types*.

FIGURE A-17   Auger Drill

Automatic Fire Sprinkling System A fire protective sprinkler system consisting of a network of overhead pipes with spaced outlets called "heads," which open at a predetermined temperature to discharge water onto the fire area.

There are four basic types of systems: the WET PIPE SYSTEM where the pipes are always filled with water under pressure and the water discharge begins immediately when any head opens. This is the type of system most often used. A second type is the DRY PIPE SYSTEM which is primarily used in areas subjected to freezing. In this system the pipes are filled with air under pressure and when a head opens it causes a drop in air pressure which activates a valve allowing water to flow into the pipes and discharge from the open heads. The third type is the PRE-ACTION SYSTEM which is used to counteract the time delay disadvantage of the dry pipe system and give an early alarm. In this system the pipes are also "dry" but water flow is actuated by an independent heat detection system rather than by the opening of a head. The fourth method is the DELUGE SYSTEM wherein all heads are open all the time and the piping system is without water until a detection device opens the water supply valve and discharges a maximum amount of water from all sprinklers. This system is used in areas of extra high hazard, so water can be discharged ahead of a fire, rather than waiting for the fire to reach and open a head.

Water supply for automatic fire sprinkler systems is supplied independently from the building supply system, either from a city water main or an independent supply system such as an elevated water tank. In all types of systems, an alarm sounds whenever the system is activated.

Spacing of sprinkler heads and the number of heads which may be served from given sizes of pipes are regulated depending upon the hazard involved. Sprinkler systems are divided into three hazard classifications: LIGHT HAZARD (apartments, office buildings, schools, etc.); ORDINARY HAZARD (most industrial plants and warehouses); EXTRA HAZARD (certian chemical plants, oil refineries, paint shops, etc.). In some cases of very high hazard, such as high piled storage, insurance requirements specify that the system be hydraulically designed (also called a CALCULATED SYSTEM or DENSITY SYSTEM), that is, the en-

tire piping system is designed to provide a specified minimum gallons-per-minute discharge from any specific head, over a given fire area.

The sprinkler piping system consists of an underground supply main, a secondary supply, controls and alarms, a vertical *Riser* either just within or without the building, *Main* lines supplied from the riser, and *Branch* lines which contain the heads and which are connected to the main lines. Mounted on the riser are pressure gages, the alarm system, and shut-off valve called a POST INDICATOR VALVE outside the building is provided to give an above ground means of indicating whether the valve is open or closed.

Sprinkler heads are basically of three types: UPRIGHT HEADS which extend above the supply pipe; PENDANT HEADS which extend down from the supply pipe (these are used where the piping is above a ceiling and only the sprinkler heads project below the ceil-

Courtesy Grinnell Corp.

FIGURE A-18   Automatic Fire Sprinkler Head (Upright type)

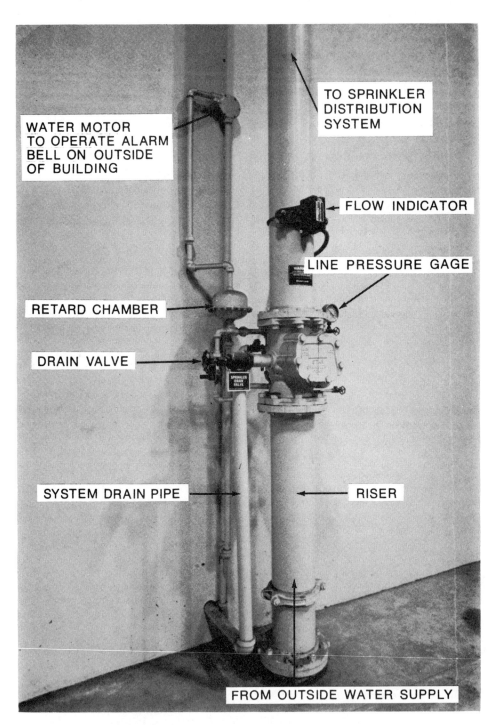

WATER MOTOR TO OPERATE ALARM BELL ON OUTSIDE OF BUILDING

TO SPRINKLER DISTRIBUTION SYSTEM

FLOW INDICATOR

LINE PRESSURE GAGE

RETARD CHAMBER

DRAIN VALVE

SYSTEM DRAIN PIPE

RISER

FROM OUTSIDE WATER SUPPLY

FIGURE A-19   Automatic Fire Sprinkler Riser

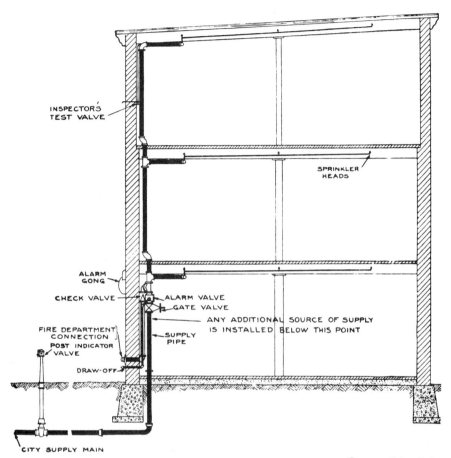

FIGURE A-20   *Automatic Fire Sprinkler System*

ing); and FLUSH HEADS which, for appearance reasons, are virtually flush with a ceiling.

Due to the fire protective value of sprinkler systems, most *Building Codes,* when such systems are installed, permit greater areas between *Fire Walls,* greater distances between exits, and other loosening of fire safety requirements. (See Figure A-18, A-19, and A-20.)

Automatic Level   See *Level.*

Automatic Transfer Switch   In electrical work, an emergency type of switch which, in case of an interruption of a normal power source, automatically switches over to an auxilliary source of power such as batteries or a gas driven *Generator.* When normal power returns, it automatically switches back to that source.

Automatic Welding   A term used to indicate machine controlled movement of the welding rod as opposed to manual welding.

Average   In mathematics, the "aver-

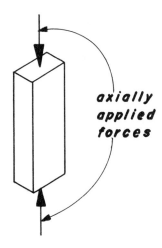

FIGURE A-21   Axial Load

age'' of two or more numbers is obtained by finding the sum of the numbers and dividing by the number of them. It is the same as ARITHMETIC MEAN or "mean.'' For example, the average cost of homes selling for \$24,500, \$29,000, and \$32,800 =
$$\frac{24{,}500 \; + \; 29{,}000 \; + \; 32{,}800}{3} \; =$$
\$28,767.

Award (of a Contract)   An indication by an owner to a contractor that his *Bid* or proposal is accepted and that a contract will be entered into.

Awning   A lightweight, and often temporary, or adjustable, exterior, roof-like sun shade over a window or door, attached to the building wall. Formerly of canvas, they are now usually of aluminum or plastic.

Axial Flow Fan   See *Fans*.

Axial Load   A load (force) applied along the longitudinal *Axis* of a structural member. If the load is parallel to the axis, but not coincident with it, the load is eccentric (see *Eccentricity*) rather than Concentric as in the former case. (See Figure A-21.)

Axis   An imaginary straight line about which something rotates or is symmetrically arranged.

Azimuth   In surveying, the defining of the direction of a line in terms of its *Angular Measure* clockwise from north. (''North,'' although usually meaning true north, might also mean grid north, *Magnetic North,* or an arbitrarily assumed north point.) Some agencies use south as a reference point rather than north.

B

Bachelor Apartment See *Apartment Nomenclature*.

Backfill Earth replaced after having been excavated, such as to refill utility trenches or behind retaining walls. (See Figure B-1.)

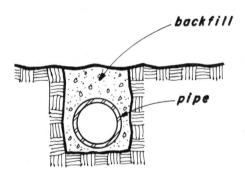

backfill

pipe

FIGURE B-1 *Backfill*

Backflow In plumbing, a flow of water or other liquid into the distributing pipes of a *Potable* (drinkable) supply of water from any source other than that intended, such as could occur if water for an industrial process were piped directly from the potable supply and contaminated process water could, through failure of a valve, flow back into the supply. *Air Gaps* or *Backflow Valves* are used to prevent this backflow.

Backflow Valves See *Valves*.

Backhoe An excavating machine, cable or hydraulically operated, which excavates in a hoe-like manner, with the digging action toward the operator. Its chief use is for trenching and pit-like excavations. (See Figure B-2.)

Back Plastering *Plaster* applied to both sides of a single layer of *Metal Lath,* usually to give a greater fire rating. Also, the application of plaster to the "inside" side of the lath, this is, the side of the lath which attaches to its supports.

Back Pressure In plumbing, an air pressure buildup in drainage pipes greater than *Atmospheric Pressure*.

Back Priming A coat of paint applied to the back, unexposed side of woodwork to prevent moisture from getting into the wood and causing the grain to swell.

Backsaw See *Saws*.

Back Siphonage In plumbing, the flowing back of used or contaminated water from a plumbing fixture into a water supply pipe, due to a negative pressure in the supply pipe. It is also caused by an incorrect *Air Gap*. See also: *Backflow*.

Back-Up In *Sealant* terminology, any material recessed into a joint to act as a

FIGURE B-2   Backhoe Excavator

backing, and to control the depth during application of the sealant.

Baffle   A partition to divert the flow of gas or liquid either directionally or by excluding a portion of it, such as in *Interceptors*.

Baked Enamel   A hard, glossy finish used on metals (such as refrigerators). It involves the use of synthetic resins and is baked in a recirculating gas oven, usually at about 300°F. See also: *Porcelain Enamel*.

Balancing   See *Air Balancing*.

Balcony   Any above-ground walkway or platform connected to the outside of a building.

Ballast

1. In railroad work, the *Gravel* or crushed rock between, and underneath, the ties.

2. For *Fluorescent Lamps* and *High Intensity Discharge Lamps,* a transformer-like device to limit the current flow through the *Arc* inside the lamp.

Ball Cock   See *Valves*.

Ball Joint   See *Pipe Joints*.

Balloon Framing   In carpentry, a two story wall frame system whereby the

Studs are in one piece from foundation to roof and intervening floor *Joists* are nailed to the sides of the studs and supported on a let-in *Ribbon*. The other basic wall framing system is called *Platform Framing*.

Balsam   See *Softwood*.

Balusters   The closely spaced vertical members in a stairway or balcony railing.

Balustrade   A stairway or balcony railing consisting of *Balusters*.

Bandsaw   See *Saws*.

Banker Partitions   A general term for floor supported partitions of less than ceiling height used to divide floor areas. Since not all such partitions are used in banks, they are collectively called *Freestanding Partitions*.

Bar

1. In concrete work, a *Reinforcing Bar*.

2. In structural *Steel*, a rectangular or round shape (in cross section) generally defined as being 6″ or less in width and .203″ or more in thickness, or 6″ to 8″ in width and .23″ and over in thickness. Bars are specified in ¼″ and ⅛″ thickness increments. Widths exceeding the above are generally termed *Plates*.

Barbed Nail   See *Nails*.

Bar Chart   A frequently used form of chart (usually for time schedules) where items or events are represented by "bars," the lengths of which indicate quantity or time.

Bare Conductor   See electrical *Conductor*.

Bargeboard   A finish, fascia-like, board along the overhanging edge of the *Gable* end of a roof.

Barge Rafter   See *Verge Rafter*.

Bar Joist   See *Open Web Joist*.

Bark   In lumber terminology, the outer layer of a tree, comprising the inner bark, or thin, inner living part (phloem), and the outer bark composed of dry, dead tissue.

Barrel (of Cement)   A unit of *Cement* measurement equal to four standard 94 pound bags of cement, or approximately four cubic feet.

Barrel Bolt   A door locking device consisting of a metal cylinder sliding in a guide mounted on the face of a door which engages a receiving hole on the *Jamb*. (See Figure B-3.)

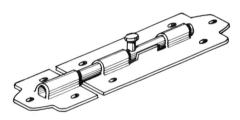

FIGURE B-3   Barrel Bolt

Barrel Pin   A tool used to erect structural steel, consisting of a round pin tapered on each end. It is inserted into one of several matching holes of two members to be joined and driven into the hole to force alignment such that a *Bolt* can be placed into an adjoining hole. It is tapered on both ends to facilitate removal by being able to drive it on through the hole in the same direction as it was inserted.

Barrel Shell Roofs   *Thin Shell Concrete* roofs having the form of a series of half-cylinders joined at their edges and supported at their ends by the building walls or *Columns*. Structurally, they can be thought of as a series of side-by-side "beams." (See Figure B-4.)

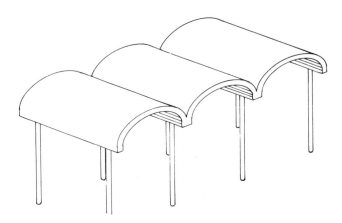

FIGURE B-4   Barrel Shell Roof

Bar Size Angles See *Angles*.

Base

1. A general term meaning an underlying support for something, such as a *Gravel* base under *Asphaltic Concrete Paving*. (See also: *Crusher Run Base*.)

2. The lowest part of something, as of a wall or *Column*.

3. In painting, the term refers to the basic make-up or *Vehicle* of a paint, such as *Oil Base Paint*.

4. In chemistry, an Alkali (the opposite of an *Acid*) which, when mixed with an acid, forms a salt.

5. Abbreviated from *Baseboard* but also often meaning ''topset base'' or *Rubber* Base.

6. As applied to *Plastering*, see *Base Coat*.

Base Bid See *Bid*.

Baseboard A wood finish piece along and flat against the bottom of a wall to hide the joint between the wall and floor and to protect the base of the wall against damage.

Base Coat In *Plastering*, the coat or coats beneath the *Finish Coat*.

Base Line In government survey *Land Description*, an east-west reference line which, at its intersection with a principal *Meridian*, forms a reference point for locating and legally describing parcels of land.

Basement A usable floor area of a building which is partly below ground level. Some *Building Codes* define it as a ''basement'' if the distance from its floor to grade is, on the average, less than the distance from grade to its ceiling. If such a distance is greater it is termed a *Cellar*.

Base Metal See *Welding*.

Base Plate A steel end plate at the base of a *Column* used to distribute the end load over a larger area. See also: *Bearing Plate, Leveling Plate*. (See Figure B-5.)

Base Screed In plastering, a formed metal straight edge fastened to and projecting from a wall a distance equal to the thickness of the plaster (usually ½"). It is used to control the thickness of the plaster coat as it is applied and to insure straight walls.

Base Shoe A wood *Molding*, usually

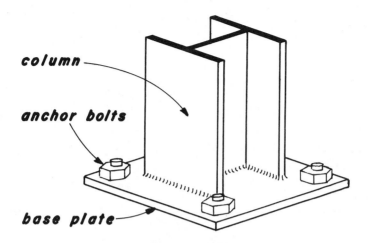

FIGURE B-5   Base Plate

one-quarter round, nailed into the joint between a floor and a *Baseboard*.

Basic Building Code Widely used performance-type *Building Code* prepared by Building Officials and Code Administrators International (BOCA International) and adopted for use by a number of counties and cities, principally in the midwestern area. A new edition is issued every five years.

Bastard Sawn *Hardwood* lumber cut from a log such that the annual rings make 30° to 60° angles with the surface of the piece.

Bat A broken, but usually usable, piece of a masonry unit.

Batch A quantity of any mixed material, such as *Concrete, Paint,* etc., made in one operation.

Batch Plant See *Asphalt Batch Plant* or *Concrete Batch Plant*.

Bathroom Classification Bathrooms are often classified according to the number and type of fixtures; called a half bathroom if it has a *Water Closet* and *Lavatory*; a three-quarter bathroom if it has a water closet, lavatory, and shower; or a full bathroom if it has a water closet, lavatory, shower and bathtub, or bathtub-shower combination.

Batt Abbreviation for *Batten*.

Batten Thin strips which cover the joints between vertical boards on a wall. See also: *Board and Batten Siding*.

Batt Insulation See *Insulation*.

Batter The slope of a wall or other surface away from vertical; inclined.

Batter Boards A system of boards set up just outside the corners of a building before construction begins and between which strings are run to indicate the lines of the building. The purpose is to have reference lines (the strings) which can remain in place during excavation.

Batter Pile See *Piles*.

Battleship Linoleum See *Resilient Flooring*.

Bay The spacing, or space, between rows of columns.

Bayonet Saw See *Saws*.

Bay Window A window placed in an outward protusion of a wall.

Bead

1. As used in *Plastering,* see *Corner Bead.*

2. In carpentry, a small *Molding* strip, usually a half-round shape.

3. A thin toothpaste-like line of *Glue, Caulking Compound,* or similar material placed on a surface, such as from a *Caulking Gun.*

See also: *Welding.*

Beam A structural member, usually horizontal, used to support loads (forces) applied transversely to it (perpendicular to its long axis) by its ability to resist *Bending* and *Shearing* forces.

A CANTILEVERED BEAM is one extending beyond a last point of support. CONTINUOUS BEAMS span over three or more supports as opposed to SIMPLE SPAN BEAMS which span between two supports, each of which is free to rotate. FIXED END BEAMS are those connected rigidly to a support in such a manner that they are not free to rotate.

Bearing In surveying, a means of describing the direction of a line by stating its *Angular Measure* from a north-south *Meridian* line. By convention, such angles do not exceed 90 degrees—a line having an angular measure clockwise from north of 100 degrees would instead by equivalently stated as "south 80 degrees east". With customarily used abbreviations, an example of the bearing of a line might be: S4°-13′-43″W (meaning 4 degrees, 13 minutes, and 43 seconds west of south).

Bearing Capacity (of Soil) The load, usually expressed in pounds per square foot or tons per square foot, which a given soil strata can support without excessive settlement or other adverse effect on the sturcture being supported. This value for most buildings and an average range of soil conditions will most often be between 1,000 and 3,000 pounds per square foot. The term SOIL BEARING VALUE is synonymous but the term *Soil Pressure* is not. The latter states the actual rather than the permissible bearing pressure under a foundation.

Bearing Pile See *Piles.*

Bearing Plate A plate under the end of a *Beam* or *Column* (when used with the latter, it is more correctly called a *Base Plate*) used to spread the *Reaction* load over a larger area.

Bearing Stratum The layer of *Rock* or *Soil* upon which a foundation rests.

Bearing Stress See *Stress.*

Bearing Wall See *Walls.*

Bed Joint In *Masonry,* a horizontal joint upon which the next *Course* is placed.

Bedrock The solid *Rock* lying below loose, unconsolidated *Soil.* If the rock is shattered or otherwise distorted or broken, it can still be classified as bedrock.

Bell and Spigot Joint See *Pipe Joints.*

Belled Pier A *Pile*-like concrete foundation made by drilling a hole into the earth, belling out (enlarging) the bottom and filling with concrete. Its purpose is to penetrate poor soil to permit improved bearing on a deeper strata. It is also called a CAISSON, particularly in the west. (See Figure B-6.)

Belt Course In *Masonry,* a continuous horizontal band of masonry, such as at window sill heights. It is also called a SILL COURSE or STRING COURSE.

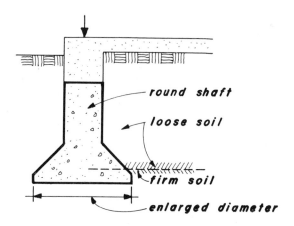

FIGURE B-6   Belled Pier

Belt Sander See *Sanders*.

Bench Mark In surveying, a reference point from which other elevations are measured and computed. A bench mark may be any relatively permanent object.

Bend See *Pipe Fittings (Ell)*.

Bending The application of a force or forces to a structural member causing such member to *Deflect* (move) transversely to its long axis. Although the term "bending" usually applies to a *Beam,* it can also apply to *Columns,* when a force is applied sidewise, or if an *Axial Load* is applied eccentrically (off-center). The term "flexure" means the same.

Bending Moment In structural design, the amount (forces multiplied by distances) of *Bending* or *Rotation* imposed upon a structural system or component by externally applied forces. When the bending moment has been mathematically determined, a determination can be made for a properly designed structural member to resist stresses caused by such bending. In engineering calculations, bending moment is abbreviated

"M" and its unit of quantity is foot-pounds or inch-pounds.

Bending Stress See *Stress*.

Bent A framework which is tranverse to the length of a building, is in one plane, and is used to carry vertical loads and to brace the building by also resisting *Lateral Loads*. The term is usually synonymous with *Rigid Frame*.

Bent Plate A steel plate which is bent into a configuration for connecting two or more members, such as an angle or "U" shape used as a bolted connector between timbers, as opposed to a similar configuration made by welding separate flat plates together.

Benzene See *Solvents*.

Benzine See *Solvents*.

Berm A smaller dike-like earth or paved embankment to divert the flow of run-off water.

Bevelled Washer See *Washer*.

Bibb A faucet having a threaded connection such as for a garden hose. It is sometimes called a *Cock,* sillcock, or *Spigot*.

Bicycle Wheel Roof   Similar in structural principle to a *Cable Roof* but round in shape, somewhat resembling a bicycle wheel; a double system of *Tendons* is used to eliminate or reduce the effects of flutter.

Bid   An offer, generally in writing (also called a proposal) to supply an indicated quantity of materials or labor or both for a stated price.

When a bid is submitted by a *Subcontractor* to a *Contractor* it is called a SUB BID (i.e., when his contract would be with the principal contractor, or general contractor, rather than directly with the owner.)

A BASE BID is a bid where alternate prices are asked for, as for different qualities or scopes of work, in which case the base bid is for the work as shown or specified with cost of the alternates indicated as additions to, or deductions from, the base bid.

Bidder   A contractor who has been asked to submit a *Bid* for a specific project. There are usually at least three bidders selected for any project, but seldom more than six or seven.

Bidding   The process whereby two or more *Contractors* compete for a contract by each submitting a *Bid* with the expectations that the one having the lowest price will receive the contract. In practice, and usually for larger projects, the procedure includes an invitation to bid by an owner to several contractors. Each of the selected contractors, after having been given *Plans* and *Specifications,* will then usually solicit *Sub-Bids,* for the various specialized portions of the work which he may not be equipped to do economically himself. Each contractor will then sum the total of the lowest sub-bids, add his estimate of that portion of the work he intends to do himself plus *Overhead* and

Profit, and submit the grand total as his final bid, usually in a sealed envelope to be opened at the same time as those received from the other contractors. See also: *Bid, Bid Opening, Bid Shopping.*

Bidet   Toilet-like bathroom fixture used for hygienic washing of the genital area.

Bid Form   A prepared document on which each *Bidder* fills in the amount of his *Bid* and delivers it, usually in a sealed envelope, to the owner at the time of the *Bid Opening.*

Bid Opening   Time, place, and date set for receiving and opening *Bids* for a specific project. (It is a formality often not used for smaller projects.) Bid openings are either public or private. For public bid openings all bidders assemble to hear each other's bids as they are opened and read. For private bid openings, the bids are opened by the owner in private and the contractors may or may not be informed as to each other's bids.

Bid Shopping   The practice, generally considered unfair, whereby a *Contractor* who is low bidder for a project, "shops" among *Subcontractors* for a price lower than that upon which his estimate was based, thereby increasing his potential profit.

Bi-Fold Door   See *Door Types.*

Billet   A semi-finished piece of *Steel* which has resulted from rolling a *Bloom* into a smaller cross section. It may be square but is never more than twice as wide as it is thick. Its cross sectional area is usually not more than 36 square inches. It, together with blooms and slabs, is further rolled or shaped into *Structural Shapes, Bars, Wire, Sheet,* and *Plates.*

Binder   A term synonymous with *Cement* meaning any material used to

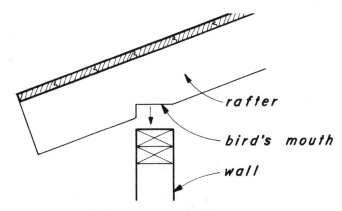

rafter

bird's mouth

wall

FIGURE B-7   Bird's Mouth

bind or cement other materials together. For example, *Asphalt* is a binder for sand and gravel to produce *Asphaltic Concrete Paving*.

Birch   See *Hardwood*.

Bird Bath   Slang term for a localized ponding of water on a paved area caused by surface irregularities which prevent complete drainage.

Bird's Mouth   Slang term for a small notch at the lower end of an inclined rafter to provide level bearing on the top of a wall. (See Figure B-7.)

Birdseye   An extreme and special case of curly-grained wood. It is a slang term for a small localized area in wood having the fibers indented and otherwise contorted to form small circular or elliptical figures remotely resembling a bird's eye. They are common in sugar maple and are used for decorative purposes, but are rare in other *Hardwood* species.

Bitumen   Solid or semisolid hydrocarbons from petroleum, native mineral waxes, and naturally occurring *Asphalts*. They occur, in varying degrees, in almost every part of the world. Distillation of petroleum yields a residue similar to many native solid bitumens.

Bituminized Fiber Pipe   See *Pipe*.

Bituminous   Descriptive of a material containing *Asphalt, Bitumen,* or *Tar*.

Bituminous Paint   See *Paint*.

Blackboards   See *Chalkboards*.

Black Spruce   See *Softwood*.

Blacktop   Slang for *Asphaltic Concrete Paving*.

Black Wire   Wire that is not galvanized or insulated.

Blade   Slang for a *Grader* (also called a MOTOR GRADER). Also, in earthmoving, an implement on the front of a *Tractor* or *Grader,* used to push earth.

Blanket Insulation   Similar to *Batt Insulation* but manufactured in a continuous roll. (See Figure B-8.)

Blast Furnace   A tall, cylindrical furnace extracting *Iron* from iron ore. The iron ore (most common is Hematite), coke (which is made from coal and serves as a fuel), and limestone (used as a *Flux* to help remove impurities) are put into the top of the furnace. A blast of hot air is circulated through the furnace to elevate the temperature to about 3,000°F. at which level the iron separates from the ore and is drawn off at the bottom of the

FIGURE B-8   *Installation of Insulation*

furnace. *Slag,* lighter than the iron and containing the impurities, is drawn off at a higher level.

Blast Heater   A device consisting of a *Fan* and heating coils whereby a blast of hot air is discharged. An example of their use would be to provide an *Air Curtain* effect at loading dock truck doors.

Bleaching Agent   An *Acid* (usually oxalic acid) or other material applied to a wood surface to permanently lighten its color or restore discolorations.

Bleed   In *Plumbing* or *Air Conditioning,* to remove unwanted air or fluids from pipes, tubing, or other passageways.

Bleeding   In painting, the action whereby the color of an undercoat comes through the finish coat producing an undesirable spotty effect.

Blind Flange   See *Pipe Flange.*

Blind Nailing   Nailing in such a way that the nail head is concealed from view by other pieces after final assembly.

Blind Rivet A special type of rivet (one trade name is "Pop" Rivet, United Shoe Machinery Corp.) which can be inserted into a predrilled hole from one side only to fasten two or more materials together. A special type of tool is used, which, by a pulling action, withdraws a mandrel (core) through the hollow rivet which compresses the rivet end on the far side for a crimp-like sealing action. Blind rivets are available in ALUMINUM, STEEL, MONEL METAL, COPPER, and STAINLESS STEEL.

Blistering In painting, the formation of bubbles or pimples on the surface of finished work.

Blocking Short, wooden members put between *Rafters, Joists* or *Studs* for added stiffness, or as a backing for attaching other materials. When they are used to cut off passage of fire within the wall they are called FIRE BLOCKING.

Blocks See *Concrete Blocks.*

Block Tin Pure *Tin* used as a coating to prevent corrosion in pipes or tanks.

Bloom In *Steel* and *Aluminum* production, a bar rolled from an *Ingot* and further rolled or formed into structural shapes, *Bars,* or *Wire.* See also: *Billet.*

Blow (or Throw) In *Air Conditioning,* the distance that air travels from an outlet to a position at which its motion along the axis reduces to a velocity of 50 f.p.m. For unit heaters, it is the distance air travels from a heater without a perceptible rise due to temperature difference and loss of velocity.

Blower See *Fan.*

Blowing Wool Insulation See *Loose Fill Insulation.*

Blowoff The controlled discharge of water, steam, vapor, sludge, or other fluid waste from a vessel or pipeline.

Blowtorch A device used to provide an intense flame for melting *Solder* or for *Brazing,* achieved by burning alcohol or gasoline under pressure. They are now largely replaced by *Butane* and *Propane* torches.

Blue Line Print A *Blueprint* made with sensitized paper producing blue lines on a white backround, the advantage being that it is easier to make pencil notations on the white background.

Blueprint A print or copy obtained by exposing a sensitized paper overlain with a drawing made on tracing paper (or *Vellum*) to light. It results in white lines on a blue background. The *Blue Line Print* is in more general use now (although it is still called a "blueprint") since readability and the advantage of making pencil notations is easier on the white background. (See Figure B-9.)

Blue Top Stake Surveyor's stakes used for final grading which are driven into the ground with the top surface, colored blue, indicating the desired grade elevation.

Bluing A surface treatment for *Ferrous* alloys forming a thin blue film of oxide to improve the appearance and corrosion resistance. It is accomplished by subjecting the scale free surface to the action of air, steam, or other agents at a suitable temperature.

Blushing

1. A term applied to *Lacquer* when, upon application or drying, it becomes flat and in the case of a clear lacquer, white or partially opaque. It is usually caused by improper drying conditions.

2. Condensate formation on the surface of a contact *Adhesive,* caused by improper drying conditions.

Board and Batten Siding A method

FIGURE B-9 *Blueprint Machine (White-printer)*

of applying exterior wood siding whereby the joint between vertical boards is covered with a narrower vertical board called a *Batten*.

Board-Foot The unit of measurement of lumber. One board-foot is a quantity of lumber 1″ thick, 12″ wide and one foot long. For computing, divide the cross sectional area of a piece by 12 and multiply by the number of feet in length. For example, the number of board-feet in a 6″ × 12″ × 20 ft. long piece = $\frac{6 \times 12 \times 20}{12}$ = 120 board-feet (abbreviated bd.-ft. or B.F.). Nominal rather than actual dimensions are used (a 2″ × 4″ is considered 2″ × 4″ rather than its actual mill surfaced size of 1-½″ × 3-½″). Lumber unit prices are quoted in dollars per M.B.F. (thousand board feet).

Boards See *Lumber*.

Boat Nail See *Nails*.

Bobtail Truck See *Truck Types*.

Boiler A closed vessel in which a liquid is heated or vaporized, usually to provide steam or hot water.

Bolts Threaded metal connecting devices having a head on one end and usually requiring a nut on the other end. Most bolts have a square or hexagonal head for tightening with a wrench; however, some bolts have rounded or flattened heads with slots for a screwdriver (See Figure B-10). Bolt types include:

ANCHOR BOLT—A bolt, usually a *Machine Bolt,* embedded in concrete or masonry to provide a means for attaching another member.

BARREL BOLT—Not truly a bolt; a door locking device described under *Barrel Bolt*.

CANE BOLT—Not truly a bolt; a locking device used at the base of a door or gate described under *Cane Bolt*.

CARRIAGE BOLT—A bolt used in

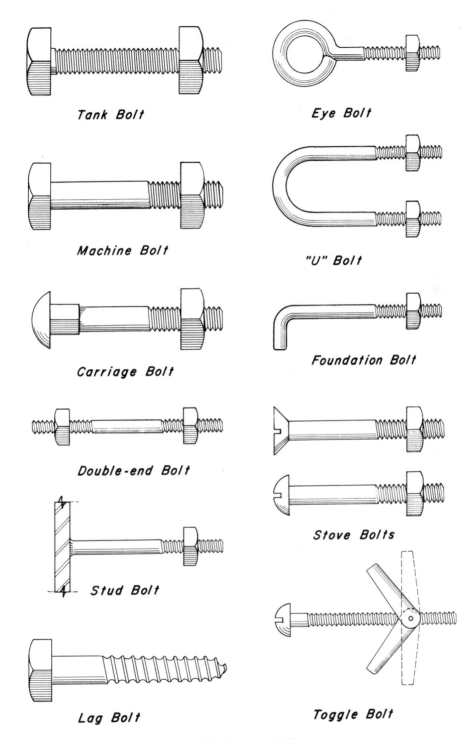

Tank Bolt

Eye Bolt

Machine Bolt

"U" Bolt

Carriage Bolt

Foundation Bolt

Double-end Bolt

Stud Bolt

Stove Bolts

Lag Bolt

Toggle Bolt

FIGURE B-10   Bolts

wood connections which has a rounded head and a short portion of square shank to prevent the bolt from turning when the nut is tightened.

DEAD BOLT—A door locking device, described under *Dead Bolt*.

DOUBLE END BOLT—See *Stud Bolt*.

ELEVATOR BOLT—A bolt having a large flat head and the top portion of the shank squared to reduce loosening.

EXPANSION BOLT—A bolt used to make an attachment to masonry or concrete. A special split sleeve is placed in a drilled hole and when the bolt is tightened into the sleeve, the sleeve expands and the wedging action or expansion against the sides holds it in place.

EYE BOLT—A bolt having a closed loop at one end instead of a head.

FOUNDATION BOLT—An *Anchor Bolt* embedded in a footing or floor slab, usually to fasten down a *Mudsill* and generally having a right angle bend at the bottom for better anchorage.

FRICTION BOLT—See *High-Strength Bolt*.

HIGH-STRENGTH BOLT—A type of hardened steel bolt used to join structural members by clamping action. A *Torque Wrench* is used to tighten the nut to a specified tension to effect the clamping force needed to resist slipping when a load is applied in *Shear*. The nut can also be tightened by the *Turn-of-the-Nut Method*.

"J" BOLT—Similar to a *Foundation Bolt* but having the hook at the end in the form of a semi-circle resembling the letter "J".

LAG BOLT—A large *Screw* having a square head requiring a wrench rather than a screwdriver.

MACHINE BOLT—A standard type of bolt widely used for connecting structural

steel and wood members, and as *Anchor Bolts*. They have either a square or hexagonal head and nut; most common diameters used are ½" to 1-¼".

MOLLY BOLT—A type of bolt similar in function to a *Toggle Bolt*.

STOVE BOLT—A type of bolt having a round or flat head which is turned with a screwdriver, and a threaded shank to receive a nut.

STUD BOLT—A bolt having each end threaded to receive a nut and *Washer*. It can also mean such a bolt welded at one end to a structural steel member to provide a means for attachment of another member.

TOGGLE BOLT—A bolt having collapsible, wing-like nut mechanisms such that it may be inserted through a hole and tightened when the nut mechanism unfolds, usually by spring action.

"U" BOLT—A bolt in the form of a "U" with threads on each end; usually used to make a connection to a round pipe.

UNFINISHED BOLT—Same as *Machine Bolt*.

Bond

1. The adherence of one material to another. This use of the term usually refers to either the *Adhesion* or holding together of *Masonry* units by means of *Grout* or *Mortar* or to the "grip" between *Concrete* and embedded *Reinforcing Bars*. The adhering force may also be by *Mechanical Bond*, as by *Anchor Ties*.

2. The pattern by which masonry units are laid up in a wall. For illustration see *Masonry Bond Patterns*.

3. A guarantee by a *Bonding Company* (also called *Surety*) that an obligation described in a *Contract* will be fulfilled regardless of any misfortune to the *Contractor*, such as his going bankrupt.

Common types of bonds are:

BID BOND—A guarantee by a *Bonding Company,* submitted with a *Bid,* that the contractor submitting the bid will enter into a contract if his bid is accepted.

COMPLETION BOND—Same as a *Contract Bond.*

CONTRACT BOND—A guarantee by a bonding company to an owner that a contract will be fulfilled regardless of a lack of performance on the part of the contractor.

LIEN BOND—A guarantee by a bonding company to the owner and to all persons who perform labor upon, or furnish materials to, a construction project, that payment for all such labor and materials will be made by the general contractor. It is also called PAYMENT BOND. If it is written in conjunction with a performance bond, it is called a performance and payment bond.

PAYMENT BOND—See *Lien Bond.*

PERFORMANCE BOND—Same as *Contract Bond.*

ROOFING BOND—A guarantee by a bonding company to an owner that a roof installed by a specific contractor will perform as represented for a stated period, usually either 10, 15, or 20 years, depending upon the specified quality and make-up of the roof.

Bond Beam In masonry construction, a specially reinforced horizontal strip of a wall designed to act as a *Beam,* usually to resist wind or *Seismic* forces.

Bond Beam Block See *Concrete Blocks.*

Bond Breaker A material (or coating) placed between concrete members which are cast in contact with each other to prevent their sticking when separated, such as between *Precast Concrete (Tilt-Up)* wall panels and the concrete slab upon which they are cast.

Bonderizing Trade name (Parker Rustproof Corporation, a division of Oxy-Metal Finishes) for a lightweight zinc phosphate coating process on *Iron, Steel, Zinc,* or *Aluminum* surfaces. Although the process is similar to *Parkerizing,* bonderizing is used primarily for paint adhesion.

Bonding Capacity The maximum dollar value of current construction contracts for which a bonding company will agree to issue a *Contract Bond* to a particular contractor. Bonding capacity is usually a good indication of a contractor's credit standing and financial condition.

Bonding Company See *Surety.*

Bonnet

1. The portion of a *Gate Valve* into which the disk rises when the valve is opened.

2. The upper enclosure of a *Furnace.*

Bonus A sum of money given in excess of a contractual amount in payment for special performance. A bonus —usually agreed upon in advance—may be given a contractor for doing a project in less time than that in which he had contracted to do it.

Booster Pump An auxillary pump used to increase the pressure in a piping system such as for a water supply for an *Automatic Fire Sprinkler System.*

Bore The inside diameter of a cylinder or pipe.

Boring See *Soil Boring.*

Boring Log In soil testing, a sequential description indicating soil classification, moisture content, color, etc., of soil retrieved from a *Soil Boring.* (See Figure B-11.)

SUMMARY — BORING NO. 3

DATE DRILLED: 4-4 —

DEPTH IN FEET	SAMPLES	SYMBOL	ELEVATION: 4.2' M.S.L.				DRIVE ENERGY FT.-KIPS/FT.	FIELD MOISTURE % DRY WEIGHT	DRY DENSITY LB./CU. FT.	SHEAR RESISTANCE KIPS/SQ. FT.
0 FILL	1		moist	medium loose & firm	lt.gr.brn. & — gray	SAND, fine	<1.3	16.8	98.2	–
						alt. SANDY SILT & silty sand streaks & occ. fine roots				
	2		very moist	soft	gray & black	SILTY CLAY & freq. fine roots	<1.6	80.7	53.9	–
5	3		wet	medium loose	light gray	SAND, fine, micaceous & occasional fine roots		29.5	93.0	0.37
	4			soft		SILTY CLAY w/org. stain & occ. fine roots	0.6	57.4	67.0	0.06
	5				dark gray	SANDY SILT, micaceous	<1.3	30.0	95.3	0.44
10				firm		SILTY CLAY				
	6					SANDY SILT, micaceous	<0.8	29.8	93.4	0.27
15	7			medium dense		SILTY	micaceous	2.4	28.3	95.4	0.65
				dense	gray	SAND	micaceous & 5-20% shells				
				& — stiff	dark gray	fine	& alt. sandy silt & clayey silt streaks & 5-20% shells				
20	8							1.6	17.1	116.4	0.81
	9		very	very stiff			slightly	5.0	14.8	120.3	1.33
				hard	brown	CLAY					
25	10		moist	very stiff			porous	5.5	19.9	110.0	1.27
				& — —							
	11		wet	stiff	gray ←silt		& freq. alkali streaks & fine concretions	5.0	34.7 27.8	87.8 96.0	1.39 top 1.00 btm.
30											

* Water at 4 feet – varies with tide

Courtesy Converse, Davis and Associates, Cal.

FIGURE B-11   *Sample Boring Log*

Borrow See *Import Fill*.

Borrow Pit A source for *Borrow*. See *Import Fill*.

Bottom-Dump Truck-Trailer See *Truck Types*.

Boulders Individual rocks or rock particles which, by general agreement, are over 12 inches in diameter.

Boundary Survey Establishing the perimeter (boundary) of a property by a surveyor, which is in accordance with the distances, *Bearings,* etc., contained in the *Land Description*. Permanent markers are usually set at all corners or changes in direction.

Bow See *Warping* (of *Lumber*).

Bowstring Truss See *Trusses*.

Box See *Electric Box*.

Box Girder (or Box Beam) A girder having more than one vertical *Web* such that an enclosed space is formed between the top and bottom *Flange* members. (See Figure B-12.)

Box Nail See *Nails*.

Box Union See *Union*.

Brace An inclined member adding rigidity to a structural system.

Braced Framing A seldom used term meaning the same as *Platform Framing*.

Bracket A seat-like device attached to a wall or column to support something, such as a bracket to support the end of a *Beam*.

Brad See *Nails*.

Branch Generally, a division or off-shoot from any distribution system, as a branch is from the trunk of a tree, as opposed to a *Main* which connects directly to a primary source and, using the same analogy, corresponds to the trunk of a tree.

Branch Vent See *Drainage and Vent Piping*.

Brashness

1. In *Wood* terminology, a condition causing some pieces of wood to be relatively low in shock resistance for the species and, when broken by bending, to fail abruptly without splintering and at comparatively small *Deflections*.

2. Brittleness or loss of flexibility of a

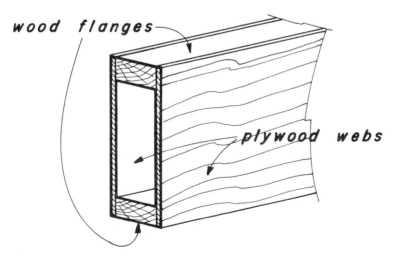

FIGURE B-12 Plywood Box Beam

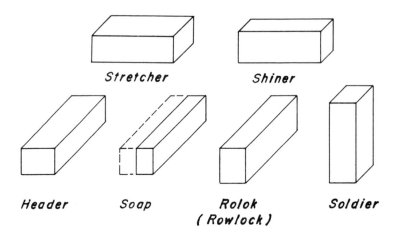

Stretcher   Shiner

Header   Soap   Rolok   Soldier
(Rowlock)

FIGURE B-13   Brick Placement Positions Within a Wall

Paint surface or *Glue* film which is generally due to certain drying conditions.

Brass An *Alloy* of *Copper* and *Zinc* used for pipe, fittings, hardware, and decorative accessories.

Brazed Joint See *Pipe Joints*.

Brazing Similar to *Soldering* but using a higher temperature as provided by an oxyacetylene torch and used usually for *Welding* of *Cast Iron,* or for conditions where the temperature of *Arc Welding* would be too high. The *filler metal,* usually brass, has a lower melting point than the material being joined and is melted and bonded into the space between the parts to be joined. Brazing is also called "hard soldering."

Breakdown See *Cost Breakdown*.

Breaker See *Circuit Breaker*.

Breathing In painting, a term used to indicate that a film has sufficient *Porosity* to permit it to expel ("breathe out") moisture vapor without any detrimental effect such as blistering, cracking, *Peeling*, etc.

Breezeway A covered passageway, usually between a house and garage.

Brick Joinery Terminology for brick placement within a wall is illustrated in Figure B-13.

Brick Solid masonry units of clay or shale formed into a rectangular prism while plastic, and burned or fired in a kiln. Bricks are manufactured in three grades: Grade SW (severe weather) where a high degree of freezing-thawing resistance is required; Grade MW (moderate weather) for moderate resistance to freezing; and NW (no weather) for use as back-up or interior masonry.

Types of brick include the following:

ACID RESISTING BRICK—Brick suitable for use in contact with chemicals, usually in conjunction with acid-resistant *Mortars*.

ADOBE BRICK—A roughly molded, unfired, clay brick dried in the sun.

ANGLE BRICK—Any brick shaped to an angle to fit a projecting corner.

ARCH BRICK—Wedge shaped brick for use in an *Arch*.

BUILDING BRICK—Brick for building purposes not especially treated for texture, or manufacturing tolerances (uniformity). This is a term replacing "common brick."

BULLNOSE BRICK—Brick having one rounded corner.

CLINKER BRICK—A very hard, burned brick whose shape is distorted or bloated due to incomplete vitrification.

COMMON BRICK—See *Standard Common Brick*.

CONCRETE BRICK—Building Brick made from concrete.

CORED BRICK—Brick having interior holes (*Cores*).

DRY-PRESS BRICK—Brick formed in molds under high pressures from relatively dry clay (5% to 7% moisture content).

ECONOMY BRICK—A brick larger than standard size.

FACE BRICK—Brick made especially for exterior use with special consideration of color, texture, and close manufacturing tolerances (uniformity).

FIRE BRICK—Brick made of refractory ceramic material which will resist high temperatures.

FLOOR BRICK—Smooth, dense brick, highly resistant to abrasion, used as finished floor surfaces.

GAGED BRICK—Brick which has been sorted or otherwise produced to accurate dimensions.

GLAZED BRICK—A brick prepared by fusing a ceramic glazing material on the surface such that one or several faces have a glassy surface.

JUMBO BRICK—A brick larger in size than the "standard." Some producers use this term to describe oversized brick of dimensions specifically manufactured by them.

NORMAN BRICK—A brick whose dimensions are $3\frac{1}{2}''$ × 2-3/16 × $11\frac{1}{2}''$.

ROMAN BRICK—A brick whose dimensions are $3\frac{1}{2}''$ × $1\frac{1}{2}''$ × $11\frac{1}{2}''$.

SALMON BRICK—Relatively soft, underburned brick, so named because of its color.

SCR BRICK—A solid brick of greater width than normal. Its nominal dimensions are $6''$ × 2-2/3$''$ × $12''$.

SEWER BRICK—Low absorption, abrasion resistant brick intended for use in drainage structures.

SOFT MUD BRICK—Brick produced by molding relatively wet clay. Often a hand process. When insides of molds are sanded to prevent sticking of clay, the product is "sand molded brick." When molds are wetted to prevent sticking, the product is "water struck" or "slick brick."

STANDARD COMMON BRICK—Same as *Building Brick* and usually $2\frac{1}{2}''$ × $3\frac{7}{8}''$ × $8\frac{1}{4}''$.

STIFF MUD BRICK—Brick produced by extruding a stiff but plastic clay through a die.

USED BRICK—Brick which has been reclaimed from a demolished building and reused in a new building, usually for rustic effect.

WIRE CUT BRICK—Brick formed by forcing plastic clay through a rectangular opening designed for the purpose of shaping clay into bars. Before burning, wires pressed through the plastic mass cut the bars into brick units.

Bridge Crane A moving overhead crane which travels along two parallel rails between which a framework (bridge) spans. One or more hoists are suspended on trolley units that travel on bridge runway rails and may be transversely movable along the "bridge" such that the entire floor area between the main rails is

Courtesy Dresser Crane, Hoist & Tower Division, Muskegon, Michigan

FIGURE B-14 Bridge Crane Installation

accessible to the crane. (See Figure B-14.)

Bridging Short sturctural members placed between adjacent *Beams, Joists, Rafters* or *Trusses* to stiffen them against sidewise or "twisting" motion. For illustration see *Wood Framing*.

Bright Shiny. It is sometimes used to mean a natural finish.

Brine In *Air Conditioning*, any liquid cooled by a *Refrigerant* and used for the transmission of heat. It is usually a salt or alcohol solution, thereby causing it to have a freezing point substantially below water so it can be piped at temperatures below that which would cause water to freeze. One example of the use of brine is to pipe chilled water to and from an *Evaporator* or *Chiller* to remote cooling *Coils,* as for a cold storage room.

Brinell Hardness A numerical measure to establish the relative hardness of various metals by measuring the diameter of an indentation made by a steel ball forced against the surface.

British Thermal Unit (B.T.U.) A unit of heat measurement. One B.T.U. is the amount of heat required to raise one pound of water one degree *Fahrenheit*. It is usually expressed relative to a period of time, for example, 1,000 B.T.U.'s per hour (abbreviated 1,000 BTUH). Some conversion factors are:

One *Ton of Refrigeration* = 12,000 BTUH

One *Horsepower hour* = 2,544 BTU

One *Kilowatt-Hour* = 3,413 BTU

One person seated at rest = 330 BTUH (average)

(= one 100 watt light bulb)

Broaching A means of making variously shaped holes in metal by removing bits of metal at a time by a reaming tool, until the desired shape is obtained. Keyholes in the cylinder of a lock are usually made by broaching.

Bronze An alloy of *Tin* and *Copper*. Bronze is primarily used for exposed architectural ornamentation, such as plaques or sign letters, due to its high corrosion resistance and attractive aging characteristics.

Bronzing A process for imparting a bronze color to metals by *Heat Treating* with chemicals or using metallically impregnated *Lacquers*.

Broom Finish A rough textured finish given concrete slabs to create a non-slip surface. It is usually done by sweeping the surface with a broom after final trowelling but before *Final Set*.

Brown and Sharpe Wire Gage Same as *American Standard Wire Gage*.

Brownstone This term usually refers to buildings or houses built, until about 1900, with a brown colored, quarried, sandstone which was laid up in mortar.

B.T.U. See *British Thermal Unit*.

Buck A rough frame around a door opening and onto which the finish *Jamb* and *Head* are attached.

Bucket In earthmoving, that part of an excavating machine which scoops up and holds earth.

Buckle The action taking place when a *Compression* member fails by bowing out sidewise.

Builder See *Contractor*.

Builder's Hardware See *Hardware*.

Builder's Risk Insurance See *Insurance*.

Building Code Published regulations and ordinances controlling design, consturction, quality of materials, use and occupancy, location and maintenance of buildings and structures within the area for which the Code has been legally adopted. Its purpose is to establish minimum standards for health, safety, property, and public welfare. In addition to specifying allowable design stresses for various materials, minimum design loads (such as wind and seismic), and minimum construction standards, they specify exiting and fire protection requirements. Most cities and counties use one of several widely adopted codes known as MODEL CODES. These are the *Uniform Building Code* published by the International Conference of Building Officials, the *Basic Building Code* by the Building Officials and Code Administrators International, Inc., the *National Building Code* prepared by the American Insurance Association, and the *Southern Standards Building Code* prepared by the Southern Building Code Congress. The groups or organizations that publish the Model Codes are nonprofit service organizations except for the American Insurance Association. They are owned and controlled by their member cities, counties, and states. Their aims and purposes are the publication, maintenance, and promotion of their Building Code and its related documents. They develop and promulgate uniformity in regulations pertaining to building construction.

Some larger cities prepare their own codes such as Los Angeles, Chicago, New York, and San Francisco. However, these too follow the guildelines of the Model Codes. Most codes are revised and updated every few years.

Building Contractor See *Contractor*.

Building Department That department of a government entity charged with establishing and enforcing building regulations within its area of jurisdiction. Such regulations usually consist of a *Building Code,* mechanical code, plumbing code, electrical code, and sometimes are supplemented by local ordinances. The head of the department is often entitled the Building Official and is charged with the administration and enforcement of the Building Code and other related ordinances.

Building Drain See *Drainage and Vent Piping*

Building Inspector An employee of the *Building Department* of a governmental entity whose function it is to inspect construction projects to verify compliance with the applicable *Building Code* and the building plans which have been reviewed and approved by the Building Department.

Building Paper A general term used to describe any heavy, water resistant paper such as is used under exterior siding. See also: *Kraft Paper, Rag Felt, Roofing Felt.*

Building Permit Authorization by the *Building Department* of a governmental entity to construct a new building, an *Addition,* or an *Alteration.* A building permit is issued after the plans have been submitted to the Building Department for checking for compliance with the *Building Code,* and any required corrections to the plans have been made and/or resolved. A fee is paid for both the plan checking and the Building Permit. These fees are to offset the cost of reviewing the plans and the field inspections after the permit has been issued.

Building Sewer See *Drainage and Vent Piping.*

Building Storm Drain See *Drainage and Vent Piping.*

Building Storm Sewer See *Drainage and Vent Piping.*

Building Trap See *Trap.*

Built-Up Beam A beam consisting of several members so connected that they work as one. For example, three 2 × 12's nailed side by side to act similarly to one 6 × 12.

Built-Up Roofing A roof covering consisting of several successive layers (each of which is called a *Ply*), usually of *Roofing Felt* with moppings of hot *Asphalt* between layers and topped by a mineral surfaced layer (*Cap Sheet*), or by *Aggregate*—such as gravel—embedded in a heavy coat of asphalt.

Bulb of Pressure In *Soil Mechanics,* the zone below *Footings* within which appreciable stresses are caused by the footing load. The bulb-shaped lines indicate "contours" of equal pressure.

Bulb Tee A slender "T" shaped steel member used in *Poured Gypsum Roof Decks* to span between *Purlins* to support gypsum planks over which the gypsum deck is poured. The "Bulb" at the top of the stem is to give better bond to the gypsum.

Bulkhead A wall used to resist pressure caused by rock or water, such as to separate land and water areas.

Bulking (of Soil) The increase in volume of *Soil* after it has been excavated, caused by the loosening of soil particles.

Bulldozer A *Tractor* having a blade, called a *Dozer,* mounted on the front such that earth can be pushed, or, by means of angling the blade, cast sideways. (See Figure B-15.)

Bullet Catch See *Catches.*

Courtesy Caterpillar Tractor Co.

FIGURE B-15   *Bulldozer*

Bullet Resistant Glass See *Glass*.

Bull-Headed Tee See *Pipe Fittings*.

Bull Nose

1. In *Plastering,* a metal *Angle* having a rounded edge used at exterior corners to form a true and even, but rounded, corner joint.

2. In *Tile* work, a rounded trim piece used at the perimeters.

Bunker A bin, usually elevated, used for storage of sand, rock or similar materials.

Burl In *Wood* and wood *Veneer,* a localized severe distortion of the grain, generally rounded in outline and swirling in appearance, from one-half inch to several inches in diameter and usually resulting from overgrowth of dead branch stubs.

Burlap A heavy, coarse cloth woven from *Jute* fibers.

Burr Uneven or jagged protrusions on metal edges where it has been cut by a saw or by a torch.

Bus Bar In electrical work, a bare bar-shaped or circular *Conductor* made of copper or aluminum which is used to carry a relatively large current, usually within an electric *Switchboard.*

Bus Duct (or Bus Way) *Bus Bars* which are enclosed in sheet metal *Ducts* along their length. (See Figure B-16.)

Bush Hammering An outdated means of texturing a *Concrete* surface by hammering. A similar effect is accomplished by *Sandblasting.*

Bushing In piping, a plug screwed into the end of a pipe which is then bored and tapped to receive a pipe of smaller diameter.

Butane A *Liquified Petroleum Gas* stored at high pressure which becomes a

Courtesy I-T-E Imperial Corp.

FIGURE B-16   Bus Duct Assembly

gas upon discharge. It is used for heating appliances and flame torches. It has the chemical formula C_4H_{10}, and is similar to *Propane* (C_3H_8).

Butterfly Reinforcement In *Plastering*, strips of *Metal Lath* reinforcement placed diagonally at corners of openings before plastering.

Butterfly Roof A roof consisting of two surfaces meeting at a *Valley* such that they resemble the wings of a butterfly.

Butterfly Valve See *Valves*.

Buttering In masonry work, the placing of *Mortar* on masonry units using a mason's *Trowel*.

Butt Hinges See *Hinge Types* (*Butt Hinge*).

Butt Joint A joint in which the end of one member connects to the end of another member. See also: *Welded Joints, Woodworking Joints*.

Button Punching In the installation of *Metal Decking*, a punch-like crimping at regular intervals along a male-female seam between adjacent decking panels for the purpose of locking them together.

Buttress A short pier-like section of masonry or concrete wall at right angles to a main wall. Its function is to give lateral stiffness to the main wall or to resist thrust from an *Arch*.

Butts See *Hinge Types* (*Butt Hinge*).

Butt Weld See *Welded Joints*.

Butyl Rubber A synthetic rubber used chiefly as a caulking *Sealant* for expansion joints in concrete work.

BX Cable Trade name for an electric *Cable* consisting of a spiral of interlocking steel within which are insulated *Conductors*. It differs from *Flexible Conduit* in that the latter is not manufactured with conductors inside.

C

Cabinet Hinge See *Hinge Types.*

Cabinetwork The making, usually in a woodworking shop, of high quality cabinets, shelving, furniture, and related wood items. It is sometimes referred to as CASEWORK with the latter more accurately meaning shelving and display cases such as for stores. See also: *Millwork.*

Cable

1. A *Rope* consisting of a number of single wires twisted or otherwise bound together.

2. In electrical work, one or more electric *Conductors* (wires) inside a protective, insulated covering. Cables are most often used for underground telephone or electric service and as electric *Feeders.* Cables are usually stranded and the term ''cable'' generally means an overall diameter larger than ½″ to ¾ inch.

Cable Roof Similar in structural concept to a suspension bridge, whereby a series of *Cables,* either parallel or crossing, support a roof structure or form an integral part of a roof structure. The resulting pulling forces at the ends of the cables must be resisted by other parts of the structure.

Cable Tray In electrical work, a continuous support, usually overhead, upon which electrical cables or *Conductors* are laid. It is of sheet metal with low sides, hence it resembles a tray in cross section.

Caisson

1. A watertight steel or concrete shell, usually circular, which can be lowered into the water or below *Ground Water* level for the purpose of constructing a foundation within its protection.

2. The term is also used, particularly in the west, to mean a *Belled Pier.*

Calcimine See *Paint.*

Calcium Carbonate The chief constituent of *Limestone* and *Marble,* having the mineral name calcite.

Calcium Chloride A white, lumpy, *Absorbent* material chiefly characterized by its ability to assimilate water. It is used on roadways to absorb dust (by becoming moist itself) and in *Concrete* to accelerate setting time and to prevent freezing.

Calculated System See *Automatic Fire Sprinkler System.*

Calculator Any high speed, electronic calculating machine. It differs from an ''adding machine'' in that it can multiply, divide, and on some models, extract square roots and perform other mathematical functions. It differs from a *Computer* in not having data storage and *Programming* capability.

Calder Coupling See *Pipe Joints.*

Caliche A type of *Soil,* usually associated with a desert environment, in which the individual grains are naturally cemented together by *Calcium Carbonate.*

California Bearing Ratio In *Soil Mechanics,* a test for determining the supporting capacity of the soil under a pavement. A measurement is made of the penetration of a cylindrical plunger into the sub-grade material and compared to the resistance to penetration of the same size plunger into a standardized sample.

California Red Fir See *Softwood.*

Camber A built-in, upward curvature of a *Beam* or *Girder* to compensate for its sag (*Deflection*).

Candle See *Illumination.*

Candlepower See *Illumination.*

Cane Bolt A locking device used at the base of a door or gate. A vertical cane-shaped *Bolt* mounted by a bracket to the door slides down into a socket in the floor.

Canopy Any roofed area projecting from a building such as to shelter the approach to a doorway or cover a storage area.

Cantilevered Beam See *Beam.*

Cant Strip In roofing, a triangularly shaped piece of wood or pre-molded fiber material placed in the corner between the roof and a *Parapet* wall. Its purpose is to eliminate what would otherwise be a sharp corner, thereby making placement of the roofing material easier and less subject to tearing. (See Figure C-1.)

Cap Block See *Concrete Blocks.*

Capillary Action The tendency of water (or other liquid) to rise in a small tube or network of spaces, such as in a *Soil,* caused by the surface tension of the liquid.

Cap Sheet On a roof, a heavy layer (*Ply*) of *Roofing Felt* used as the top sheet and usually having mineral granules embedded on the exposed surface to increase wearability.

Carbon Steel See *Steel.*

Carborundum Trade name (Carborundum Corporation) for an *Abrasive* material composed chiefly of silicon carbide (sand and carbon).

Carburizing A form of *Case Hardening.* It is a process of heating a ferrous (iron base) alloy in contact with a carbon containing solid, liquid, or gas to diffuse

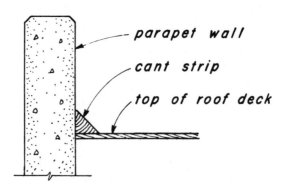

FIGURE C-1 Cant strip

carbon into the outer suface (case) of the alloy. It is usually followed by a *Heat Treating* cycle to produce a hardened "case."

Carpet Carpet is fabric constructed to serve as a soft floor covering and is generally fastened to an entire floor area as opposed to a rug which covers only a partial area, is not fastened down, and usually has bound edges. Carpet can be manufactured by the tufted, woven, or knitted method—the latter two generally connecting pile and backing yarns in one operation while the tufted method consists of machine insertion of tufts of pile into a pre-made backing. Tufted carpets are produced faster and are therefore less expensive. Carpet consists of pile yarns (made of pile fibers) which constitute the upper, wearing surface and backing yarns which interlock and hold the pile yarns in place. Pile fibers may be wool or synthetic—*Nylon, Acrylics, Polypropylene* olefin, and *Polyester*, the latter becoming the most widely used. INDOOR-OUTDOOR CARPETING utilizes synthetic pile fibers and a special backing which resists damage by moisture; generally this type of carpet is less durable.

When included in construction *Specifications,* carpeting is specified as to pile fiber, method of manufacture, backing, method of installation, fire and other resistance abilities, and other technical qualities.

Carport A covered automobile parking shelter differing from a garage in that it is open on two or more sides.

Carriage Bolts See *Bolts.*

Case Hardening A process of *Heat Treating* a ferrous alloy so that the surface layer is made substantially harder than the interior core.

Casein A material obtained from milk used in *Glues* and as a binder in *Paints.*

Casein Glue See *Glue.*

Casein Paint See *Paint.*

Casement Window See *Window Types.*

Casework See *Cabinetwork.*

Cash Flow On an income producing real estate investment, the excess of cash income over cash requirements for debt repayment, operating expenses, and overhead. It is also called "net spendable income."

Casing

1. In carpentry, a wood finish piece placed around the inside of a door or window opening.

2. In foundations, a lining on the inside of an excavated shaft to prevent the soil from caving in.

Casing Bead In plastering, a formed metal trim piece at the edge of a plastered area, such as at a door opening. It forms an even-edged stopping point for the plaster.

Casing Nail See *Nails.*

Casting The process of making an item by pouring a fluid mixture, such as *Concrete* or molten metal, into a mold and allowing it to harden.

Cast-in-Place Pile See *Piles.*

Cast Iron See *Iron.*

Cast Iron Pipe See *Pipe.*

Cast Stone Precast concrete components made with a high degree of quality and precision; for example, the sculpture-like decorative facing panels on buildings. It is also called ARTIFICIAL STONE.

Cat A trade mark (Caterpillar Tractor Co.); the term is used in reference to a track-type tractor or any of the earthmoving equipment manufactured by the Caterpillar Tractor Company.

Cat and Can A slang abbreviation referring to a Caterpillar track-type tractor used with a towed *Scraper*. See also: *Cat*.

Catch Basin A concrete, masonry, or cast iron box-like receptacle set into the ground in a parking lot or similar area into which run-off water is drained. They have underground outlet pipes and are generally covered with a metal grate.

Catches Fastening devices used on cabinet and cupboard doors, including the following basic types:

BULLET—A type of catch that depends upon "snap-in" action to hold the door closed. It is usually concealed in the cabinet or door frame and is made of metal, usually brass.

ELBOW—A closing device whereby two "L" shaped pieces of metal overlap each other. It requires one "L" to be moved to hook over the second, stationary one.

FRICTION—Operates by pressure, that is, one part is pressure caught against another. It is not a locking type catch.

MAGNETIC—Uses a magnet to hold the door closed.

Catenary The natural curve which a rope or chain will form when freely suspended between two points.

Cathodic Protection A means of reducing *Galvanic Corrosion* of iron and steel by introducing an opposing electric current.

Caulked Joint See *Pipe Joints*.

Caulking The weather resistant sealing of a *Joint* by filling the void or crack

with a permanently elastic material. See also: *Mastic, Putty, Sealant*.

Caulking Compound A plastic material used to seal joints or fill cracks around windows, chimneys, etc. It is semi- or slow-drying and generally comes in two types; a cartridge type for use with a special applicating gun or in a can for use with a putty knife.

Cavity Wall Masonry A type of masonry wall construction having two layers of *Brick*—each called a *Wythe*—separated by an air space (cavity) and using metal ties to connect and space them apart.

Cedar See *Softwood*.

Cedar Shingle See *Wood Shingles*.

Ceiling The overhead surface of a room or space; the surface opposite the floor. The term SUSPENDED CEILING means a ceiling hung by wires or wood framing, from a floor or roof above, and generally consisting of a grid of metal *Tees* into which *Acoustical Tile* panels are placed.

Ceiling Joists See *Joist*.

Ceiling Tile See *Acoustical Tile*.

Cell (Masonry) See *Core*.

Cellar A usable floor area of a building which is partly below ground level. Some building codes call it a cellar if it is mostly in the ground and a *Basement* if it is mostly out of the ground.

Cellular Metal Floor See *Metal Decking*.

Cellular Plastic See *Foamed Plastic*.

Cellulose In trees, cellulose is the chief structural component of the cell walls and makes up more than half the volume of *Wood*. Paper, made from *Pulpwood*, is almost all cellulose.

Cement The term "cement" generally means any material used to bind other materials together; however, it is most often used to describe any powdery material which, when mixed with water into a plastic state, will harden in place and in so doing act as a binder between other materials such as between particles of sand and gravel in *Concrete*. *Portland Cement* is the most widely used type of cement and unless a specific type of cement is indicated (see description below), the word "cement" nearly always means "Portland Cement." Types of cement include the following, some of which refer—as indicated—to types of Portland Cement:

AIR ENTRAINING CEMENT—See *Portland Cement*.

EXPANSIVE CEMENT—A type of cement which, when mixed with water, forms a paste that, during and after setting and hardening, significantly increases in volume to compensate for drying shrinkage. It uses essentially the same raw materials as *Portland Cement* with a slight change in formulation during manufacture.

GRAY CEMENT—Same as *Portland Cement*.

GYPSUM CEMENT—Refers to *Gypsum* used as the cementing agent, such as for *Gypsum Plaster* and other *Gypsum* products.

HIGH EARLY STRENGTH CEMENT—See *Portland Cement*.

HYDRAULIC CEMENT—A general term meaning any cement which will harden under water (as well as in the air) and principally includes *Portland Cement, Natural Cement,* and *Pozzolan Cement*.

KEENE'S CEMENT—A type of *Gypsum Cement* used as a plaster mix which is harder and more resistant to moisture than ordinary *Gypsum Plaster*.

MAGNESIA CEMENT—An insulating type of cement consisting of magnesium oxide mixed with *Asbestos* fibers and water and primarily used to coat pipes.

MAGNESITE—See *Magnesium Oxychloride Cement*.

MAGNESIUM OXYCHLORIDE CEMENT—A cementing material consisting of magnesium chloride and magnesium oxide mixed with fibers of mineral filler material, such as sawdust or sand, and which can be spread and finished similar to concrete. It is usually used as a resilient and sound deadening floor covering and is applied ½" to 1½" thick. It is only moderately weather resistant. It is also known as MAGNESITE, OXYCHLORIDE, or SOREL CEMENT.

MASONRY CEMENT—Also called *Mortar Cement* or *Plastic Cement,* it contains a *Plasticizer,* usually *Hydrated Lime,* such that it is more easily workable as *Mortar* mix. Its composition varies slightly form *Portland Cement*.

MORTAR CEMENT—Same as *Masonry Cement*.

NATURAL CEMENT—Historically, the first cement used. It is made by heating and pulverizing *Limestone*. Commercially, it is rarely used today.

NEAT CEMENT—A mixture of *Cement* and water only, without any sand or *Aggregate*.

OXYCHLORIDE CEMENT—See *Magnesium Oxychloride Cement*.

PLASTIC CEMENT—Same as *Masonry Cement*.

PORTLAND CEMENT—This is the type of *Cement* most widely used in construction and the one which is usually meant when the term "cement" is used. It derives its name from its similarity to a stone quarried on the Isle of Portland, England. It is a type of *Hydraulic Cement* (can harden under water) and when hardened is resistant to moisture as opposed, for example, to *Gypsum Cement*. The chief raw materials used in its manufacture are *Limestone* (or similar materials rich in *Lime*) and *Silica* (sand). Other ingredients are added to provide specific qualities. Portland cement (also called *Gray Cement* due to its color) is produced in five types, as described below, plus several special types.

TYPE I—The usual type used for general construction and the one which is usually provided unless another type is specified.

TYPE II—A modification of *Type I* which, by chemical formulation, gives off less heat during *Hydration,* consequently a lessened tendency for shrinkage cracks in bulkier strictures. It is a moderatley sulfate-resistant cement.

TYPE III—A formulation to provide earlier strength attainment, also called *High Early Strength Cement,* it gives in three days the strength attained by *Type I* cement in seven days, and a strength in seven days equivalent to 28 days for Type I.

TYPE IV—A slow setting cement with a very low *Heat of Hydration*; it is principally used for massive structures, such as dams, where the heat build-up during hydration could adversely affect the sturcture.

TYPE V—A high sulfate-resistant cement used for exposure to alkaline soils or sea water.

WHITE CEMENT—Portland cement (usually Type I or Type III) containing negligible amounts of iron and manganese oxide which produce a gray color. When used with a light colored *Aggregate,* a "whiter" effect is achieved than using regular cement.

POZZOLAN CEMENT—A material consisting of *Silica* and *Alumina* which is not in itself cementitious, but will react with *Slaked Lime* to form cementing properties. Wide varieties of formulated pozzolans have various effects as an additive to *Concrete* mixes, such as improved workability, less heat generation, reduced shrinkage, and economy (since it can replace some of the cement in a mix). It is marketed as an *Admixture* under several different trade names.

SOREL CEMENT—See *Magnesium Oxychloride Cement.*

SULFUR CEMENT—A mixture of molten sulfur with mineral fillers such as lead particles. It is commonly known as *Lead and Sulfur* and is used like a *Grout,* usually to embed anchors, pipe railings, etc., into holes in concrete and masonry.

Cement Asbestos Board Sheet-like boards made from a mixture of Portland cement, water, and *Asbestos* fibers which are pressure formed into flat or corrugated sheets. They are weather and corrosion resistant and non-combustible. Installation is by nailing or screwing to support framing. Flat sheets are available in thicknesses of ⅛, ¼, and ½ inches; the standard sheet size is 4' × 8'. Corrugated cement asbestos board comes in sheets 42″ wide with 10 corrugations per sheet and lengths to 12', in 6″ increments.

Cement Base Paint See *Paint.*

Cement Coated Nail See *Nails.*

Cement Finishing The process of floating, trowelling, and otherwise leveling and "finishing" of a *concrete* surface. (See Figure C-2.)

Cement Gun See *Gunite.*

Cement Mason This trade classification includes cement finishers, plasters, and related trades represented by the Operative Plasterers and Cement Masons International Association of the United

FIGURE C-2 *Cement Finishing Machine*

States and Canada, affiliated with the AFL-CIO.

Cement Plaster A *Plaster* mix using *Portland Cement* rather than *Gypsum Cement*. When used for exterior work it is called *Stucco*. It is used where weather and moisture would be harmful to *Gypsum Plaster*.

Cement Tamper A *Cement Finishing* tool used on concrete slabs to tamp down the coarser aggregate such that a smoother surface finish can be obtained.

Centering In masonry work, the *Falsework* (supports) over which an *Arch* is formed.

Center of Gravity A single imaginary point about which an object or shape would "balance" if freely suspended.

Centigrade Scale See *Temperature Scales*.

Centimeter Metric system unit of length measurement. 100 centimeters = 1 meter. One inch = 2.54 centimeters. It is abbreviated cm.

Central Air Conditioning An assemblage of heating and cooling (see also *Refrigeration Equipment*) equipment at one location serving remote rooms. Such assembly of equipment can be in one complete unit, called PACKAGED AIR CONDITIONING, or separate units such as a *Compressor, Condenser, Coils,* and *Evaporator*, called UNITARY AIR CONDITIONING. Where such components are located apart, such as the compressor and condensor outdoors or on a roof and the evaporator with the building, connected by *Refrigerant* lines, it is called a SPLIT SYSTEM. Generally, a central conditioning provides heating and cooling to remote spaces by one of three methods: a SINGLE DUCT SYSTEM, whereby the heated or cooled air, as controlled by a

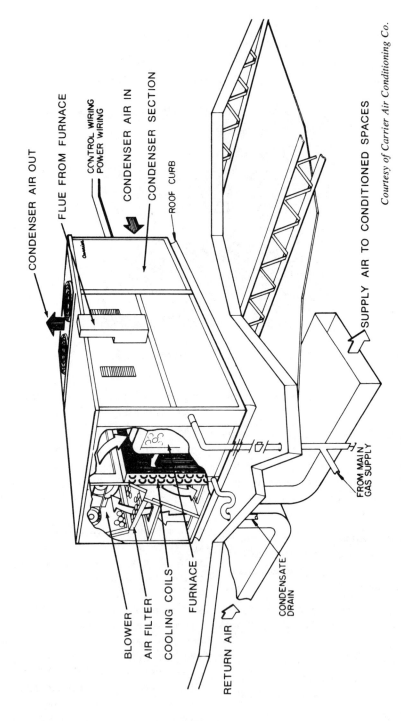

Courtesy of Carrier Air Conditioning Co.

CONDENSER AIR OUT

FLUE FROM FURNACE

CONTROL WIRING
POWER WIRING

CONDENSER AIR IN

CONDENSER SECTION

ROOF CURB

SUPPLY AIR TO CONDITIONED SPACES

FROM MAIN
GAS SUPPLY

BLOWER

AIR FILTER

COOLING COILS

FURNACE

RETURN AIR

CONDENSATE
DRAIN

*FIGURE C-3   Central Air Conditioning Roof Mounted "Packaged" Unit with Gas
Heating and Electric Refrigeration*

Thermostat, is forced by a fan through the *Ductwork* to the room to be served, discharging in *Diffusers*, and by a separate opening is returned by a *Return Air Duct* for a continuous recirculating cycle; a DOUBLE DUCT SYSTEM whereby heated and cooled air is supplied in separate ducts and at each room a *Mixing Box* controls their discharge; or by the use of *Fan Coil Units* at each room, where air from the ducts passes through the coils within the units, the coils contain hot or cold water supplied from the central equipment. (See Figure C-3.)

Centrifugal Compressor See *Compressor*.

Centrifugal Fan See *Fan*.

Centroid Same as *Center of Gravity*.

Ceramic A term descriptive of the materials and processes whereby moist *Clay* is molded and fired to form end products.

Ceramic Tile *Clay* tiles used over floor and wall surfaces to give an attractive, durable, easily cleaned, and relatively waterproof finish. They are machine made by pressing and firing clay and are available either glazed (gloss finish) or unglazed. Most common glazed tile sizes are 4½″ square and 4″ × 6″; thickness is usually 5/16 inch (see Figure C-4). They are set against a surface by

Courtesy of American Olean Tile Co.

FIGURE C-4   *Ceramic Wall Tile*

FIGURE C-5   Ceramic Mosaic Floor Tile

one of three methods: on a *Cement Mortar* setting bed; the "thin set" Portland cement method which is a cementitious mixture of cement, sand, and water; or by the use of an organic adhesive. Smaller size, usually unglazed, tiles (1″ square is most common) are called CERAMIC MOSAIC TILE (see Figure C-5). ·Such tiles are usually factory mounted face down on paper such that a sheet of tile can be placed against a surface upon which adhesive or cement mortar has been applied and allowed to set, after which the paper is peeled off the finished surface.

Certificate of Occupancy   A term used in *Building Codes* meaning an approval issued by the Building Department of a governmental entity to permit a building to be occupied for a particular purpose. It is issued at the completion of construction when the Building Department is satisfied that the structure complies with the Building Code and is safe to occupy.

Cesspool   A temporary underground storage chamber for *Sewage,* usually round and masonry lined with relatively porous joints so the liquid portion of the sewage can seep into the surrounding soil. A cesspool usually is from 3 ft. to 5 ft. in diameter and from 10 ft. to 25 ft. deep depending upon the soil porosity. It differs from *Seepage Pit* in that sewage is emptied directly into a cesspool, whereas a *Septic Tank* retains solids and only *Effluent* passes to the seepage pit.

Chaining In surveying, measuring horizontal distances by means of a steel tape, usually 100 ft. long. For measuring longer distances, pins or other markers are set at 100 ft. intervals by the man at the lead end of the chain, from which the rear man can hold the 100 ft. mark as they progress. In measuring distances over sloped ground the tape is held level so that only the horizontal measure is made; a *Plumb Bob* is held from the tape at one or both ends for precise positioning over a point on the surface.

Chain Link Fencing A diamond patterned interlocked system of wires stretched between pipe, wood, or other supports. The usual heights are 3, 3½ (popular residential size), 4, 5, 6, and up to 12 feet for commercial fencing. Wire size can be 6, 9, or 11 gage and is galvanized after fabrication—some are vinyl clad.

Chain Saw See *Saws*.

Chain Wrench A plumber's wrench similar in operating principle to a *Pipe Wrench* but which uses a sprocket-like chain to grip the pipe. It is used to turn large diameter pipes.

Chair Rail A wooden protective strip placed horizontally along a wall to prevent damage from chairs striking the wall.

Chairs In *Concrete* construction, saddle-like bent wire devices used to support reinforcing bars off the bottom of a form so concrete can get under the bars.

Chalkboards Chalkboards are manufactured in three basic ways: pigmented monolithic synthetic stone, enameled steel, and hardboard with a baked on finish. Standard sizes are 4' × 6' and 4' × 8', ¼" thick, and standard available colors are green and charcoal gray. BLACKBOARDS consisted of quarried slate and are no longer manufactured.

Chalking In painting, the breakdown of the surface of a paint film to a loose powder.

Chamfer The bevelling of a corner. In *Concrete* construction, such bevelling prevents a jagged, rough edge and is accomplished by putting a triangularly shaped wood strip into the corner of the *Formwork*.

Change of Air See *Air Changes*.

Change Order A directive to the *Contractor* issued during the course of construction, authorizing him to deviate from the *Drawings* and/or *Specifications* in order to effect certain changes.

Channel A metal member having, in cross section, a shape resembling a channel. They are available in a range of standardized sizes in extruded *Aluminum*, formed *Sheet Metal*, and roll-formed structural *Steel*. For illustration see *Structural Shapes*.

Channel Block See *Concrete Blocks*.

Charcoal Charcoal is made by burning wood with not enough air for complete combustion but enough to drive off fluid, sap, etc., and leave almost pure carbon, weighing about one-third of its original weight.

Charcoal Filters See *Activated Carbon*.

Charge In *Air Conditioning*, and *Refrigeration*, the amount of *Refrigerant* in a system; also, to put in the refrigerant charge.

Chase A groove or slot, cut or built into a concrete or masonry wall to accomodate pipes, electric *Conduit,* etc.

Check In *Wood,* a lengthwise separation of fibers causing longitudinal splits on the surface or a split on the end of a piece. A check usually extends across the annual rings and commonly is caused by stresses produced in *Seasoning*.

Checkered Plate Steel or aluminum plate having raised ribs to provide a non-slip surface; used primarily for industrial decking.

Check Valve See *Valves.*

Cherry See *Hardwood.*

Cherry Picker Slang term for a relatively small crane mounted on a truck.

Chicken Wire Slang term for a light gage *Stucco Mesh* of a type having hexagon shaped openings 1" wide and using 20 gage wire.

Chiller In *Air Conditioning,* a *Refrigeration* machine used to cool water. Such "chilled" water is then piped to cooling *Coils* where air to be cooled is blown over the coils.

Chimney A vertical shaft to conduct gaseous products of combustion to the outside air, as from a *Fireplace, Furnace,* or *Boiler.* It usually encloses a *Flue* (a term which is sometimes used synonymously with chimney).

Chipboard Same as *Particleboard.*

Chord (of a Circle) A straight line connecting two points on the *Circumference* of a circle. For illustration see *Circle Nomenclature.*

Chord (of a Truss) See *Trusses.*

Chromium An important alloying element in steel production; it provides

corrosion resistance and increases strength at high temperatures. It is obtained from the ore chromite.

Chromium Plating A type of metal plating used to prevent corrosion (by excluding oxygen) and to provide a hardened and decorative surface. It is accomplished by *Electroplating* from a chromic acid solution. Screws, for example, may be chromium plated.

Cinder Blocks See *Concrete Blocks.*

Cinders *Slag* or slag-like residue, most often from a *Blast Furnace,* and frequently used as an *Aggregate,* as in "cinder blocks."

Circle Nomenclature For terminology used for various parts of a circle, see Figure C-6.

Circuit One complete run of a set of electric *Conductors* from a power source to various electrical devices (appliances, lights, etc.) and back to the same power source. Generally, electric circuits originate from a distributing electric *Panelboard* where they are connected to an *Overcurrent Protective Device (Fuse* or *Circuit Breaker).* Required wire sizes and circuit load capacities are established by *Codes,* of which the most frequently used is the National Electrical Code. The term BRANCH CIRCUIT is applied to a circuit beyond a final overcurrent protective device (usually at a panelboard) and to the device being served. The most widely used type of circuit for general wiring is the PARALLEL CIRCUIT where the devices are connected between two conductors such that current is divided—if one device has an open circuit current will still flow to the others—as opposed to a SERIES CIRCUIT in which the devices are connected one after another with the same current going through each—if the current is broken, all current stops flowing.

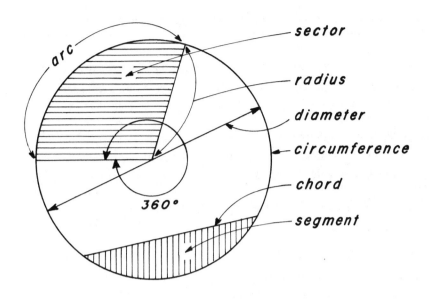

FIGURE C-6   *Circle Nomenclature*

Circuit Breaker A mechanical *Overcurrent Protective Device,* placed in an electric *Circuit.* It is designed to open the circuit automatically on a predetermined overload of current and can be reset when the overload condition has been corrected. See also: *Fuse.*

Circular Mil A unit of measurement of the area of a wire. One circular mil is the area of a wire .001 inches (one thousandth of an inch) in diameter. The abbreviation for 1,000 circular mils is MCM.

Circular Sander See *Sanders.*

Circular Saw See *Saws.*

Circumference The distance around a circle; the perimeter of a circle. For illustration see *Circle Nomenclature.*

Cistern A tank or similar receptacle, usually underground, used to store rain water.

Civil Engineer See *Engineers.*

Clamshell A jaw-like excavating bucket used with a *Truck Crane* or *Crawler Crane.* It has cables, independent from the lifting cables, to open and close the jaws and is primarily used for handling loose material.

Clapboard Horizontal wood siding consisting of boards thicker at their bottom edges and laid in an overlapping manner.

Class (of Fires) See *Fire.*

Claw Hammer Standard carpenter's hammer having a "claw" end for pulling nails.

Clay Very fine grained (under .002 mm in diameter) earthy material (the particles having a flat, plate-like shape) which is powdery when dry but plastic and moldable when moist. When molded and burned in a kiln it becomes hard and weather resistant. It is used for a variety of building materials such as *Bricks, Tiles,* and *Pipe.* Generally the base of clay is hydrous aluminum silicate, but clays may differ greatly both chemically and mineralogically, which accounts for differences in physical properties.

Clay Soil *Soil* in which *Clay* is the principal constituent. Clay soil expands when wet and shrinks upon drying, but properties can vary widely with locality.

Cleanout In plumbing, a pipe fitting having a removable threaded plug to provide access to a run of pipe or a plumber's snake for cleaning or un-clogging.

Cleanout Holes In *Masonry* wall construction, openings left in the bottom course, before *Grouting*, to remove any debris accumulated within the *Cores* or other empty spaces within the wall.

Clean Room A room used for the assembly of precision parts, or similar uses, where airborne dust and lint particles must be excluded. Such a room has very smooth wall and ceiling surfaces so particles cannot collect and may, in addition, have a laminar flow of air through the room such that it can be filtered and recirculated to continually "clean" the room.

Clear and Grub An earthwork term meaning to clear land and remove roots (GRUB) such as preparatory to constructing a building, roadway, or parking lot.

Clear Span An open area within a building without intermediate columns.

Cleat In *Plywood* construction, a square piece of wood placed at intervals underneath an unsupported joint to prevent differential movement across the joint. The are usually held in place by wood screws.

Clerestory (Also but not preferred: "clearstory") A wall with windows which is between two roofs at different levels.

Clevis A metal fitting to connect a round rod to a plate. (See Figure C-7.)

Climbing Crane A type of crane used on the construction of tall buildings. It is supported on the building framework and is continually raised (climbs) as the building is erected.

Clinch To bend over the pointed end of a nail.

Clinch Nail See *Nails*.

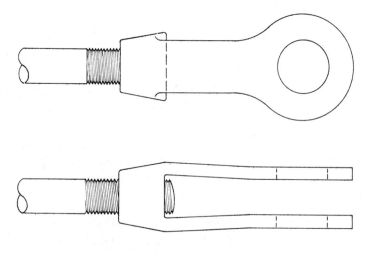

FIGURE C-7 Clevis

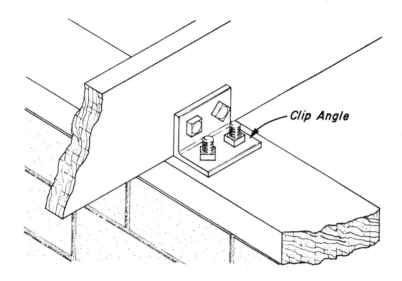

FIGURE C-8   *Clip Angle*

Clip Angle A short piece of metal angle used to connect two members together. It usually has a hole (or holes) in each leg for bolting to the members being joined. (See Figure C-8.)

Closed Shop A company employing only union members.

Closer See *Door Closer.*

Closet Bolt A *Bolt* used for fastening the bowl of a *Water Closet* to the floor.

Clout Nail See *Clinch Nails.*

Cluster Housing A high density type of housing development where units are grouped (clustered) and separated by open areas such as parks, green belts, and similar common-use space.

CO₂ Fire Extinguisher See *Fire Extinguishers.*

Coal Tar A black, gummy liquid which is a by-product of the manufacture of coke. It is used to make roofing *Pitch* and *Creosote.*

Coarse Aggregate See *Aggregate.*

Coat A thickness of *Paint, Plaster,* or other material applied in a single application.

Coatings Thin films of material applied to the surface of other materials for a specific purpose such as corrosion prevention, color (paints), prevention of electrolysis, or waterproofing.

Cobbles A classification of rocks or rock particles ranging in diameter from 3 to 12 inches as opposed to *Boulders* (larger) and *Gravel* (smaller).

Cock See *Valves.*

Code Any published regulatory material. See also: *Building Code.*

Coefficient In mathematics, a number representing a fixed value which is used as a multiplying factor before another quantity which is usually a variable or unknown. For example, the circumference of a circle = 3.14 × D, where 3.14 is the coefficient and "D" represents the diameter.

Coefficient of Expansion The

amount by which a given material will lengthen or shorten with a change in temperature. It is expressed in inches per inch per degree *Fahrenheit*. For steel the coefficient is .000006 (which is equivalent to a ½ inch change in a 100 foot length of steel for a 35° temperature change). (.000006 × 35° × 100 ft. × 12 inches per foot = 0.50 inches).

Coefficient of Friction A measure of the resistance to sliding of one material resting on another. It is the ratio of the sliding ("pushing") force required, to the weight of the material being "pushed." For example, the coefficient of friction for wood on a *Concrete* surface is about 0.40.

Coefficient of Heat Transmission Also called a "U-Value," it is a measure of the insulating value of building materials, expressed in B.T.U.'s per hour per square foot per degree Fahrenheit temperature difference between each side. The lower the U-Value the better the insulation.

Coefficient of Permeability See *Permeability*.

Cofar Floors A trade name (Granco Steel), being a contraction of "combined Form and Reinforcement." It is a means of using a corrugated steel sheet as the form for a *Concrete Slab*, which is not removed after the concrete has been placed so that it bonds to the concrete and acts as tension reinforcing.

Cofferdam A temporary dam designed to keep water and soil out of a below ground construction site.

Cohesion In *Soil Mechanics*, the forces causing soil particles to adhere to each other. The binding forces are caused by molecular attraction and the surface tension of the water in the spaces between the particles.

Coil In *Air Conditioning*, a cooling or heating device made of pipe or tubing through which hot or cold fluids circulate to heat or cool air blown past the coils.

Cold Forming Bending or otherwise shaping structural members without heating. The term usually applies to *Sheet Metal*. See also: *Cold Working*.

Cold Joint In concrete construction, a joint between portions of the work placed at different times. See also: *Contraction Joint*.

Cold Mix See *Asphaltic Concrete Paving*.

Cold Working Strengthening of *Steel* by rolling, pressing, pounding, drawing, etc.—at ordinary temperature—causing a slippage of the internal grains and their re-alignment into a stronger oriented plane. It is also called WORK HARDENING. The effects of cold working can be removed by heat (*Annealing*).

Collar Beam In carpentry, a horizontal framing member tying rafters together below the *Ridge* but above the *Plate* line. (See Figure C-9.)

Collector In electrical work, a spring loaded, moving contact rubbing or rolling along a fixed supply *Bus Bar*. It is used to supply current to a moving, electrically operated machine, such as a *Bridge Crane*.

Collusion A secret agreement between two or more parties to take unfair advantage of a party with whom they anticipate doing business.

Color A term referring to all the colors of the spectrum as well as black and white. Terms relating to color include HUE, which designates the name of a color such as red, blue, etc.; SHADE, which is a term describing the difference of lightness between colors but frequently

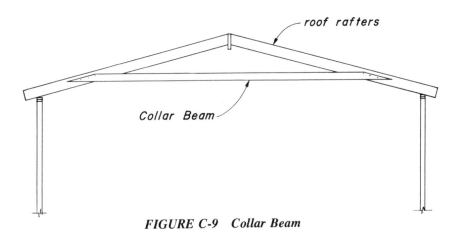

roof rafters

Collar Beam

FIGURE C-9   **Collar Beam**

used only to indicate a darker color, and TINT meaning lightening a color by adding white.

Color Finish (Plastering) See *Plaster Finishes*.

Column A slender vertical structural member used to support loads parallel to its length. The design of columns primarily involves their resistance to *Buckling* which, generally, is a function of their length and *Radius of Gyration*. Resistance to such buckling varies inversely as the square of the ratio of the laterally unsupported length of the column to its radius of gyration. A column is said to be *Concentrically Loaded* if an *Axial* load is applied as its *Neutral Axis*, and *Eccentrically Loaded* if the load is applied at a distance from the neutral axis (called *Eccentricity*). Eccentric loads cause bending in the column. Bending can also be imposed upon a column by loads applied perpendicular to their length, such as from wind on a column at the exterior of a building.

Combination Sewer See *Sewer*.

Combined Stress See *Stress*.

Combustible Generally, any material or structure which can burn, but since any material will "burn" if raised to a sufficiently high temperature, combustible is usually defined as a material which will ignite below a specified temperature. In the *Uniform Building Code* this temperature is 1382° Fahrenheit. Wood ignites at about 500°.

Combustion The chemical union of a *Combustible* material with oxygen with the resulting generation of light and heat. In the combustion of *Fuel*, each particle of fuel, previously heated to kindling temperature, is brought into contact with the proper amount of oxygen causing the fuel to oxidize as completely as possible.

Combustion Air Air required for the burning (combustion) of fuel.

Commercial Group Name In *Lumber* terminology, an identification of a group of various wood species which may have varying characteristics but for marketing purposes are designated under the group name. For example, the commercial group name "Southern Pine" includes the species Shortleaf Pine, Loblolly Pine, Longleaf Pine, and Slash Pine.

Commercial Projected Window See *Window Types*.

Commercial Standards See *Product Standard*.

Common Bond See *Masonry Bond Patterns*.

Common Brick See *Bricks*.

Common Labor See *Labor*.

Common Nail See *Nails*.

Common Wall See *Party Walls*.

Compaction Artificial densification of soil. By depositing *Fill* soil in layers, controlling its moisture content, and compacting by means such as a *Sheepsfoot Roller,* the soil can be made safe to support a building.

Compaction Test A test for determining the degree of *Compaction* of filled soil. The most commonly used field test is the SAND DISPLACEMENT METHOD whereby the soil removed from a small test hole is weighed, its volume determined, and its density then compared to the laboratory established maximum density for that particular soil.

Compactors In earthmoving, any of several types of equipment used for *Compaction.* Most common types are *Sheepsfoot Rollers* which are pulled by *Tractors,* and self-propelled rubber tire types having many tires and large built-in water tanks to provide additional weight. Some models have tires that are so mounted that they wobble when moving to assist compaction. There is also the self-propelled *Roller,* used to roll or smooth an already levelled grade prior to paving.

Compass A drawing instrument for making circles.

Compass Saw See *Saws*.

Composite Construction A construction technique whereby two different but connected materials act together ("compositely") to resist loads. A concrete slab on a steel beam will act compositely if stud bolts placed on the top flange of the beam project into the slab to effect bonding between the two materials. Concrete poured over *Metal Decking* will act compositely if the concrete bonds properly to the metal. A wooden beam with steel side plates will act compositely if the two materials are firmly bolted together.

Composition Roofing A general term for any roofing consisting of *Asphalt* base materials.

Composition Shingles See *Shingles*.

Compression A structural member is said to be in compression if the forces acting upon it tend to cause shortening of the particles in the member.

Compression Joint See *Pipe Joints*.

Compressive Stress See *Stress*.

Compressor In *Air Conditioning,* a machine in the *Refrigeration Cycle* in which vapor at a low temperature is taken from the *Evaporator,* compressed to a high pressure and then discharged to the *Condenser.* The most common types of compressors are:

CENTRIFUGAL COMPRESSOR—A nonpositive displacement compressor which depends, at least in part, on centrifugal effect for pressure rise.

HERMETIC COMPRESSOR—A compressor in which the motor and the compressor are an integral unit and the *Refrigerant* is introduced into the motor windings.

RECIPROCATING COMPRESSOR —A positive displacement compressor with a piston or pistons moving in a straight line but alternately in opposite directions.

ROTARY COMPRESSOR—One in which compression is attained in a cylinder by rotation of a positive displacement member.

SEALED REFRIGERANT COMPRESSOR—A mechanical compressor combination consisting of a compressor and a motor, both of which are sealed in the same housing, with no external shaft or shaft seals, the motor operating in the refrigerant atmosphere. It generally is smaller than a *Hermetic Compressor*.

Computer An essentially electronic device for high speed calcualtion. It differs from a *Calculator* by its capability to store data into a memory and by its ability to retrieve and mathematically manipulate such data in accordance with a devised series of steps, called a PROGRAM, such that a planned computational result is obtained.

Computer Program A sequential list of instructions that directs the operation of a computer. Such instructions can be stored on magnetic tape, punched cards, etc. Computer programs are also called *Software*.

Concave Curving inward; the inside of a bowl is concave. It is the opposite of *Convex*.

Concentrated Load See *Load*.

Concrete A mixture of *Portland Cement, Sand,* and *Gravel* to which water is added which reacts chemically with the *Cement* by a process called *Hydration*. The resultant hardening of the water-cement paste binds the sand and gravel into a solid mass by coating and filling the spaces between particles. The sand and gravel are called *Aggregate* and make up 60% to 80% of the total volume. Strength of concrete is related to the *Water-Cement Ratio*; the lower the water content per volume of cement the higher the strength. Concrete weighs about 150 pounds per cubic foot, hardens to *Initial Set* in about 45 minutes after adding water, and has a *Compressive* strength of usually between 2,000 and 4,000 lbs. per square inch. It

has very low value in *Tension*. When *Reinforcing Bars* are embedded in the work (to increase tension resisting capability) it is called *Reinforced Concrete*. *Admixtures* are sometimes added to give certain qualities.

Concrete Batch Plant A centralized plant for mixing *Concrete*. Weighed proportions of *Portland Cement,* varying sizes of *Aggregate* and water, and sometimes *Admixtures,* are dispensed into a mixer, then discharged into *Ready-Mix Concrete* trucks for delivery to the job site. (See Figure C-10.)

Concrete Blocks Concrete blocks are *Masonry* units which are press-molded by machine from a *Cement-Sand-Aggregate*-water mix (Figure C-11). Concrete blocks are either standard weight (using *Gravel* as the aggregate), or LIGHT-WEIGHT BLOCKS, which most often use expanded shale or expanded slag as the aggregate. CINDER BLOCK is now a slang term, since blocks are no longer made with cinders as the aggregate. Concrete blocks are generally rectangular in shape, are laid up with *Mortar* between them, and have inner openings called *Cores* both for the purpose of decreasing the weight of the block and to provide a space for *Grouting* and/or *Reinforcing Bars* (called *Grouted Reinforced Masonry*). Standard nominal block widths (perpendicular to the face of the wall) are 4″, 6″, 8″, 10″, 12″ and 16″. Heights are 4″, 6″, 8″ and lengths are 16″ or 24″. Actual dimensions are ⅜″ or ½″ less than nominal dimensions, such that with *Mortar* in the joints a modular block spacing results. Block sizes are stated by three dimensions, as 8 × 6 × 16, meaning 8″ wide by 6″ high by 16″ long. Types of concrete blocks include:

BOND BEAM BLOCK—A block with the transverse webs depressed at the top to

Courtesy of Conrock Co.

FIGURE C-10   Concrete Batch Plant

permit a channel-like path for horizontal reinforcing bars within the wall.

CAP BLOCK—A solid (without cores) block to close and "cap" the top of a wall.

CHANNEL BLOCK—Same as *Bond Beam Block*.

CINDER BLOCK—See general description above.

DOUBLE OPEN END BLOCK—See *"H" Block*.

"H" BLOCK—A block open at each end, also called a *Double Open End Block*.

JAMB BLOCK—Blocks having a vertical groove in one end for attachment of a window assembly.

LIGHTWEIGHT BLOCK—See general description above.

LINTEL BLOCK—Similar to a *Bond Beam Block* but without the cores running through the block, so the solid bottom

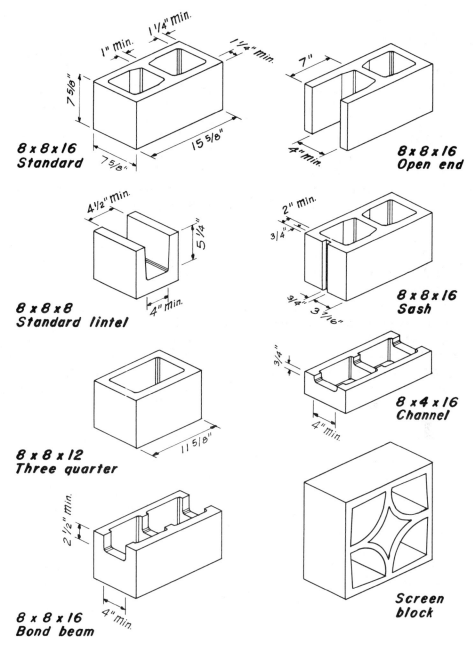

By convention, dimensions for concrete block are stated in the following order: width x height x length.

FIGURE C-11   Concrete Blocks

can be used exposed, such as over door and window openings.

OFFSET FACE BLOCK—Block having part of the front face recessed (off-set) for a decorative purpose.

OPEN END BLOCK—A block having one open end to facilitate placing it around reinforcing bars.

PATTERN BLOCK—Blocks having offset or recessed patterns on the front surface for a decorative effect.

PILASTER BLOCK—Blocks of various shapes to form *Pilasters* within the wall.

SASH BLOCK—Same as *Jamb Block*.

SCORED BLOCK—Blocks manufactured with vertical grooves resembling additional vertical joints.

SCREEN BLOCK—Blocks having variously shaped holes in a decorative pattern which, when laid in a wall, gives a screen-like effect.

SHADOW BLOCK—Same as *Pattern Block*.

SHEAR FACE BLOCK—Same as *Split Face Block*.

SILL BLOCK—Block without cores having a sloped top surface for use at window sills.

SPLIT FACE BLOCK—Blocks having one or more faces sheared off (fractured) to give a rough textured surface.

SLUMPED BLOCK—Blocks having a rough surface resembling Adobe Brick.

"U" BLOCK—Same as *Lintel Block*.

Concrete Brick Similar to a clay *Brick* but made of concrete.

Concrete Columns Concrete columns are almost always reinforced with longitudinal bars to increase load carrying capacity. Lateral ties are placed around the vertical bars to hold them in place and to keep them from "spreading" under load. TIED COLUMNS are those in which the lateral ties are square or rectangular in shape, and SPIRALLY REINFORCED COLUMNS are those in which the ties are in the form of a con-

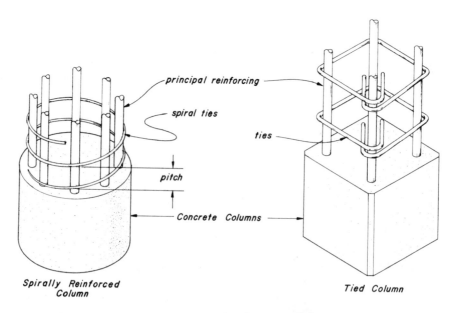

Spirally Reinforced Column

principal reinforcing

spiral ties

ties

pitch

Concrete Columns

Tied Column

FIGURE C-12 Concrete Columns

tinuous spiral—the spacing between such adjacent bars being called the PITCH. In computing the compression supporting strength of a concrete column, credit is taken for both the compressive resistance of the concrete and of the longitudinal reinforcing bars. (See Figure C-12.)

Concrete Fill See *Lightweight Concrete Fill*.

Concrete Formwork See *Formwork*.

Concrete Mix Proportions The ratio designation 1: 2½: 3½ means the number of parts by volume of *Cement, Fine Aggregate*, and *Coarse Aggregate*, in that order. Water content (which is not indicated in this type of mix description) is designated by the number of gallons per sack of cement. Mix proportions are also expressed in ratios of weights of the constituent materials.

Concrete Nail See *Nails*.

Concrete Piles See *Piles*.

Concrete Pipe See *Pipe*.

Concrete Pumping The placing of concrete by pumping through a flexible rubber or steel hose to the point of discharge. The concrete mix is first dumped into a *Hopper*, usually directly from a *Ready Mix Concrete* truck, then fed into the hose by a screw or piston type pump. Hose sizes are usually 2 to 6 inches in diameter. Use of pumping facilitates placing of concrete to otherwise inaccessible areas vertically or horizontally and can be pumped to considerable heights. At the end of the pumping operation the hose and hopper are cleaned by flushing with water. (See Figure C-13.)

Concrete Saw See *Saws*.

Courtesy of Burger Bros. Concrete Pumping, Pasadena, Ca.

FIGURE C-13   **Concrete Pumping**

FIGURE C-14   *Concrete Slab-on-Grade Being Placed*

Concrete Slabs A level surface of concrete either on the ground or supported above ground. The former are usually 3½″ to 6″ thick and, depending upon soil condition or loading, may or may not be reinforced with either *Welded Wire Fabric* or *Reinforcing Bars*. Slabs on the ground, also called Slabs-on-Grade (See Figure C-14), are often underlain with a gravel *Base* to improve supporting capacity, or by a plastic *Membrane* to eliminate dampness (also called a *Vapor Barrier*) due to water coming through the slab by *Capillary Action*. Slabs supported above ground are either ONE-WAY SLABS spanning in each direction between supporting beams, TWO-WAY SLABS spanning in each direction to beams on all sides, or FLAT SLABS, which are similar to a two-way slab but

do not have beams between supporting columns; they are slabs reinforced to span like a flat plate between supporting columns. See also: *Lift Slab, Pan Joist Concrete Floor, Waffle Slab*.

Condemnation The legal process by which property of a private owner is taken for public use, without his consent, but upon payment of *Fair Market Value* as determined by the courts based upon appraisals.

Condensate The liquid formed by *Condensation* of a vapor such as the water forming on the *Coils* of an *Air Handling Unit* or water formed from steam vapor.

Condensation The moisture producing effect which results when warm, moist air comes in contact with a colder surface and deposits moisture onto that surface. The resulting water is called *Condensate*.

Condenser That component of an *Air Conditioning* system which changes gases or vapors to a liquid state. It dissipates heat in the porcess.

Condominium A building (residence, apartment, townhouse, etc.) owned by a number of persons, each of whom holds title to an undivided interest of a portion of land and of a structure as a Tenant-in-Common with others, and a Title-in-Fee (absolute ownership) to the air space encompassed by the walls of his unit. The owners of a condominium usually pay a prorated share of common area maintenance, generally on a monthly basis.

Conductance In electricity, the ability of a material to conduct an *Electric Current* through it. It is the reciprocal of *Resistance*.

Conduction The movement of heat through a material by molecule to molecule contact, as through a spoon in hot coffee, as opposed to heat transfer by *Convection* or *Radiation*.

Conductive Floor A concrete floor which has metal particles embedded in the surface to prevent sparking due to static electricity. It is an essential requirement for rooms containing explosive materials, such as hospital operating rooms.

Conductor

1. In plumbing, a vertical pipe for carrying water from the roof, or from a gutter, to the ground. It is the same as a *Leader* or *Downspout*.

2. In electrical work, any material used to conduct electricity but the term usually refers to a current carrying wire. Conductors are *Aluminum Wire* or *Copper Wire* and are either insulated or non-insulated, the latter being called a *Bare Conductor*.

Conduit Any round, tubelike enclosure to protect insulated electric *Conductors* from external mechanical damage. After conduit has been installed, the conductors are "pulled" in place by *Fish Tape* and connected between *Panelboards* and *Electric Boxes*. Conduit is classified into the following categories:

RIGID CONDUIT—Straight sections of aluminum, steel, or plastic pipe joined by fittings. Although it is relatively "rigid" it can be bent to facilitate placement.

FLEXIBLE CONDUIT—A spiral of flexible, interlocking steel or aluminum used for continuous runs. (See' Figure C-15.)

FIGURE C-15 Flexible Conduit Entering Outlet Box

THIN WALL CONDUIT—Also called Electric Metallic Tubing (EMT). This is usually of steel tubing having a thinner wall than rigid conduit. Thin wall conduit is generally used where there is less likelihood of mechanical injury to the conduit.

ASBESTOS CEMENT CONDUIT—Consisting of *Asbestos Cement Pipe* and usually used for underground applications.

Conical Cone shaped.

Consortium A temporary partnership among several business firms for the development of a particular building project.

Constant In mathematical formulas, a number which remains fixed throughout the specific conditions of a given mathematical statement, in contrast to a variable. For example, the quantity $\pi = 3.14$ (the ratio of the *Circumference* to the *Diameter* of a circle) is a constant.

Construction Associations For construction related associations, agencies, and professional organizations see Figure C-16.

FIGURE C-16

Construction Related Organizations
Classified in accordance with the Uniform Construction Index

Division 1—General Requirements:

American Arbitration Association
140 W. 51st St., New York, N.Y. 10020

American Association of Cost Engineers
308 Monongahela Building,
Morgantown, West Virginia 26505

American Building Contractors Association
3345 Wilshire Blvd., Los Angeles, Ca. 90010

American Consulting Engineers Council
1155 15th St., NW, Washington, D.C. 20005

American Federation of Labor and
Congress of Industrial Organizations
815 16th St., NW, Washington, D.C. 20006

American Industrial Real Estate Association
5670 Wilshire Blvd., Suite 850, Los Angeles, Ca. 90036

American Institute of Architects
1735 New York Ave., NW, Washington, D.C. 20006

American Institute of Building Design
839 Mitten Rd., Suite 128, Burlingame, Ca. 94010

American Institute of Consulting Engineers
345 E. 47th St., New York, N.Y. 10017

American Institute of Industrial Engineers
25 Technology Park/Atlanta, Norcross, Ga. 30071

American Institute of Interior Designers
730 Fifth Ave., New York, N.Y. 10019

American Institute of Plant Engineers
1021 Delta Ave., Cincinnati, Ohio 45208

American Insurance Association, Engineering and Safety Service
85 John St., New York, N.Y. 10038

American National Standards Institute
1430 Broadway, New York, N.Y. 10018

American Society of Civil Engineers
345 E. 47th St., New York, N.Y. 10017

American Society of Heating, Refrigerating, and
Air Conditioning Engineers, Inc.
Engineering Center, 345 E. 47th St., New York, N.Y. 10017

American Society of Mechanical Engineers
345 E. 47th St., New York, N.Y. 10017

American Society for Nondestructive Testing
914 Chicago Ave., Evanston, Ill. 60202

American Society of Safety Engineers, Inc.
850 Busse Highway, Park Ridge, Ill. 60068

American Society for Testing and Materials
1916 Race St., Philadelphia, Pa. 19103

American Subcontractors Association
815 15th St., NW, Suite 902, Washington, D.C. 20005

Apartment Owners and Managers Association of America
65 Cherry Plaza, Watertown, Conn. 06795

Associated Builders and Contractors
P.O. Box 8733, International Airport, Md. 21240

Associated General Contractors of America
1957 "E" St., NW, Washington, D.C. 20006

Building Officials and Code Administrators International, Inc.
1313 E. 60th St., Chicago, Ill. 60637

Building Officials Conference of America
1313 E. 60th St., Chicago, Ill. 60637

Building Owners and Managers Association, International
224 S. Michigan Ave., Chicago, Ill. 60604

Building Research Advisory Board, National Research Council
2101 Constitution Ave., NW, Washington, D.C. 20418

Building Research Institute
2101 Constitution Ave., NW, Washington, D.C. 20418

Construction Industry Manufacturers Association
111 E. Wisconsin Ave., Suite 1700 Marine Plaza,
Milwaukee, Wis. 53202

Construction Specification Institute
1150 17th St., NW, Washington, D.C. 20418

Earthquake Engineering Research Institute
429 40th St., Oakland, Ca. 94609

Engineers Joint Council
345 E. 47th St., New York, N.Y. 10017

Environmental Protection Agency
401 M Street SW, Washington, D.C. 20460

Factory Insurance Association
85 Woodland St., Hartford, Conn. 06102

Factory Mutual System
1151 Boston-Providence Turnpike, Norwood, Mass. 02062

Federal Housing Administration (FHA)
c/o Dept. Housing & Urban Development,
451 7th St., SW, Washington, D.C. 20410

Federal National Mortgage Association
811 Vermont Ave., NW, Washington, D.C. 20005

General Building Contractors Association, Inc.
#2 Penn Center Plaza, Suite 1212, Philadelphia, Pa. 19102

Global Engineering Documentation Services, Inc.
3950 Campus Drive, Newport Beach, Ca. 92660

Illuminating Engineering Society
345 E. 47th St., New York, N.Y. 10017

International Association of Plumbing and Mechanical Officials
5032 Alhambra Ave., Los Angeles, Ca. 90032

International Conference of Building Officials
3560 S. Workman Mill Rd., Whittier, Ca.

National Association of Building Manufacturers
1619 Massachusetts Ave., N.W., Washington, D.C. 20036

National Association of Home Builders of the United States
15th and M Streets, NW, Washington, D.C. 20005

National Association of Realtors
155 E. Superior St., Chicago, Ill. 60611

National Association of Women in Construction
2800 Lancaster Ave., Fort Worth, Texas 76107

National Building Material Distributors Association
221 N. LaSalle, Chicago, Ill. 60601

National Bureau of Standards, U.S. Department of Commerce
Building Research Division, Washington, D.C. 20234

National Construction Industry Council
1625 "I" Street NW, Room 606, Washington, D.C. 20006

National Constructors Association
1133 15th St. NW, Washington, D.C. 20005

National Environmental Systems Contractors Association
1501 Wilson Blvd., Arlington, Va. 22209

National Society of Professional Engineers
2029 "K" St., NW, Washington, D.C. 20006

Occupational Safety and Health Administration (OSHA)
Contact regional offices in: New York, Philadelphia,
   Atlanta, Chicago, Dallas, Kansas City, Denver,
   San Francisco, and Seattle.

Producers Council, Inc.
1717 Massachusetts Ave., NW, Washington, D.C. 20036

Redwood Inspection Service
617 Montgomery St., San Francisco, Ca. 94111

Seismological Society of America
P.O. Box 826, Berkeley, Ca. 94701

Society of American Registered Architects
600 Michigan Ave., Chicago, Ill. 60605

Society of Construction Superintendents, Inc.
150 Nassau, New York, N.Y. 10038

Society of Fire Protection Engineers
60 Batterymarch St., Boston, Mass. 02110

Society of Industrial Realtors
925 15th St., NW, Washington, D.C. 20005

Society of Real Estate Appraisers
7 South Dearborn St., Chicago, Ill. 60603

Underwriters Laboratories, Inc.
207 E. Ohio St., Chicago, Ill. 60611

Division 2—Site Work

American Concrete Paving Association
1211 W. 22nd St., Suite 727, Oak Brook, Ill 60523
American Concrete Pipe Association
1501 Wilson Blvd., Arlington, Va. 22209
American Public Works Association (APWA)
1313 E. 60th St., Chicago, Ill. 60637

American Society of Landscape Architects
2000 "K" St., NW, Washington, D.C. 20006

Associated Landscape Contractors
700 Southern Building, Washington, D.C. 20005

Scaffolding and Shoring Institute
2130 Keith Building, Cleveland, Ohio 44115

Division 3—Concrete

American Concrete Institute
P.O. Box 19150, Redford Station, Detroit, Mich. 48219

The American Society for Concrete Construction
P.O. Box 85, Addison, Ill. 60101

Architectural Precast Concrete Association
825 E. 64th St., Indianapolis, Ind. 46220

Cellular Concrete Association
715 Boylston St., Boston, Mass. 02116

Concrete Reinforcing Steel Institute
108 N. LaSalle St., Chicago, Ill. 60601

Lightweight Aggregate Producers Association
B & B Building, 546 Hamilton St., Allentown, Pa. 1801

National Lime Association
5010 Wisconsin Ave., NW, Washington, D.C. 20016

National Precast Concrete Association
825 E. 64th St., Indianopolis, Ind. 46220

National Ready Mixed Concrete Association
900 Spring St., Silver Spring, Md. 20910

National Sand and Gravel Association
900 Spring St., Silver Spring, Md. 20910

National Slag Association
300 S. Washington St., Alexandria, Va. 22314

Perlite Institute
45 W. 45th St., New York, N.Y. 10036

Portland Cement Association
Old Orchard Rd., Skokie, Ill. 60076

Prestressed Concrete Institute
20 N. Wacker Dr., Chicago, Ill. 60606

The Vermiculite Association, Inc.
52 Executive Park South, Atlanta, Ga. 30329

Western Concrete Reinforcing Steel Institute
1499 Bayshore Highway, Suite 113, Burlingame, Ca. 94010

Wire Reinforcement Institute
7900 Westpark Dr., McLean, Va. 22101

Division 4—Masonry

The Barre Granite Association, Inc.
51 Church St., P.O. Box 481, Barre, Vt. 05641

Brick Institute of America
1750 Old Meadow Rd., McLean, Va. 22101

Building Stone Institute
420 Lexington Ave., New York, N.Y. 10017

Expanded Shale, Clay & Slate Institute
1041 National Press Building, Washington, D.C. 20004

Indiana Limestone Institute of America, Inc.
Stone City National Bank Building, Suite 400, Bedford, Ind. 47421

International Masonry Institute
823 15th St., NW, Suite 1001, Washington, D.C. 20005

Marble Institute of America, Inc.
1984 Chain Bridge Rd., McLean, Va. 22101

Mason Contractors Association of America
208 S. La Salle St., Chicago, Ill. 60604

Masonry Institute of America
2550 Beverly Blvd., Los Angeles, Ca. 90057

Mo-Sai Institute, Inc.
110 Social Hall Ave., Salt Lake City, Utah

National Association of Brick Distributors
1750 Old Meadow Rd., McLean, Va. 22101

National Concrete Masonry Association
1800 N. Kent St., Arlington, Va. 22209

National Crushed Stone Association
1415 Elliot Place, NW, Washington, D.C. 20007

National Limestone Institute, Inc.
1315 16th St., NW, Washington, D.C. 20036

Structural Clay Products Institute
NAME CHANGED TO: Brick Institute of America

Division 5—Metals

Aluminum Association
750 Third Ave., New York, N.Y. 10017

American Hot Dip Galvanizer Assocication
1000 Vermont Ave., NW, Washington, D.C. 20005

American Institute of Steel Construction
1221 Ave. of the Americas, Room 1580, McGraw-Hill Bldg.,
New York, N.Y. 10020

American Iron and Steel Institute
150 E. 42nd St., New York, N.Y. 10017

American Welding Society
2501 N.W. 7th St., Miami, Fla. 33125

Architectural Aluminum Manufacturers Association
410 N. Michigan Ave., Chicago, Ill. 60611

Copper Development Association, Inc.
405 Lexington Ave., New York, N.Y. 10017

Lead Industries Association, Inc.
292 Madison Ave., New York, N.Y. 10017

Metal Treating Institute
Box 448, Rye, N.Y. 10580

National Association of Architectural Metal Manufacturers
1033 South Blvd., Oak Park, Ill. 60302

Steel Deck Institute
9836 W. Roosevelt Rd., Westchester, Ill. 60153

Steel Joist Institute
2001 Jefferson Davis Highway, Suite 707, Arlington, Va. 22202

Welded Steel Tube Institute
522 Westgate Tower, Cleveland, Ohio 44126

Wire Reinforcement Institute
7900 Westpark Dr., McLean, Va. 22101

Zinc Institute, Inc.
292 Madison Ave., New York, N.Y. 10017

Division 6—Wood and Plastics

American Forest Institute
1619 Massachusetts Ave., NW, Washington, D.C. 20036

American Forest Products Association
2740 Hyde St., San Francisco, Ca. 94119

American Hardboard Association
20 N. Wacker Dr., Suite 1452, Chicago, Ill. 60606

American Institute of Timber Construction
1757 "K" St., NW, Washington, D.D. 20006

American Lumber Standards Committee
20010 Century Blvd., Suite 204, Germantown, Maryland 20767

American Plywood Association
1119 "A" St., Tacoma, Wash. 98401

American Wood Council
1619 Massachusetts Ave., NW, Washington, D.C. 20036

The American Wood Preservers Bureau
P.O. Box 6085, Arlington, Va. 22206

Appalachian Hardwood Manufacturers, Inc.
NCNB Bldg., Room 408, High Point, N.C. 27260

Architectural Wood Institute
5055 S. Chesterfield Rd., Arlington, Va. 22206

California Redwood Association
617 Montgomery St., San Francisco, Ca. 94111

Fine Hardwoods—American Walnut Association
666 N. Lake Shore Drive, No. 1730, Chicago, Ill. 60611

Forest Products Research Society
2801 Marshall Court, Madison, Wisc. 53705

Hardwood Dimension Manufacturers Association
3613 Hillsboro Rd., Nashville, Tenn. 37215

Hardwood Plywood Manufacturers Association
2310 S. Walter Reed Dr., P.O. Box 6246, Arlington, Va. 22208

National Forest Products Association
1619 Massachusetts Ave., NW, Washington, D.C. 20036

National Hardwood Lumber Association
332 S. Michigan Ave., Chicago, Ill. 60604

National Lumber and Building Material Dealers Association
1990 M Street NW, Washington, D.C. 20036

National Particleboard Association
2306 Perkins Place, Silver Spring, Md. 20910

National Woodwork Manufacturers Association, Inc.
400 W. Madison St., Riverside Plaza, Chicago, Ill. 60606

Northeastern Lumber Manufacturers Association, Inc.
13 South St., Glens Falls, N.Y. 12801

Northern Hardwood and Pine Manufacturers Association
Northern Bldg., Suite 501 Green Bay, Wis. 54301

Pacific Lumber Inspection Bureau
1411 4th Ave. Building, Suite 1130, Seattle, Wash. 98101

Society of the Plastics Industry, Inc.
250 Park Ave., New York, N.Y. 10017

Southern Cypress Manufacturers Association
805 Sterick Bldg., Memphis, Tenn. 38103

Southern Forest Products Association
P.O. Box 52468, New Orleans, La. 70150

Southern Hardwood Lumber Manufacturers Association
805 Sterick Bldg., Memphis, Tenn. 38103

Southern Pine Association
NAME CHANGED TO: Southern Forest Products Association
Southern Pine Inspection Bureau
P. O. Box 846, Pensacola, Fla. 32594

Timber Engineering Co. Subsidiary of National Forest Products Association
1619 Massachusetts Ave. NW, Washington, D.C. 20036

USDA Forest Products Laboratory, check with Office of Communications,
U.S. Dept. of Agriculture—14 St. and
Independence Ave. SW, Washington, D.C. 20250

West Coast Lumber Inspection Bureau
6980 S.W. Varns Rd., P.O. Box 23145, Portland, Ore. 97223

Western Red Cedar Lumber Association
Yeon Bldg., Portland, Ore. 97204

Western Wood Products Association
Yeon Building, Portland, Ore. 94204

Division 7—Thermal and Moisture Protection

The Adhesive and Sealant Council, Inc.
1410 Higgins Rd., Park Ridge, Ill. 60068

Asphalt Institute
Asphalt Institute Bldg., College Park, Md. 20740

Asphalt Roofing Manufacturers Association
757 Third Ave., New York, N.Y. 10017

Building Waterproofer's Association
60 E. 42nd St., New York, N.Y. 10017

Gypsum Roof Deck Foundation
5820 N. Nagle, Chicago, Ill. 60646

National Asphalt Pavement Association
6811 Kenilworth Ave., Riverdale, Md. 20840

National Insulation Contractors Association
8630 Fremont St., Suite 506, Silver Springs, Md. 20910

National Mineral Wool Insulation Association
211 E. 51st St., New York, N.Y. 10022

National Roofing Contractors Association, Inc.
1515 N. Harlem Ave., Oak Park, Ill. 60302

Red Cedar Shingle and Handsplit Shake Bureau
5510 White Bldg., Seattle, Wash. 98101

Shingle Inspection Service, Inc.
P.O. Box 272, Anacortes, Wash. 98221

Division 8—Doors and Windows

American Hardware Manufacturers Association
2130 Keith Bldg., Cleveland, Ohio 44115

American Society of Architectural Hardware Consultants
77 Mark Dr., P.O. Box 3476, San Rafael, Ca. 94902

Builders Hardware Manufacturers' Association
60 E. 42nd St., New York, N.Y.,10017

Door Operator and Remote Controls Manufacturers Association
110 N. Wacker Dr., Chicago, Ill. 60606

Flat Glass Marketing Association
1325 Topeka Ave., Topeka, Kansas 66612

National Builders Hardware Association
1815 N. Ft. Myer Dr., Suite 412, Rosslyn, Va. 22209

National Sash & Door Jobbers Association
20 N. Wacker Dr., Chicago, Ill. 60606

Sealed Insulating Glass Manufacturing Association
200 S. Cook St., Suite 209, Barrington, Ill. 60010

Stained Glass Association of America
1125 Wilmington Ave., St. Louis, Mo. 63111

Steel Door Institute
2130 Keith Bldg., Cleveland, Ohio 44115

Division 9—Finishes

Acoustical and Insulating Materials Association
205 W. Touhy Ave., Park Ridge, Ill. 60068

Acoustical Society of America
335 E. 45th St., New York, N.Y. 10017

Asphalt and Vinyl Asbestos Tile Institute
NAME CHANGED TO: Resilient Tile Institute

Carpet and Rug Institute
310 S. Holiday Ave., Box 2048, Dalton, Ga. 30720

Ceilings and Interior Systems Contractors
1201 Waukegan Rd., Glenview, Ill. 60025

Ceramic Tile Institute
700 N. Virgil Ave., Los Angeles, Ca. 90029

Decorating Products Association
9334 Dielman Ind. Dr., St. Louis, Mo. 63132

Facing Tile Institute
111 E. Wacker Dr., Chicago, Ill. 60601

Gypsum Association
201 N. Wells St., Chicago, Ill. 60606

Gypsum Drywall Contractors International
2010 Massachusetts Ave., NW, Suite 600,
Washington, D.C. 20036

International Association of Wall and Ceiling Contractors
1775 Church St., NW, Washington, D.C. 20036

Jute Carpet Backing Council, Inc.
30 Rockefeller Plaza, 23rd floor, New York, N.Y. 10020

Maple Flooring Manufacturers Association, Inc.
424 Washington Ave., Oshkosh, Wis. 54901

Metal Lath Association
221 N. LaSalle St., Chicago, Ill. 60601

National Paint and Coatings Association
1500 Rhode Island Ave., NW, Washington, D.C. 20005

National Paint, Varnish, and Lacquer Association
NAME CHANGED TO: National Paint and Coatings Association

National Oak Flooring Manufacturing Association
814 Sterick Bldg., Memphis, Tenn. 38103

National Terrazzo and Mosaic Association, Inc.
2A W. Loudoun St., Leesburg, Va. 22075

Painting and Decorating Contractors of America
7223 Lee Highway, Falls Church, Va. 22046

Paint and Wallpaper Association of America, Inc.
NAME CHANGED TO: Decorating Products Association

Resilient Tile Institute
26 Washington St., East Orange, N.J. 07017

Steel Structures Painting Council
4400 Fifth Ave., Pittsburgh, Pa. 15213

Tile Contractors Association of America
112 N. Alfred St., Alexandria, Va. 22314

Tile Council of America
P.O. Box 326, Princeton, N.J. 08540

Wallcovering Industry Bureau
969 Third Ave., New York, N.Y. 10022

Wallcovering Manufacturers Association, Inc.
969 Third Ave., New York, N.Y. 10022

Wood Flooring Institute of America
201 N. Wells St., Chicago, Ill. 60606

Division 10—Specialties

National Association of Mirror Manufacturers
1225 19th St., NW, Washington, D.C. 20036

Division 11—Equipment

Gas Appliance Manufacturers Association, Inc.
1901 N. Ft. Myer Dr., Arlington, Va. 22209

National Kitchen Cabinet Association
334 E. Broadway, Louisville, Ky. 40202

Division 12—Furnishings

American Canvas Institute
1918 N. Parkway, Memphis, Tenn. 38112

Furniture Manufacturers Association
220 Lyon St., NW, Grand Rapids, Mich. 49502

Division 13—Special Construction

Metal Building Manufacturers Association
2130 Keith Bldg., Cleveland, Ohio 44115

Mobile Home Manufacturers Association
14650 Lee Rd., P.O. Box 201, Chantilly, Va. 22021

National Swimming Pool Institute
2000 "K" St., NW, Washington, D.C. 20006

Division 14—Conveying Systems

International Material Management Society
Monroe Complex, Bldg. 3, Suite 1B, 2520 Mosside Blvd.,
Monroeville, Pa. 15146

The Material Handling Institute, Inc.
1326 Freeport Rd., Pittsburgh, Pa. 15238

National Elevator Industry, Inc.
600 Third Ave., New York, N.Y. 10016

National Elevator Manufacturing Industry
(NAME CHANGED TO: National Elevator Industry, Inc.)

Truck Trailer Manufacturers Association
2430 Pennsylvania Ave., NW, Washington, D.C. 20037

Division 15—Mechanical

Air-Conditioning and Refrigeration Institute
1815 N. Ft. Myer Dr., Arlington, Va. 22209

American Gas Association
1515 Wilson Blvd., Arlington, Va. 22209

American Water Works Association
6666 W. Quincy Ave., Denver, Colo. 80235

Bituminous Pipe Institute
8 S. Michigan Ave., Chicago, Ill. 60603

Cast Iron Pipe Research Association
1301 W. 22nd St., Room 509, Oak Brook, Ill. 60521

Cast Iron Soil Pipe Institute
2029 "K" St., NW, Washington, D.C. 20006

Cooling Tower Institute
3003 Yale St., Houston, Texas 77018

Mechanical Contractors Association of America
55530 Wisconsin Ave., NW, Suite 750, Washington, D.C. 20015

National Clay Pipe Institute
350 W. Terra Cotta Ave., Crystal Lake, Ill. 60014

National Fire Protection Association
470 Atlantic Ave., Boston, Mass. 02210

National Utility Contractors Association
815 15th St. NW, Washington, D.C. 20005

Plastics Pipe Institute
250 Park Ave., New York, N.Y. 10017

Plumbing-Heating-Cooling Information Bureau
35 E. Wacker Dr., Chicago, Ill. 60601

Sheet Metal and Air Conditioning National Association, Inc.
1611 N. Kent St., Suite 200, Arlington, Va. 22209

Division 16—Electrical

Electric Energy Association
90 Park Ave., New York, N.Y. 10016

Electronic Industries Association
2001 Eye St., NW, Washington, D.C. 20006

National Electrical Contractors Association
7315 Wisconsin Ave., Washington, D.C. 20014

National Electrical Manufacturers Association
155 E. 44th St., New York, N.Y. 10017

Construction Joint In concrete construction, a joint made between concrete portions placed at different times. The term is used interchangeably with *Cold Joint*. See also: *Contraction Joint, Expansion Joint.*

Construction Loan See *Interim Financing.*

Construction Management This term currently means a unified way of performing construction whereby one person (or company) called the "Construction Manager" has coordination and administrative control of both the design and construction of a project, as opposed to the traditional method whereby the *Architect* prepares the plans without *Contractor* involvement, and after the plans are completed general contract bids are taken. In the CM method, the Construction Manager works closely with the architect, preparing cost budgets and analyzing alternate construction types while the plans are being prepared. When the plans are complete the CM takes subcontract bids and coordinates and otherwise administers the job during construction, thus replacing the usual function of the general contractor. The CM may be a firm specializing in such services, or it may be a general contractor, architect, or consulting engineer. This is a general current description; however, variations do and will occur as use of the method increases.

Consulting Engineer See *Engineers.*

Contact Cement A glue-like substance distinguished by its ability to immediately stick on contact without the necessity of "sliding" the material around and in place.

Contingency Fund A sum of money set aside to provide for any unknown expense.

Continuous Beam See *Beam.*

Continuous Footing See *Footings.*

Continuous Inspection Special inspections required by a *Building Code* when higher than usual stresses are used in design or when unusual or unsafe hazards may exist during construction. An approved inspector is to be constantly present while the work is being performed to assure quality (e.g., placing high strength concrete) and/or safety. Continuous inspection may also be specified for a construction process (such as *Welding* or *Masonry* work) to assure quality work. The person performing the inspection is often called a *Deputy Inspector.*

Contour A line on a *Topographic Map* or *Grading plan(s)*, each point of which represents the same *Elevation* (height above a reference point).

Contour (or Belt) Sander See *Sanders.*

Contract An agreement between two or more persons to do or not to do a particular thing in exchange for an agreed upon payment.

Contract Documents In addition to the *Agreement,* other documents which are usually a legal part of a construction contract include the *Drawings, Specifications, Instructions to Bidders,* and *Addendum.*

Contraction

1. Decrease in size of a material due to a lowering of its temperature.

2. Shrinkage of *Concrete, Masonry,* or *Plaster* due to loss of moisture content during hardening.

Contraction Joint In concrete construction, a means of controlling where a *Shrinkage Crack* will occur by providing an artifically weakened joint, usually by embedment of a strip of plastic or sheet metal, or by a saw cut.

Contractor An individual or company offering construction services. An ENGINEERING CONTRACTOR (also: BUILDING CONTRACTOR, or BUILDER) is one specializing in works such as highways, flood control projects, etc., as contrasted to GENERAL CONTRACTORS who are usually thought of as constructing buildings and related structures. A general contractor has the responsibility for a complete project and may in turn engage SUB-CONTRACTORS who are usually specialty contractors, such as for plastering, roofing, etc., and who do not have contracts with the owner.

Conversion Tables See Figure C-17.

Cooler Nail See *Nails.*

Cooling Coils See *Coil.*

Cooling Tower A device to cool water by *Evaporation* so it can be reused for *Air Conditioning* equipment or industrial processes. (See Figure C-18.)

Figure C-17

Conversion Factors

Multiply	by	to obtain
acres	43,560	square feet
British thermal units	778.2	foot-pounds
British thermal units	3.930×10^{-4}	horse-power-hours
British thermal units	2.930×10^{-4}	kilowatt hours
centimeters	3.281×10^{-2}	feet
centimeters	0.3937	inches
centimeters	6.214×10^{-6}	miles
circular mils	7.854×10^{-7}	square inches
cubic centimeters	3.531×10^{-5}	cubic feet
cubic centimeters	6.102×10^{-2}	cubic inches
cubic centimeters	2.642×10^{-4}	gallons
cubic centimeters	10^{-3}	liters
cubic centimeters	2.113×10^{-3}	pints (liq.)
cubic centimeters	1.057×10^{-3}	quarts (liq.)
cubic feet	2.832×10^{4}	cubic cms.
cubic feet	1728	cubic inches
cubic feet	0.02832	cubic meters
cubic feet	0.03704	cubic yards
cubic feet	7.481	gallons
cubic feet	28.32	liters
cubic feet	59.84	pints (liq.)
cubic feet per minute	0.1247	gallons per sec.
cubic inches	16.39	cubic centimeters
cubic inches	5.787×10^{-4}	cubic feet
cubic inches	4.329×10^{-3}	gallons
cubic inches	0.01732	quarts (liq.)
cubic yards	27	cubic feet
cubic yards	46,656	cubic inches
cubic yards	0.7646	cubic meters
cubic yards	202.0	gallons
cubic yards	764.6	liters
degrees (angle)	60	minutes
degrees (angle)	0.01745	radians
degrees (angle)	3600	seconds
feet	30.48	centimeters
feet	12	inches
feet	0.3048	meters
feet	1.894×10^{-4}	miles
feet	⅓	yards

Multiply	by	to obtain
feet per second	30.48	centimeters per second
feet per second	1.097	kilometers per hour
feet per second	0.01136	miles per minute
foot-pounds	1.285×10^{-3}	British thermal units
foot-pounds	5.050×10^{-7}	horse-power-hours
foot-pounds	3.766×10^{-7}	kilowatt-hours
gallons	3785	cubic centimeters
gallons	0.1337	cubic feet
gallons	231	cubic inches
gallons	3.785×10^{-3}	cubic meters
gallons	4.951×10^{-3}	cubic yards
gallons	3.785	liters
gallons	8	pints (liq.)
gallons	4	quarts (liq.)
gallons per minute	2.228×10^{-3}	cubic feet per second
gallons per minute	0.06308	liters per second
grams	10^3	milligrams
grams	0.03527	ounces
grams	2.205×10^{-3}	pounds
horse-power	42.40	B.t. units per min.
horse-power	33,000	foot-pounds per minute
horse-power	1.014	horse-power (metric)
horse-power	745.7	watts
horse-power-hours	2544	British thermal units
horse-power-hours	1.98×10^6	foot-pounds
inches	2.540	centimeters
inches	8.333×10^{-2}	feet
kilograms	1.102×10^{-3}	tons (short)
kilometers	10^5	centimeters
kilometers	3281	feet
kilometers	3.937×10^4	inches
kilometers	10^3	meters
kilometers	0.6214	miles
kilometers	1094	yards
kilowatts	56.88	B.t. units per minute
kilowatts	4.427×10^4	foot-pounds per min.
kilowatts	737.8	foot-pounds per sec.
kilowatts	1.341	horse-power
kilowatts	14.33	kg. calories per min.
kilowatt-hours	3413	British thermal units
kilowatt-hours	2.656×10^6	foot-pounds
kilowatt-hours	1.341	horse-power-hours
knots (speed)	1.152	miles per hour

Conversion Factors

Multiply	by	to obtain
liters	10^3	cubic centimeters
liters	0.03531	cubic feet
liters	61.02	cubic inches
liters	10^{-3}	cubic meters
liters	1.308×10^{-3}	cubic yards
liters	0.2642	gallons
liters	2.113	pints (liq.)
liters	1.057	quarts (liq.)
lumens per sq. ft.	1	foot-candles
meters	100	centimeters
meters	3.281	feet
meters	39.37	inches
meters	10^{-3}	kilometers
meters	6.214×10^{-4}	miles
meters	10^3	millimeters
meters	1.094	yards
miles	5280	feet
miles	1.609	kilometers
miles per hour	88	feet per minute
miles per hour	1.467	feet per second
miles per hour	26.82	meters per minute
milligrams	10^{-3}	grams
millimeters	0.1	centimeters
mils	2.540×10^{-3}	centimeters
mils	10^{-3}	inches
minutes (angle)	2.909×10^{-4}	radians
minutes (angles)	60	seconds (angle)
ounces	28.35	grams
ounces	0.0625	pounds
pints (liq.)	473.2	cubic centimeters
pints (liq.)	1.671×10^{-2}	cubic feet
pints (liq.)	28.87	cubic inches
pints (liq.)	4.732×10^{-4}	cubic meters
pounds	453.6	grams
pounds	16	ounces
pounds per cubic foot	0.01602	grams per cubic cm.
pounds per cubic foot	5.787×10^{-4}	pounds per cubic inch
pounds per square foot	6.944×10^{-3}	pounds per square inch
pounds per square inch	144	pounds per square foot
quarts (liq.)	0.25	gallons
quarts (liq.)	0.9463	liters

Multiply	by	to obtain
radians	57.30	degrees
square centimeters	1.973×10^5	circular mils
square centimeters	1.076×10^{-3}	square feet
square centimeters	0.1550	square inches
square centimeters	10^{-4}	square meters
square feet	2.296×10^{-5}	acres
square feet	144	square inches
square feet	0.09290	square meters
square feet	3.587×10^{-8}	square miles
square feet	0.1111	square yards
square inches	1.273×10^6	circular mils
square inches	6.452	square centimeters
square inches	6.944×10^{-3}	square feet
square kilometers	247.1	acres
square meters	2.471×10^{-4}	acres
square meters	10.76	square feet
square miles	640	acres
square miles	27.88×10^6	square feet
square miles	2.590	square kilometers
square yards	9	square feet
square yards	1296	square inches
square yards	0.8361	square meters
tons (short)	907.2	kilograms
tons (short)	2000	pounds
watts	0.05688	B.t. units per min.
watts	44.27	foot-pounds per min.
watts	1.341×10^{-3}	horse-power
watts	10^{-3}	kilowatts
watt-hours	3.413	British thermal units
watt-hours	2656	foot-pounds
yards	91.44	centimeters
yards	3	feet
yards	36	inches
yards	0.9144	meters

Cooperative Ownership Ownership of property divided among several persons. It can take the form of a corporation, partnership, *Limited Partnership*, etc. See also: *Condominium*.

Cope To cut out an interfering part of a member so it will fit snugly against, or into, an adjoining member.

Coping

1. A sheet metal covering over the top of a wall. (See Figure C-19.)

2. The top finish course of a masonry wall.

Coping Saw See *Saws*.

Copper A reddish-brown metal

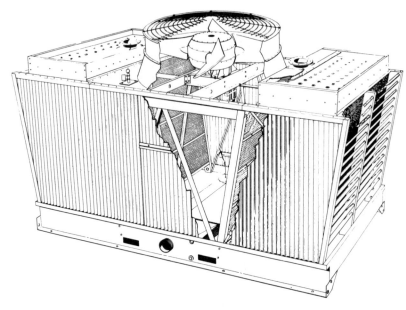

Courtesy of Marley Company

FIGURE C-18   Cooling Tower

which is relatively *Ductile* and resistant to corrosion by air and salt water. Its chief use is for *Electric Wire* due to its high electrical conductivity. As tubing, it is used for water and refrigeration piping due to its ease of bending and installation.

Copper Pipe   See *Pipe*.

Copper Wire   The largest use of copper wire is for electrical *Conductors*. It is higher in cost than aluminum wire but has excellent conductive properties. It is designated in *American Standard Wire Gage* (A.W.G.) sizes. See also: *Electric Wire*.

Coordinate   A number or other scaled reference used in a system for locating points in a geometric, geographic, or space relationship; typical coordinate systems include Cartesian, polar, bipolar, spherical, cylindrical, etc.

A distance or other quantitative measurement which locates the position of a point by its distance from each of two perpendicular reference lines (each called an *Axis*). Such a system, using fixed reference lines at right angles, is called a Cartesian coordinate system.

Cord (of Wood)   A measure of cut wood, such as firewood, which is equivalent in quantity to a pile 8' long, 4' high, and 4' deep.

Core

1. In *Masonry,* a hollow opening through a masonry unit; also called a CELL.

2. A central vertical space of a building which may contain elevator shafts, stairways, and/or mechanical equipment.

Core Drill   A machine for drilling into rock or concrete to retrieve a continuous cylinder (*Core*) as the bit progresses. The cores are used for visual examination and physical testing.

Core Sample   A sample of material obtained from a *Core Drill*.

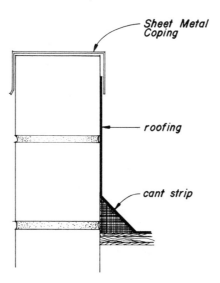

FIGURE C-19   *Coping*

Cork A lightweight, relatively spongy material harvested from the *Bark* of the cork oak tree. Over 50% of its volume consists of small air pockets. It is used for insulation, bulletin boards, flooring, and as an ingredient in *Linoleum*.

Corner Bead (Plastering) See *Metal Corner Bead*.

Cornice A decorative element projecting from a wall, at or near the roof line.

Corrosion A surface deterioration of metal caused by the chemical reaction of the metal with moisture and oxygen, such as when *Rust* forms.

Corrugated Cement Asbestos See *Cement Asbestos Board*.

Corrugated Metal Pipe See *Pipe*.

Corrugated Steel Sheets Also called corrugated iron sheets, they are steel sheets corrugated for stiffness and used as a roof and wall covering, generally on industrial type buildings. The most commonly used type has corrugation ½″ deep and spaced about 2½″ apart. These sheets are generally 27½″ wide, to give a net coverage 24″ wide, allowing for overlapping. Lengths available are in 2′ increments, from 6′ to 14′ and longer. They are available galvanized, uncoated, or with a variety of other finishes. See also: *Metal Decking*.

Cost-Plus Contract A means of performing construction work where the contractor is paid on the basis of the actual cost plus an agreed upon fee or percentage.

Coulomb In electricity, a quantity of electric charge, being equivalent to that transported by a current of one *Ampere* flowing for one second.

Counterflashing See *Flashing*.

Countersink To drill or ream a hole into which a screw or bolt can be placed so that its head will not project above the surface.

Counter Top The finished cover surface over a counter, usually of *Ceramic Tile* or *Laminated Plastic*.

County Recorder (or Clerk) That department of a county charged with keeping records showing legal ownership of land within the county.

Couple In structural design, two equal forces acting parallel to each other but in opposite directions and tending to rotate about an imaginary point midway between them. The *Bending Moment* caused by a couple is equal to one of the forces multiplied by the distance between them. See also: *Thermocouple.*

Coupling See *Pipe Fittings.*

Course In *Masonry,* one horizontal layer of masonry units.

Coved Base A rounded corner at the junction of a wall and floor to facilitate cleaning.

Cover Plate A steel plate connected to the *Flanges* of a *Beam* or *Column* to increase its strength.

C.P.M. See *Critical Path Method.*

Crack Control Joint See *Contraction Joint.*

Crane Any of a number of types of overhead machines to lift and move materials. This term is also used interchangeably with *Overhead Crane.* See also: *Bridge Crane, Crawler Crane, Derrick, Jib Crane, Monorail Crane, Self-Climbing Crane, Tower Crane, Truck Crane.*

Crawler Describes construction equipment such as a *Crane* or *Tractor* using, instead of tires, a continuous track-like belt of interlocking steel treads for better traction.

Crawler Crane Similar to a *Truck Crane* but using *Crawler* type motive power. Such cranes are used on terrain where greater traction and maneuverability are needed, and for "flotation" where

soil may be too soft to support a truck crane.

Crawler Tractor See *Tractor.*

Crawl Hole An access opening into an otherwise enclosed space.

Crazing

1. The formation of random, very thin, hairline cracks on the surface of a *Concrete Slab.* They are usually caused by too rapid drying of the surface.

2. In painting and/or gluing, minute cracks in the film due to extreme cold, lack of elasticity, or shrinkage.

Creep The lengthening or shortening of a structural member caused by a *Load* applied over a long period of time.

Creep (of Soil) A very slow, down-hill movement of soil caused by a combination of gravitational and weather induced forces.

Creosote A wood preservative manufactured from *Coal Tar* which is in the form of a brownish liquid. It seals the surface of wood in contact with earth and poisons fungi.

Cribbing A box-like stack of interwoven steel beams or timbers used to temporarily support a heavy load; a framework of timbers to support the walls and roof of a mine shaft or tunnel.

Cricket A specially raised or built-up portion of roof to facilitate drainage. (See Figure C-20.)

Cripple Studs *Studs* less than a story high, such as those used under or over a window opening.

Crippling See *Web Crippling.*

Critical Path Method A graphical method for planning, scheduling and controlling the interrelated events for a construction project such that time losses are minimized and controlled.

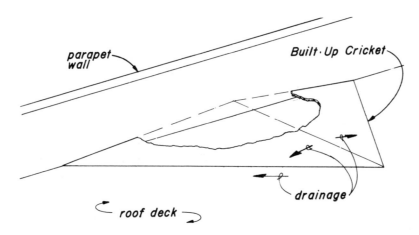

FIGURE C-20   **Cricket**

Crook   See *Warping (of Lumber)*.

Cross   See *Pipe Fittings*.

Cross Bracing (or X-Bracing) Bracing consisting of two intersecting diagonals, usually in the form of round rods, but can also be *Angles* or other *Structural Shapes*.

Cross Brake   In *Lumber* terminology, a separation of *Wood* cells perpendicular to the direction of the grain. Such breaks may be due to internal strains resulting from unequal longitudinal shrinkage or to external forces.

Cross Connection   In plumbing, an erroneous connection whereby a water supply pipe is inadvertently connected to a *Waste Line* causing contaminated water to mix with the supply water.

Crosscut Saw   See *Saws*.

Cross-Hairs   In the telescope of a surveying instrument, thin intersecting horizontal and vertical lines used for sighting.

Cross Section   In drafting terminology, an imaginary slice, cross-wise, through an object in order to better visualize or illustrate it.

Crown   The highest point of an *Arch,* roadway, or of a curved structural member.

Crushed Rock   Gravel-like material obtained by crushing larger rocks. Crushed rock has angular shaped particles in contrast to the rounded shape of *Gravel*.

Crusher Run Base   A *Base* course consisting of fragmented *Crushed Rock* as it comes from a rock crusher.

Cube (of a Number)   In mathematics, a number multiplied by itself three times; as the cube of $3 = 3 \times 3 \times 3 = 27$.

Cube Root   A number which, when multiplied by itself three times, will equal the number for which the cube root is desired. The cube root of $27 = 3$ $(3 \times 3 \times 3 = 27)$.

Cubic Yard   The standard unit of measurement for *Earthwork* and *Concrete*. One cubic yard contains 27 cubic feet.

Cul-De-Sac   A dead-end street having a turn-around at the end.

Culvert   A pipe or similar covered channelway under a road or railroad to provide drainage from one side to the other.

Cup See *Warping (of Lumber)*.

Cupola A relatively small dome-like structure on a roof.

Curing The time, or process, during which *Concrete,* or other cementitious mixture, hardens from its plastic state, and chemical reaction (*Hydration*) of cement and water is taking place.

Curing Compound In *Concrete* work, a liquid which is sprayed onto concrete surfaces after final trowelling to seal the surface against too rapid drying.

Current See *Electric Current*.

Curtain Wall Any of several types of prefabricated finished wall panels attached to the exterior structural frame of a building to form a finished wall surface. The panels may or may not incorporate windows.

Cut Removal of earth; opposite of *Fill.*

Cut Nails See *Nails.*

Cut Washer See *Washers.*

Cutting Torch See *Gas Cutting and Welding.*

Cyaniding A form of *Case Hardening.* It is the process of heating a ferrous alloy in contact with cyanide to diffuse carbon and nitrogen simultaneously into the outer surface of the alloy. It is usually followed by a heat treating cycle that produces a thinner but harder case than that produced by *Carburizing.*

Cycle

1. A time-related series of events that progress through an orderly sequence and return to the initial state.

2. In electricity, the progression of *Alternating Current* from zero to a positive maximum, back to zero, then to a negative maximum, then back to zero.

Cyclone Collector A sheet metal device for collecting and separating dust and other airborne particles from the air. The incoming particle laden air passes at high velocity around the inside of the cylinder. By centrifugal force, the particles are thrown against the side and slide down into a collecting bin below. The air passes out the top. A cyclone collector is one type of *Dust Collector.*

Cylinder (of a Lock) The tubular mechanism into which a key is inserted and which contains tumblers which can only be activated by the correct key.

Cylinder Locks and Latches This term includes locks and latches designed to be inserted into a door through a bored hole (through the door) with a separate hole into the edge to receive the latch bolt. It is the most common type of *Lockset* used, as opposed to a *Mortise Lock* or *Rim Lock.*

Cylinder Test A standard test to determine the compressive strength of concrete. Following a specified procedure, a standard size cylinder (usually 6″ round and 12″ high) is filled with concrete, stored for *Curing* for a certain period of time, then placed in a testing machine to determine the force required to crush it. Specifications usually call for such tests to be made at 7 and 28 days after pouring. Cylinders are generally metal but may be of other materials and sizes for different situations.

D

"d" Abbreviation for *Penny* (nail size).

Dado Joint See *Woodworking Joints*.

Damper Adjustable blades within an air duct to regulate the flow of air.

Dap In woodworking, a notch, or the act of cutting a notch.

Dap Joint See *Woodworking Joints*.

Darby A plasterer's tool for levelling the *Brown Coat* and leaving it slightly rough to receive the *Finish Coat*. It is a board about 42″ long and 4″ wide with two handles.

Dash Coat See *Plaster Finishes*.

Datum In surveying, an imaginary surface, such as mean sea level, used as a reference for measurements of vertical distances.

Dead Bolt A door locking bolt operated by a key or thumb turn rather than by a knob. When locked, the bolt cannot be forced back by end pressure on the bolt. (See Figure D-1.)

Dead Latch A type of *Latch Set* that locks automatically when the door is closed.

Dead Load See *Load*.

Dead Lock A door lock consisting of a *Dead Bolt* only (no knob). It can be keyed from either one or both sides.

Deadman Any kind of anchoring device set into the ground, such as a block of concrete to anchor the lower end of a *Guy* wire.

Decay The decomposition or rotting of *Wood* caused by certain types of fungi. Moisture, air, and temperature imbalances enhance such action by the fungi which feed on the wood, causing a softening and often a discoloration.

Decibel A unit of measurement of the intensity of sound. It is a nonlinear scale where zero decibels is just audible, the average speaking voice is about 60, a riveting machine at 30 feet is about 100, and a sense of pain to the ear occurs at about 120 decibels.

Declination See *Magnetic Declination*.

Decomposed Granite A granular soil consisting of *Quartz,* feldspar and mica. It results from the decomposition of *Granite*.

Deflection The amount of movement of the *Axis* of a beam caused by loads acting upon it.

Deformed Bars See *Reinforcing Bar*.

Degree See *Angular Measure*.

Dehumidifier An *Air Conditioning* device to lower the moisture content of air, either for human comfort or as re-

Courtesy Schlage Lock Co.

FIGURE D-1 Dead Bolt

quired by an industrial process. Dehumidifiers are of two basic types: they either remove moisture by chemical means such as passing air over a moisture absorbing material such as *Silica Gel*, or by a physical means whereby moisture in the air is condensed out by passing the air over cold coils which reduces its temperature to below its *Dew Point*.

Dehumidify The process of reducing the amount of moisture in the air, usually by a *Dehumidifier*.

Delamination In *Glued Laminated Lumber*, the separation of layers in an assembly becuase of failure of the *Adhesive*.

Deluge Sprinkler System See *Automatic Fire Sprinkler System*.

Demand Factor In electricity, the ratio or percentage of the actual use, or demand, of an electric system to the total connected load. For example, the demand factor of a 100 watt light which is on half the time would be 50%; the same factor for two 100 watt lamps on 25% of the time each would be 25%.

Density Weight per unit of volume. For example, the "density" of water is 62.4 pounds per cubic foot.

Deputy Inspector A person approved by the Building Department and employed by the builder to independently inspect the certain types of work specified under *Continuous Inspection*.

Derrick A pivoted type of *Crane*, usually of timber construction with *Guy* wires, for lifting and moving materials within a limited radius; sometimes called a STIFF-LEG DERRICK. It is also a non-movable tower-like structure for supporting a lifted load.

Design Stress See *Stress*.

Detailer A draftsman who is primarily engaged in making detailed drawings for parts of a building. A SHOP DETAILER is a draftsman who primarily makes *Shop Drawings*.

Detector Check In an *Automatic Fire Sprinkler System*, a device to prevent *Backflow* of water from the sprinkler system back into a public, *Potable* water supply.

Developer A company or individual who, for investment purposes (usually of a speculative nature), conceives of a building project, determines its conceptual use, organizes the design and construction services required for its completion, and arranges the financing and sale of the completed project, or may retain a part interest or the entire ownership for his or its own investment portfolio.

Dew Point The temperature of air at which its moisture content will begin to condense (see *Condensation*). For example, when air comes into contact with a cooler surface, droplets of water are formed. The dew point temperature varies with the moisture content (*Relative Humidity*) of the air.

Diagonal Bracing A structural member installed in a diagonal direction in a wall to resist horizontal forces such as from wind or *Earthquake*.

Diagonal Sheathing See *Sheathing*.

Diagonal Tension A tensile *Stress* that has a tendency to produce inclined or diagonal cracks in concrete members; special reinforcement is usually required to resist diagonal tension.

Diamond Mesh See *Metal Lath*.

Diaphragm A sheathed floor, roof, or section of wall which acts as a bracing element against wind or seismic forces. A plywood roof diaphragm, for example, can be thought of as a plate girder laid on its side, whose depth is the width of the building and whose length is the "span" between the end walls. The plywood "web" resists shearing forces like the web of a plate girder, and tension and compression members at the perimeter of the diaphragm act as the "flanges" of the girder. An analogy would be a solid cover nailed onto a box, stiffening the box by "diaphragm" action. (See Figure D-2.)

Die

1. In *Extrusion* forming of metal or plastic members, the die is the piece having the opening through which the material is squeezed, the opening giving shape to the item being formed.

2. A tool used for cutting threads on a pipe or bolt.

3. A tool used in a punch press for forming, punching out a shape from flat stock, or piercing metal.

Differential Settlement In foundation design, the settlement of different parts of a building at different rates. It is caused by varying compressibility of the underlying soil for the same *Soil Bearing Pressure,* or by differing soil bearing pressures on the same type of soil.

Diffuser See *Air Diffuser*.

Dike An earth embankment to confine water.

Diluent A liquid generally blended with a *Solvent* of *Varnish* or *Lacquer*. It is used to reduce the cost or increase the bulk. In gluing, it is an additive to an *Adhesive* to minimize bonding material (*Glue*) concentration.

Dimension The distance between two points.

Dimmer Switch In electrical work, a *Solid State* type device used to vary voltage to a light fixture and thereby change its intensity. It can be used on incandescent lamps and fluorescent lamps having special *Ballasts*.

Dipper Shovel See *Shovel*.

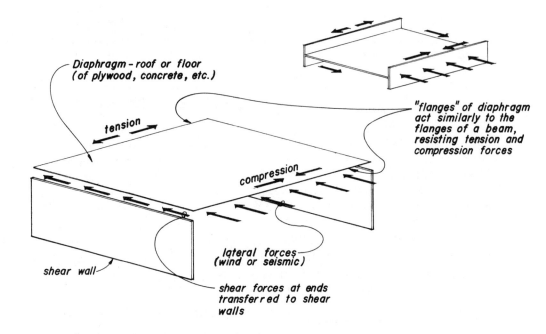

Diaphragm - roof or floor
(of plywood, concrete, etc.)

tension

"flanges" of diaphragm
act similarly to the
flanges of a beam,
resisting tension and
compression forces

compression

shear wall

lateral forces
(wind or seismic)

shear forces at ends
transferred to shear
walls

FIGURE D-2 *Diaphragm*

Direct Current See *Electric Current*.

Direct Shear Test In *Soil Mechanics*, a test for determining the strength of soil. A carefully controlled lateral load is applied to the top half of a disk of sampled soil, and the force causing slippage, or *shearing* of the upper and lower halves is used to determine its *Soil Bearing Valve*.

Disconnect In electrical work, a *Switch* or circuit breaker adjacent to a piece of electrical equipment (such as a motor) for the purpose of disconnecting power to the equipment for servicing.

Disposal Field See *Leach Field*.

Distribution Curve See *Grain Size Distribution Curve*.

Distribution Panel See *Panelboard*.

Ditcher An excavating machine for digging ditches. They usually operate by using a series of scoop-like buckets arranged like a continuous conveyor belt, discharging the excavated earth alongside the ditch (or trench).

Division Wall See *Walls*.

Dock See *Loading Dock*.

Dockboard A portable steel plate, or similar device, used on the edge of a *Loading Dock* to form a bridge between a truck bed and the loading dock.

Dock Leveler A manufactured ramp-like device placed on, or built into, the edge of a *Loading Dock* to provide an adjustable bridge-like transition between the dock and varying truck bed, or rail car, heights. (See Figure D-3.)

Dome Any structural shape, the surface of which, in plan view, is a circle at any horizontal section.

Domestic Sewer See *Sewer*.

Door Bucks A rough wood frame

FIGURE D-3 *Dock Leveler*

around a door opening in a masonry or concrete wall and to which the finished door *Jamb* and *Head* are attached.

Door Casing See *Casing*.

Door Check A device to limit the swing of a door—also called a DOOR HOLDER or *Door Stop*.

Door Closer A device attached to the bottom or top of a door to cause it to close automatically. It generally operates pneumatically.

Door Holder Same as *Door Check*.

Door Jamb The sides of a door opening.

Door Pull A handle or any similar device to provide a means for grasping and opening a door. It is most commonly used on cabinet doors and doors that close automatically but do not require a locking device.

Door Stop A small wood or metal strip around the inside of a *Door Jamb* against which a door closes. Also, a device synonymous with *Door Check*.

Door Types

ACCESS DOOR—A small door provided to either enter, or as a means to visually inspect, what would otherwise by an inaccessable area.

ACCORDIAN DOOR—A system of hinged door panels hung from a head rail in such a way that they fold flat against each other when in the open position.

AUTOMATIC CLOSING DOOR—On *Fire Doors*, a means of having a normally open door close automatically in the event of fire. It is usually accomplished by a system of weights and pulleys—or a *Door Closer*—each of which would have a fusible link which, when it melts from heat, would cause the door to close. There are also magnetic holding devices which operate electrically through a *Smoke Detector*.

BI-FOLD DOOR—A type of door similar to an *Accordian Door* but which consists of just four panels. In the open position the two panels on each side fold flat against each other.

DOUBLE ACTING DOOR—A door hinged such that it can swing in or out.

DUTCH DOOR—A door separated horizontally near mid-height such that the top or bottom halves may be opened separately (or together).

FIRE DOORS—Doors intended to prevent the passage of fire from room to room. They are classified and "labeled" by Underwriters Laboratories according to their time-rated degree of fire resistance. Class A doors are rated for three hours protection and are used in four-hour *Division Walls*. Class B doors are rated one or one and one-half hours and are usually used in two-hour division walls, stairwells, and elevator shafts. Class C doors are rated three-quarter hours and usually used for doors opening into corridors and through one-hour fire separation walls. Class D doors are rated one and one-half hours and used for exterior doors subject to exposure from an outside fire. Class E and F doors are rated three-quarter hours and are used as for Class D doors but in areas having less exposure. Due to an increased range of available hour ratings (3, 1½, 1, ¾, ½ hour and 20 minutes), there is a trend toward using the hour rating system rather than the letter classification above.

FRENCH DOOR—A pair of doors having glazed window panels.

GUILLOTINE DOOR—An overhead door consisting of a single plane which raises and lowers vertically in its entirety, guided in tracks at each side.

HOLLOW CORE DOOR—A wood door having smooth, flush surfaces and a hollow or honeycombed interior.

HOLLOW METAL DOOR—An all metal passageway door having light gage metal front and back panels with metal ribs to form a hollow interior.

KALAMEIN DOOR—A wood door covered with light gage steel, usually for fire protection.

LABELED DOOR—A door that has been approved, and carries a label, by *Underwriters Laboratories* to indicate its fire resistive time classification.

MAN DOOR—A door used by people as opposed to a larger door such as an overhead door which is also used for passage of materials, trucks or equipment.

METAL CLAD DOOR—A wood door covered with sheet metal, also called a *Kalamein Door*.

OVERHEAD DOOR—Any of a number of types of door which open by lifting vertically. They include *Roll-Up Doors* and *Sectional Doors*.

PANEL DOOR—A door with panels or lights (glass panes).

PILOT DOOR—A *Man Door* set within an *Overhead Door* or *Sliding Door* to provide passage without having to open the larger door.

ROLL-UP DOOR—Overhead truck type door having a slat-like construction such that it can be rolled up into a receptacle over the door opening.

SECTIONAL DOOR—Overhead type door which lifts vertically and is constructed in panel-like sections which move on tracks mounted at the sides of the door opening.

SELF-CLOSING DOOR—A door which closes automatically after opening, usually by means of a *Door Closer* mounted on or in the door, or concealed in the head or floor, but sometimes by a system of weights and pulleys.

SLIDING DOOR—A door which slides to open. They are usually hung on wheels within a track.

SOLID CORE DOOR—A wood slab type flush surface door whose interior core is solid as opposed to a *Hollow Core Door.*

Dope See *Pipe Dope.*

Double Acting Door See *Door Types.*

Double Duct System See *Central Air Conditioning.*

Double End Bolt See *Bolts.*

Double Headed Nail See *Nails.*

Double-Hung Window See *Window Types.*

Double Shear See *Shear.*

Double Strength Sheet Glass See *Glass.*

Douglas Fir-Larch See *Softwood.*

Dovetail Joint See *Woodworking Joints.*

Dowel

1. In concrete or masonry construction, a short length of *Reinforcing Bar* acting as a splice (by overlapping with another bar) between adjoining areas of concrete or masonry.

2. In wood construction, a round, peg-like, short connecting piece fitting into matching holes in parts to be joined.

Dowel Pins See *Nails.*

Downspout The vertical portion of a rain-water drainage pipe. It is also called a conductor or leader.

Dozer

1. Short for *Bulldozer.* Also, the term "to doze" means to push material by means of a bulldozer.

2. A blade mounted on the front of a *Tractor* such that material can be pushed. Some types can be angled or tilted so the pushed material is cast sideways.

Draft (of Air) See *Natural Draft* or *Forced Circulation.*

Draft Curtain A vertical partition extending downward from a roof (usually six feet) for the purpose of slowing the spread of fire along the underside of the roof. It must be noncombustible. Where draft curtains are used, *Smoke Vents* are also used to relieve the entrapped smoke. See also: *Draft Stops.*

Drafting Drawing by means of tools such as triangles, *Parallel Rules, Drafting Machines,* etc., as opposed to free-hand sketching; the art of providing information visually by means of illustrations and details. See also: *Drafting Tools.*

Drafting Machine A unitized drafting tool which incorporates the functions of a *"T" Square* or *Parallel Rule,* *Adjustable Triangle,* and scales. See also: *Drafting Tools.*

Drafting Service A individual or firm offering drafting services, usually to supplement an architect's or engineer's staff; however, they sometimes provide complete plans for simpler structures —residences, carports etc. Since they are not licensed architects or engineers their design services are usually legally restricted.

Drafting Tools For commonly used drafting tools, see Figure D-4.

Draft Stop A vertical partition closing off or separating areas of an *Attic* for fire protective purposes. See also: *Draft Curtain.*

From left to right, on a Hamelton drafting table with a Vemco drafting machine: scale, electric eraser, adjustable triangle, masking tape, french curve, template, brush.

FIGURE D-4 Drafting Tools

Drag In foundations, a downward force or "negative friction" on a *Pile* or *Caisson* (pier) shaft due to consolidation of the compressible soil layer penetrated. The soil tends to cling to the pile or pier shaft and pull it down as consolidation occurs.

Dragline A scoop shaped *Bucket* rigged onto a *Crane* by cables connected to the end of the boom and to a drum near the cab. It excavates by being pulled along the ground toward the operator. (See Figure D-5.)

Drag Strut A horizontal structural member in a roof or floor framing system which is used to transfer ("drag") wind and seismic forces from a *Diaphragm* into a *Shear Wall* or other bracing element.

Drainage and Vent Piping In plumbing, the system of piping whereby waste products are collected and removed from the building, to a public *Sewer* or an on-site sewage disposal system. *Vent Piping* is connected to the drainage piping to let in air so siphonage will not occur, and to discharge gases. The vent pipes are connected at certain locations and extend up through the roof and are open at the top. Pipes carrying wastes from sinks, lavatories, showers and similar fixtures are termed WASTE PIPES. When carrying discharge from *Water Closets* and uri-

Courtesy Northwest Engineering Co.

FIGURE D-5   *Dragline*

nals, it is called SOIL PIPE. Horizontal runs of pipe (actually they are sloped slightly to drain properly) are called BRANCHES. Vertical pipes are called STACKS. Accordingly, there can be BRANCH SOIL PIPE, BRANCH WASTE PIPE and BRANCH VENT PIPE, also SOIL STACK, WASTE STACK and VENT STACK. One vertical pipe may be both a waste stack and vent stack, the latter applying to that portion of pipe above the highest point where waste enters it. *Cleanouts* are installed at code specified locations to provide access to any run of pipe for unclogging. The lowest horizontal pipe, which collects the discharge from all waste and soil stacks, is called the BUILDING DRAIN (also HOUSE DRAIN). At a point just outside the building—usually 5 feet—its continu-

ation is called the BUILDING SEWER (also HOUSE SEWER). The above described drainage piping is also called SANITARY DRAINAGE, as opposed to STORM DRAINAGE, which is a similar system, either separate or connected to the sanitary drainage, and which handles rainwater, as from roof drains and *Leaders.* (See Figure D-6.)

Drawbar   A hitch mounted on the rear of a *Tractor* for attaching and pulling implements such as *Rippers, Sheepsfoot Rollers,* and *Scrapers.*

Drawings   A graphic representation of the projected construction. They consist of three main elements: PLANS, which show horizontal relationships; ELEVATIONS, which show features in a side view; and *Sections* and details which

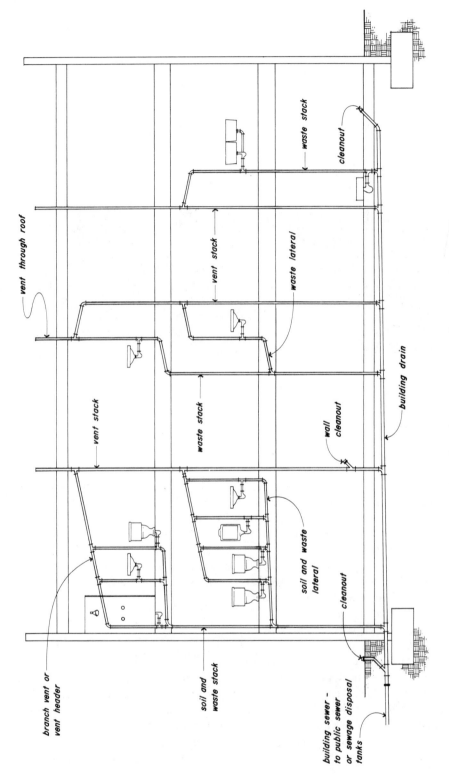

FIGURE D-6 Building Drainage and Venting System

show either horizontal or vertical relationships in a larger scale for clarity, and exhibit more detailed content than shown on the plans and elevations.

Dressed Size Lumber See *Lumber*.

Driers A material added to *Paint* or *Varnish* to accelerate drying time. They are usually compounds of certain metals such as lead, manganese, or cobalt.

Drift Pin A round spike-like tool used in the erection of structural steel to align holes in members to be joined. The round, tapered end is inserted through one hole into the hole to be matched, holding the two aligned while a bolt is put into an adjacent hole.

Drilled Caisson See *Belled Pier*.

Drilling Mud A thick slurry of water and bentonite (*Clay* powder) used in drilling and trenching to prevent caving. The pressure of the mud counterbalances the pressure of the soil at the excavation wall.

Drill Track See *Railroad Terminology*.

Drip A projecting piece of material from a wall so shaped as to throw off water running down from above to prevent its staining the surface below.

Drive Pin See *Powder Actuated Tools*.

Drive Screws See *Nails*.

Drop Panels As used in *Concrete Slab Flat Slab* floor construction, a thickened area of slab at the tops of the columns, usually for the purpose of increasing resistance to *Punching Shear*.

Dry-Bulb Temperature Temperature as indicated on an ordinary thermometer. See also: *Wet-Bulb Temperature*.

Dry Chemical Fire Extinguisher See *Fire Extinguishers*.

Dry Density The weight of the solid, dry, portion of *Soil* or rock, expressed in pounds per cubic foot.

Drying Shrinkage Shrinkage of a material caused by loss of moisture. It applies to wood, concrete, or masonry.

Dry-Pack A mixture of Portland *Cement* and sand with just enough water to form a putty-like consistency that can be worked into small spaces such as under beam or column *Bearing Plates*.

Dry Pipe System See *Automatic Fire Sprinkler System*.

Dry Rot In wood terminology, a term applied to dry, crumbly decay, especially that which, when in an advanced stage, permits the wood to be crushed easily to a dry powder. Dry rot is caused by a fungus.

Dry Standpipes See *Standpipe*.

Drywall Refers to a wall covering of *Gypsum Board* or similar ''dry'' sheets as opposed to *Plaster* which is applied ''wet.''

Drywall Accessories Any of a number of formed metal devices used in *Drywall* application such as to form true corners (by acting as a guide for puttying and taping to the adjacent sheets) or for trim around door openings or at the perimeter of drywall surfaces.

Drywall Nail See *Nails*.

Drywell A hole in the ground, usually filled with gravel, used to dissipate unwanted water into the soil.

Dual Duct System See *Central Air Conditioning*.

Duck Board Boards placed on a roof to provide walking access to roof mounted equipment without walking on the roof itself. Also, boards laid over wet or muddy ground to form a walkway.

Duct A round, oval, or rectangular pipe-like passageway for conducting air, as from an *Air Conditioner* or *Furance,* to room outlets. Ducts are made of *Asbestos Cement, Sheet Metal, Plastic, Fiberglass* or other noncombustible materials.

Ductile Frame A structural design concept applicable to rigidly connected frameworks as for a multi-story building and which can apply to either concrete or steel members, whereby the *Ductility* of the material (its ability to yield without fracturing) is taken into account in the design.

Ductility The property of a metal describing the degree to which it can deform beyond its *Elastic Limit* before breaking. It is the opposite of brittleness.

Dumbwaiter A small elevator-like means of sending food, trash, etc., from floor to floor within an enclosed shaft.

Dump Truck See *Truck Types.*

Dumpy Level See *Level.*

Duplex See *Apartment Nomenclature.*

Duplex Nail See *Nails.*

Durham Fitting In plumbing, pipe fittings used for *Drainage and Vent Piping* systems where the threads of the fittings are recessed so the inside diameter of the fitting matches the inside diameter of the pipe to provide a smooth water passageway.

Dust Collector Any of several means by which dust particles can be removed from dust-laden air. One method is by the use of a *Cyclone Collector.*

Dusting A surface condition of a *Concrete Slab* characterized by loose dust-like particles of concrete which have ground loose from the mass.

Dutch Door See *Door Types.*

Dutchman Slang term for any odd size piece of wood to fill an opening in a "make shift" manner such as to correct a mistake.

Dwarf Wall Any low wall of less than a story height.

Dry Penetrent Testing See *Weld Testing.*

Dynamic Changing.

E

"E" Abbreviation for *Modulus of Elasticity*.

Earlywood See *Annual Rings*.

Earth See *Soil*.

Earth Drill See *Auger*.

Earthmoving Equipment See *Backhoe, Bulldozer, Clamshell, Ditcher, Dragline, Grader, Loader, Shovel, Scraper, Skip Loader, Truck Types*.

Earth Pressure In structural design, pressure caused by the weight of earth or its sideways pressure. ACTIVE EARTH PRESSURE is force exerted by the earth, as against a *Retaining Wall*. PASSIVE EARTH PRESSURE is the resistance offered by a soil to its being pushed against (compressed).

Earthquake A shaking motion of the surface of the earth caused by a sudden stress relieving slippage (or rupture) inside the earth's crust. The magnitude of its shaking force, as recorded on a *Seismograph*, is classified, most commonly, by the *Richter Scale* which uses a non-linear scale from one to ten. *Building Codes* require buildings in areas subject to earthquakes to be designed with sufficient bracing to resist *Seismic Loads* of a statistically anticipated magnitude.

Earthwork A general term meaning all acts relating to the moving, shaping, and/or compacting of earth.

Easement The legal right of one party to cross land owned by another such as for roadway access or to repair underground utilities.

Eastern Hemlock-Tamarack See *Softwood*.

Eastern Spruce See *Softwood*.

Eave That part of the lower edge of a roof extending beyond the sides of a building.

Eave Strut A structural member spanning between columns at the edge of a roof. The term usually applies to such members in a *Pre-Engineered Steel Building*.

Eccentricity The distance from which a force acts away from the *Center of Gravity* of a member or connection, and which tends to cause such member of connection to rotate or bend. (See Figure E-1.)

Edge Distance The distance a *Bolt, Screw*, or *Nail* is from the edge of the member into which it is inserted, as opposed to the *End Distance*.

Efficiency The ratio of actual performance to theoretical maximum performance.

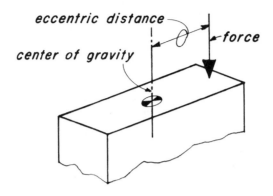

FIGURE E-1   *Eccentricity*

Efflorescence A white, powder-like deposit sometimes appearing on masonry walls. It is generally caused by soluble salts within the wall being carried to the face of the wall by moisture and deposited as a residue when the water evaporates.

Effluent In *Sewage* terminology, the liquid portion which is partially purified and relatively clear of solid matter. In general terms, it is the gas, liquid, or dust by-product of a process.

Eggshell See *Paint Finishes.*

Ejector In plumbing, a device, usually a pump, used for pumping *Sewage* from a lower to a higher elevation.

Elasticity The degree to which a material can be stretched, twisted, bent, or otherwise deformed and return to its original dimensions when the applied force is removed. See also: *Hooke's Law.*

Elastic Limit The limiting *Stress* to which a material follows *Hooke's Law.* See also: *Stress-Strain Curve.*

Elastomer Any of various *Plastic* materials resembling rubber which can be stretched and which return to their original dimensions when released.

Elastomeric A highly flexible mate-

rial which can be extruded and is used primarily as a *Sealant* and as a pre-formed joint filler. Some elastomerics also have *Adhesive* properties.

Elbow See *Pipe Fittings.*

Elbow Catch See *Catches.*

Electric Arc Welding See *Arc Welding.*

Electric Box A metal box open on one face and having round knock-out openings on the ends for *Conduit* such that electric *Conductors* (wires) can be pulled into the box. They are generally recessed into a wall flush with the finished surface and attached to *Studs* or other framing by steel straps. The three basic types of electric boxes are: an OUTLET BOX which is used to terminate conductors such as for connection to a receptacle, light fixture, motor, etc.; a SWITCH BOX containing one or more toggle type switches; and a PULL BOX (or JUNCTION BOX), which is used to connect conductors where the run through conduit would otherwise be too long or have too many bends.

Electric Circuit See *Circuit.*

Electric Current The amount of electricity flowing in a wire or other conductor as measured in *Amperes*. The two types of current are:

ALTERNATING CURRENT—Electric current which reverses its direction of flow at a regular interval—usually 60 cycles per second, abbreviated A.C.

DIRECT CURRENT—Electric current which flows in one direction only through a circuit, such as from a battery, abbreviated D.C.

Electric Elevator See *Elevators*.

Electric Generator See *Generator*.

Electricians Union electricians are represented by the International Brotherhood of Electrical Workers (I.B.E.W.), affiliated with the AFL-CIO.

Electric Measurements The four basic electrical measurements and the instrument for measuring them are: *Volts* (by a *Voltmeter*); *Amps* (by an AMMETER); *Watts (by a WATTMETER); and Resistance* (by an OHMMETER). See also: *Ohm's Law*.

Electric Metallic Tubing See *Conduit*.

Electric Panelboard See *Panelboard*.

Electric Power The "real" energy per unit of time at a given point in an electrical *Circuit,* measured in *Watts.* Electric meters indicate power consumption in *Kilowatt Hours* (1,000 watts acting for one hour). *Watts = Volts × Amps* in single phase, or *Volts × Amps ×* 1.73 in three phase power.

Electric Service The means by which an electric utility company supplies electric power to a customer. The quantity of power desired is stated in *Amperes* (e.g., a "400 amp service"), such quantity being determined by the electrical *Code* and the customer's consumption require-

ments, and is usually provided in the following choices of voltages and phases:

TWO-WIRE SERVICE—Where one *Conductor* is "hot" and the other is a *Neutral Conductor*. The voltage across the two is usually 120 volts. This system is rarely used now.

THREE-WIRE SERVICE—This consists of either two "hot" conductors and a third neutral conductor, whereby single phase power can be provided at either 120 or 240 volts, or if all three conductors are "hot," three-phase power can be provided at either 240 or 480 volts.

FOUR-WIRE SERVICE—This consists of three "hot" conductors and one neutral conductor. By a connection arrangement called FOUR WIRE WYE, three-phase power can be provided in the following pairs of voltages: 120 or 208 volts, or 277 or 480 volts. By means of a FOUR WIRE DELTA connection, three-phase power can be provided at 240 volts or single phase at 120 or 240 volts.

The utility company furnishes the power to the building including intermediary *Transformers* where required, and an *Electric Service Meter*. Service is either overhead or underground. In the former, the conductors, called a SERVICE DROP, run from the transformer on a power pole to a SERVICE HEAD at the building, the latter being a weatherproof means of the conductors entering the building and usually the point at which the customer provides subsequent distribution wiring. Underground service, either from the base of a power pole or from an underground distribution system is run in *Cable* to a utility company provided transformer (when required) which is placed on a slab outside the building or within a vault inside the building. Such transformer converts the power company primary voltage to the customer utilization voltages and phases desired. From

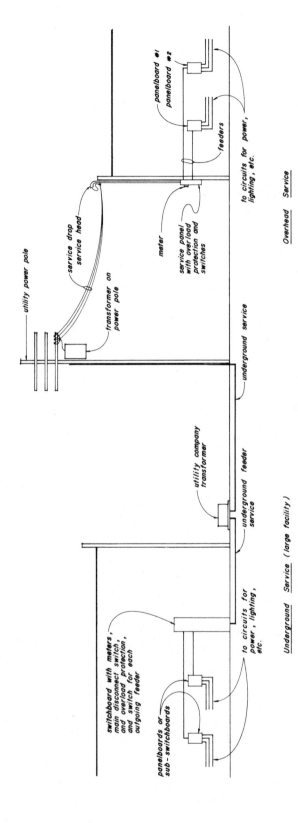

Typical Electric Service and Distribution

utility power pole

service drop
service head

transformer on
power pole

meter

service panel
with overload
protection and
switches

panelboard #1
panelboard #2

feeders

to circuits for power,
lighting, etc.

Overhead Service

underground service

utility company
transformer

underground feeder

underground
service

switchboard with meters,
main disconnect switch,
and overload protection,
and switch for each
outgoing feeder

panelboards or
sub-switchboards

to circuits for
power, lighting,
etc.

Underground Service (large facility)

FIGURE E-2 Electrical Service and Distribution

this transformer, or from the service head in the case of overhead service, the customer provides a *Feeder,* called a SERVICE ENTRANCE, to a utility company provided electric meter within or adjacent to an electric *Panelboard* or electric *Switchboard* from which subsequent distribution continues. (See Figure E-2.)

Electric Service Meter A meter installed by an electric utility company to measure *Electric Power* consumption by a subscriber and which is read periodically for billing purposes. It registers the number of *Kilowatt Hours* used (one

kilowatt hour = one thousand watts acting for one hour).

Electric Switchboard See *Switchboard.*

Electric Symbols Commonly used symbols for electrical devices are shown in Figure E-3.

Electric Transformer See *Transformer.*

Electric Wire Most electric wire is either copper or aluminum. Both copper and aluminum wire sizes are indicated by *American Standard Wire Gage* (AWG)

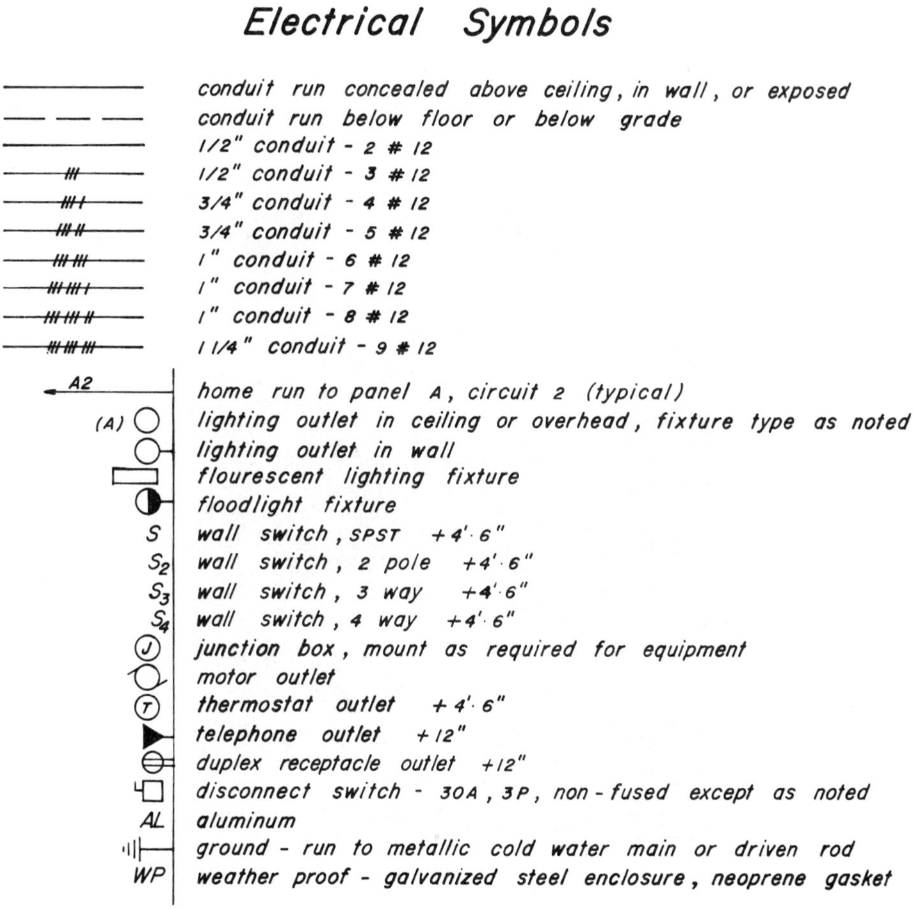

Electrical Symbols

————————	conduit run concealed above ceiling, in wall, or exposed
— — —	conduit run below floor or below grade
————————	1/2" conduit - 2 # 12
——///——	1/2" conduit - 3 # 12
——////——	3/4" conduit - 4 # 12
——//////——	3/4" conduit - 5 # 12
——/// ///——	1" conduit - 6 # 12
——/// ////——	1" conduit - 7 # 12
——//// ////——	1" conduit - 8 # 12
——/// /// ///——	1 1/4" conduit - 9 # 12
←A2	home run to panel A, circuit 2 (typical)
(A) ○	lighting outlet in ceiling or overhead, fixture type as noted
○—	lighting outlet in wall
▭	flourescent lighting fixture
◑	floodlight fixture
S	wall switch, SPST +4' 6"
S₂	wall switch, 2 pole +4' 6"
S₃	wall switch, 3 way +4' 6"
S₄	wall switch, 4 way +4' 6"
Ⓙ	junction box, mount as required for equipment
○	motor outlet
Ⓣ	thermostat outlet + 4' 6"
►	telephone outlet + 12"
⊖	duplex receptacle outlet +12"
⌐▢	disconnect switch - 30A, 3P, non-fused except as noted
AL	aluminum
⊣⊢	ground - run to metallic cold water main or driven rod
WP	weather proof - galvanized steel enclosure, neoprene gasket

FIGURE E-3 *Electrical Symbols*

sizes which is the same as the *Brown & Sharp (B & S) Gage*. The numbering system is in the opposite order to the size of the wire diameter. No. 12 wire is probably most commonly used, after which the size goes "down" to No. 40 which is about the size of a hair and "up" to Nos. 0,00,000, and 0000, and larger (also designated 1/0, 2/0, 3/0, 4/0). The latter size is approximately ½″ in diameter. Wire is insulated with various types of plastic or rubber coating with the composition and thickness related to the voltage rating and conditions of use (dry or wet location, high heat, etc.). The term CIRCULAR MIL means a unit of measurement of the area of a wire. One circular mil is the area of a wire .001 inches (one thousandth of an inch) in diameter. The abbreviation for 1,000 circular mils is MCM. Electric wire

is sizes larger than 4/0 is measured in circular mils.

Electrode See *Welding*.

Electrolysis A chemical change in a material caused by a flow of electricity. *Electroplating* is an example, as is *Galvanic Corrosion*.

Electromotive Series In electricity, a listing of metals in order of their current producing ability when in contact with metals lower on the list. Some common metals in their descending order are: aluminum, zinc, chromium, iron, nickel, tin and copper. It is also referred to as Galvanic Series.

Electronic Distance Measuring (E.D.M.) In surveying, measuring distances using an electronic instrument

FIGURE E-4 Electronic Distance Measuring Device

rather than by taping. The instruments operate on a radar-like principle—a pulse of high frequency radio waves is beamed from the instrument to a point whose distance is to be measured where a transponder bounces the beam back to the instrument. The elapsed time is converted to distance. It can measure up to 3,500′ (1 kilometer) with an accuracy of ± 1/100th foot per 1,000 feet. Recently developed models using *Lasers* and *Infra-Red* waves can be used for short range measurements from one foot to two miles with an accuracy to within one inch. (See Figure E-4.)

Electroplating A process of putting a thin layer of metal onto another metal by *Electrolysis*. The metal to be plated is dipped into a conducting solution (called the ELECTROLYTE). Electrons from the metal to be deposited (called the ANODE) flow to, and are deposited on, the metal to be plated which acts as the CATHODE.

Electrostatic Painting A spray painting process based upon the law of physics that unlike charges attract and like charges repel. Paint particles leaving a spray gun are ionized, giving them a negative charge. The product being finished is electrically grounded and is therefore "positive" in relation to the paint spray. The negative charge of the paint particles consequently pulls them onto the product. Overspray is greatly reduced because paint that misses or rebounds from the product on the first pass "turns around" under the influence of the electrostatic field and returns to the product where it is deposited on the rear side of the surface. This phenomenon—known as "electrostatic wrap around"—makes possible the finishing of many small items by spraying from one side only.

Electrostatic Precipitators In *Air*

Conditioning, a filtering device to remove dust particles from the air by electrically charging the particles so they will be attracted to oppositely charged collector plates and thereby removed from the air flow. Another method uses a dry filter media upon which an electrostatic charge is impressed. The latter combines electrostatic precipitation and mechanical filtration.

Elevation

1. A vertical distance with respect to a reference point.

2. A drawing showing a side view.

Elevator Bolt See *Bolts*.

Elevators The two basic types of elevators are:

ELECTRIC ELEVATORS—Used in taller buildings and operated by counterweighted cables driven through a grooved, electrically driven drum at the top. The elevator cab is guided on vertical rails. Most new elevator installations have fully automatic (no operator) control systems.

HYDRAULIC ELEVATORS—Used for relatively small vertical travel distances, usually less than four floors. They are operated by a vertical supporting piston which is guided in a cylindrical tube buried under the lower floor. Oil is pumped into the tube to raise and lower the cab.

Ell Same as *Elbow*. See Pipe Fittings.

Ellipse An oval-shaped geometric figure. It is the shape obtained by an inclined "slice" through a cylinder. For obtaining the area of an ellipse see *Areas of Common Shapes*.

Ellipsoid A solid, egg-like shape of which each cross section parallel or perpendicular to its long axis is either a circle or an *Ellipse*.

Elongation The lengthening of a material subjected to *Stress* or to *Thermal Expansion*.

Embankment A term usually meaning a body of *Fill* soil which has a length exceeding both width and height. Levee and highway fills are types of embankments.

Embeco Trade name (Master Builders) for a group of non-shrink *Grout* products.

Embossing A process whereby a material is raised above its surrounding surface such as to form letters or a decorative effect.

Emergency Generator Also called a STANDBY GENERATOR, it is an electric *Generator* usually driven by a gasoline or diesel engine and used to supply electricity to critical areas in buildings when normal power fails.

Emery Paper Paper which is coated with glue or other bonding material and into which a mineral (aluminum oxide) is embedded. Emery paper is available in various degrees of coarseness and is used for polishing purposes, generally for aluminum or steel. Grades are: extra coarse, coarse, medium, fine, and extra fine. See also: *Sandpaper*.

Empirical A term meaning to rely upon observation or experimentation rather than upon theoretical analysis.

E.M.T. Abbreviation for electric metallic tubing. See also: *Conduit*.

Emulsion The suspension of minute droplets in an immiscible liquid, such as oil particles in water.

Enamel See *Paint*.

End Bearing Pile See *Piles*.

End Distance The distance a bolt, screw, or nail is from the end of a member, as opposed to its *Edge Distance*. For bolts, codes specify this minimum distance in terms of bolt diameters and varies it depending upon whether the load on the bolt is toward or away from the free end.

Energy The capacity of a body to do work, gained through its position or condition, as classified below:

POTENTIAL ENERGY—The energy of a body due to its position so that it can do work if released. For example, a weight held suspended by a rope; the position of the weight is such that if the rope is released, work can be done by the weight as it drops.

KINETIC ENERGY—The energy in a body in motion and the work it can do before it is brought to rest. For example, a stream of water from a nozzle striking the paddles of a wheel. The magnitude of available kinetic energy varies as the square of its velocity (mathematically, the formula is $\frac{1}{2} MV^2$ where M = mass, and V = velocity).

Engelmann Spruce See *Softwood*.

Engineering The various branches of engineering are described under *Engineers*.

Engineering Contractor See *Contractor*.

Engineers A person especially trained, knowledgeable, and experienced in one of a number of technical areas. Engineering is subdivided into several branches which, in the building industry, are described below, and the use by a person of one or more of the titles generally requires licensing by the state, such licensing requiring a college degree or its equivalent plus a certain number of years experience and a written and/or oral examination.

CIVIL ENGINEER—An engineer involved in the design of fixed works such as bridges, highways, airports and industrial facilities. Such engineers in building design are involved with grading, drainage design and *Structural Design.* See also: *Structural Engineer* (Structural engineering is a branch of Civil Engineering.)

CONSULTING ENGINEER—A collective term for engineers in private practice offering engineering services in one of the branches of engineering.

ELECTRICAL ENGINEER—An engineer whose work includes the design of electric power distribution, lighting systems, etc. Such engineers usually act as consultants to *Architects,* providing the design, drawings and specifications for the above items for incorporation into the general building plans.

FOUNDATION ENGINEER—An engineer specializing in foundation problems. His services usually include sampling and testing soil for a proposed structure and recommending *Soil Bearing Values* and suitable types of foundations.

MECHANICAL ENGINEER—An engineer whose work deals with the design of plumbing systems, heating, ventilating and air conditioning systems, process piping, and related mechanical components of a building. Such engineers usually act as consultants to *Architects,* providing the design, drawings and specifications for the above items, for incorporation into the general building plans.

PLANT ENGINEER—A general term for a technically trained person, usually but not necessarily licensed, who is employed by a manufacturing plant to oversee functional operation and maintenance of the plant.

PROFESSIONAL ENGINEER—A term used by many states to indicate an engineer in any of the branches of engineering who is so licensed by that state.

STRUCTURAL ENGINEER—A *Civil Engineer* who specializes in the design of structures. Some states require special licensing for use of the title "Structural Engineer." Such engineers usually act as consultants to *Architects,* providing the design, drawings, and specifications for structural items incorporated into the general building plans.

Engineer's Level See *Levels.*

Engineer's Scale A measuring scale based on decimals of an inch as opposed to an *Architect's Scale* which is based on fractions of an inch. Such scales are generally used where a large reduction in scale is needed, such as for *Site Plans.* Most engineer's scales divide an inch into 10, 20, 30, 40, 50, or 60 parts; for example, a scale for a drawing might read: "1 inch = 40 feet," as opposed to an architect's scale designation which would be stated as $\frac{1}{4}'' = 1' - 0''$. (The two are not equivalent reductions.)

English Bond See *Masonry Bond Patterns.*

Environmental Protection Agency (EPA) An independent agency of the executive branch of government created to coordinate and effect governmental action in an attempt to control the nation's environment with respect to air, water, pesticides, and solid waste management, radiation and noise control. It supports research and antipollution efforts by state and local governments as well as private groups. Its major function is to make public its comments and determinations regarding environmental quality, particularly with regard to new projects. The EPA maintains regional offices with representatives commited to local programs for pollution abatement—its main office address is 401 M Street SW, Washington, D.C. 20460.

Epicenter The point of the earth's surface directly above the focus of an *Earthquake*.

Epoxy A family of *Thermosetting Plastics* used in construction, primarily as coating and binding agents. They have high weather and chemical resistance, resilience and electrical insulation properties.

Epoxy Paint See *Paint*.

Equilibrium A state of loading on a physical system in which all the forces are in exact balance.

Erection The term usually refers to the job site assembly of pre-assembled structural members or components into their final position.

Erosion The wearing away by water or wind.

Errors and Omissions Insurance See *Insurance*.

Escalator A moving stairway.

Escutcheon A decorative plate or covering àround an opening, such as that surrounding the hole for a door knob or lock.

Estimating The act of determining in advance the anticipated cost of supplying services, materials, labor, or any combination thereof to a construction project, as in the act of preparing a *Bid* (See Figure E-5). See also: *Quantity Surveying*.

Evaporator In *Air Conditioning*, that part of a *Mechanical Refrigeration* cycle where the *Refrigerant* changes from a liquid to gas (vaporizes) and absorbs heat during the process.

Evaporative Cooler An *Air Conditioning* unit using the evaporation principle whereby incoming air is forced by a fan through a spray of water, or over a wet surface, and is cooled by losing its heat in evaporation of the water. The moisture content (*Humidity*) of the outgoing air is increased.

Excavating Equipment See *Earthmoving Equipment*.

Excavation Removal of earth.

Exhaust Fan A *Fan* used to remove air overloaded with smoke, odor or heat, from a room.

Expanded metal Lath See *Metal Lath*.

Expanded Plastic See *Foamed Plastic*.

Expanded Slag *Slag* which, in a molten state, is treated with steam to produce a rapid popcorn-like expansion. It then has numerous pores and its resulting light weight makes it useful as a lightweight *Aggregate* for cinder blocks and related uses.

Expansion The increase in size or length of a material, generally referring to the effect produced by an increase in temperature, more accurately termed *Thermal Expansion*. See also: *Coefficient of Expansion*.

Expansion Anchors Same as *Expansion Shields*.

Expansion Bolt See *Bolts*.

Expansion Coil In *Air Conditioning*, the coils in an *Evaporator*, where the *Refrigerant* absorbs heat while vaporizing. See also *Refrigeration Cycle, Refrigeration Equipment*.

Expansion Joint A joint containing compressible material which will absorb movement cuased by thermal expansion. The term usually refers to such joints in concrete work.

Expansion Shields Any of a number

BROOKS CONSTRUCTION CO.

Project Name _MYERS MANUFACTURING_   Date of Estimate _APRIL 2, 1975_

COST ESTIMATE & RE-CAP SHEET

No.	Operation	Subcontractor	Amount
1.	* General Conditions		$ 8,000
2.	Earthwork	HARRIS EXCAVATING	14,168
3.	Demolition	JONES BROS.	2,107
4.	Concrete	COAST CONCRETE CONT'RS	37,980
5.	Masonry	H & R MASONRY	16,293
6.	Structural Steel & Misc. Iron	SMITH IRON WORKS	3,446
7.	Rough lumber	PACIFIC LUMBER CO.	17,975
8.	Rough Carpentry	WESTERN FRAMERS	14,570
9.	Rough Hardware	SANTA FE SUPPLY	865
10.	Finish Carpentry	J. H. FREEMAN	1,410
11.	Finish Hardware	ANDERSON CO.	2,110
12.	Millwork - Cabinets	CUSTOM CABINET CO.	1,975
13.	Drywall	BENTSON DRYWALL	4,222
14.	Plastering	TYLER PLASTERING	1,210
15.	Acoustical Ceilings	A & B CEILINGS	4,369
16.	Ceramic Tile	POTTER TILE CO.	780
17.	Resilient Flooring	BEST FLOOR COVERINGS	3,070
18.	Insulation	DYER INSULATION	1,066
19.	Metal Doors & Frames	M & S DOOR CO.	968
20.	Overhead Doors	✓	1,410
21.	Toilet Partitions	METALCRAFT	342
22.	Glass & Glazing	CROWN GLASS	1,088
23.	Store Front Work	✓	
24.	Fire Protection System	BROWN CO.	17,840
25.	Caulking	H.L. HOOPER	369
26.	Waterproofing	DONLEY CO.	890
27.	Carpets	WILLIAMS CARPET CO.	1,741
28.	Electrical	HARRIS ELECTRIC	28,428
29.	Plumbing	SANDERS PLUMBING	21,682
30.	Heating, Vent., & Air Cond.	KENNETH WARREN INC.	19,400
31.	Sheet Metal Work	C & T SHEET METAL	925
32.	Roofing	ACE ROOFING	6,100
33.	Painting	H. BROCK	3,200
34.	Asphalt Paving	SUPERIOR PAVING	12,822
35.	Fencing	NELSON FENCE	1,428
36.	Landscaping	CARO BROS.	2,500
37.	Signs	R & T SIGNS	840
38.	Parking Lot Stripping & Bumpers	BOB PETERSON	280
39.			
40.			
41.			

$ 257,869

**   Performance Bond   1,934
***   Overhead 4%   10,315
     Profit 6%   15,472

TOTAL $ 285,590

*   Includes job site related costs such as superintendent's salary, field office, temporary utilities, telephone, clean-up, etc.

**   This usually averages 3/4 of 1%.

***   This includes accounting, insurance, purchasing, project manager, etc.

FIGURE E-5   Cost Estimate

of types of devices for inserting—for the purpose of attachment—a *Bolt* or *Screw* into a concrete or masonry wall. Although the means by which they operate can vary with the manufacturer, they generally consist of a split "sleeve" (shield) which is placed into a drilled hole and expands against the sides as the bolt or screw is inserted. They are also called EXPANSION ANCHORS.

Expansive Cement See *Cement*.

Expansive Soil Soil which changes in volume with changes in moisture content, swelling when wetted and shrinking when dried. Most clay soils are expansive. A soil is generally considered to be "expansive" if its volume change potential is enough to require special design of structures supported on it.

Explosion Proof Electrical equipment built to comply with hazardous location requirements, such as areas exposed to a high concentration of dust particles or *Flammable Liquids*.

Export See *Import Fill*.

Extra Hazard (Fire-Sprinklers) See *Automatic Fire Sprinkler System*.

Extender An inexpensive *Pigment* used to increase the bulk of paint or putty. Their use does not necessarily indicate an inferior product as they occasionally improve certain film characteristics such as durability and porosity.

Extinguisher See *Fire Extinguishers*.

Extreme Fiber Stress in Bending See *Stress*.

Extrusion Any item manufactured by forcing material through a *Die*. Many plastic and aluminum products are formed by the extrusion process.

Eye Bolt See *Bolts*.

F

Facade The finished, outside face of a building. Also, an architectural element used to conceal some other, less attracitve element.

Face Brick See *Bricks*.

Factor of Safety In sturctural design, the ratio of failure load to actual load (or failure *Stress* to actual stress).

Fahrenheit Scale See *Temperature Scales*.

Fair Market Value In real estate, the highest price for which a property could be sold if exposed for sale on the open market and where neither buyer nor seller would be under pressure and where the buyer would have reasonable time to investigate the property and have reasonable knowledge of all use potential of the property.

Fallout Shelter A shelter to provide protection against harmful radiation caused by radioactive dust particles settling to earth after an atomic explosion. A fallout shelter is not intended to be protection against blast forces but its enclosure materials provide radiation protection in proportion to their density. For example, one inch of lead (weighing 700 pounds per cubic foot) will provide as much protection as seven inches of earth (weighing 100 pounds per cubic foot). Extensive and authoritative information can be obtained from the Office of Civil Defense, Department of Defense, Washington, D.C.

Falsework Temporary wood supports upon which *Concrete* or *Masonry* members are constructed and which are removed after the concrete or masonry has attained sufficient strength to be self-supporting.

Fan A device to force the movement of air. Fans are of two basic types: propeller type, also called AXIAL FLOW FANS; and CENTRIFUGAL FANS, also called *Blowers*. In the latter type a squirrel cage arrangement of blades sucks air into a housing and discharges it out perpendicularly to the axis of the impeller.

Fan Coil Unit In *Air Conditioning*, a packaged unit consists of a *Fan*, cooling and/or heating *Coils*, and *Air Filters*. Hot water, cold water, or *Refrigerant* from another source, is circulated through the coils to provide the heating or cooling effect as air is blown across them. Electric resistance coils can also be used for heating. (See Figure F-1.)

Fascia The exposed vertical edge of a roof.

Fatigue A phenomenon of metals leading to *Fracture*, whereby repeated reversals of *Stress* cause a weakening of the

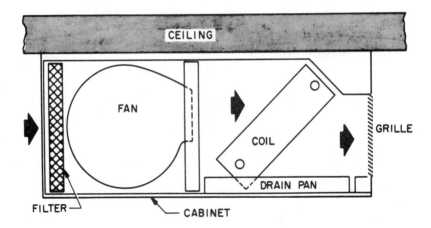

FIGURE F-1   Fan Coil Unit

metal such that failure can occur at a lower stress than would normally be expected.

Faucet See *Valves*.

Fault In geology, a fracture in the ground along which there is movement of one side relative to the other. The movement may be a few inches or several miles. *Earthquakes* are caused by stress-relieving movements along faults.

Feathering To blend or otherwise form a tapering transition between one material or surface and another.

Federal Housing Association (FHA) A government organization which does not make loans but insures them when made by supervised lending institutions. Such institutions, such as banks, life insurance companies, federal savings and loan associations, etc., and generally those whose deposits are insured by the Federal Deposit Insurance Corporation although certain approved mortgage companies and other lenders are also included.

Federal National Mortgage Associa- tion **(Fannie Mae)** An association originally created under the National Housing Act to provide a secondary mortgage market organized by private individuals and operating under federal charter and supervision. Fannie Mae is now solely owned by private stock holders but is still regulated by the Federal Government.

Federal Specification Number Federal (U.S. Government) specifications which describe the essential and technical requirements for purchase of items, materials, or services for use by the Federal Government. Such specifications are coordinated with the various governmental agencies and industries during their development. Reference to such specifications are mandatory for government procurement. Federal Speicification Numbers are made up of two nonsignificant groups of letters followed by numbers, such as MMM-A-134.

Feeder In electrical work, a primary *Conductor* generally considered as providing power between the point of connection by the electric utility company and the first distribution *Switchboard*. A

SUB-FEEDER designates such a conductor from a primary switchboard to a subsequent switchboard or *Panelboard*.

Felt See *Building Paper*.

Ferro-Cement A coined word by the originator of the process, Italian engineer Pierre Nervi, which consists of a thin shell-like sturctural surface made of alternate layers of steel mesh and cement mortar.

Ferrous Containing *Iron*.

Ferrule A metallic sleeve, joined to a *Pipe* end into which a plug is screwed which can be removed for the purpose of cleaning or examining the interior of the pipe. Also, a tube connecting device which is compressed by tightening a nut to effect the connection.

Fiberboard A flexible board-like material made from pressed fibers of *Wood*. Fiberboard panels are generally ½″ thick and 4′ × 8′.

Fiberglas A trade name (Owens-Corning Fiberglas Corp.) for a group of fiberglass related products.

Fiberglass Fiberglass is a *Plastic* material which is structurally reinforced by embedding glass fibers into a *Resin*.

Field Refers to "at the job site" as "in the field" (where the work occurs); out of the office.

Figure The pattern produced in a wood surface by *Grain* deviations.

Figured Glass See *Glass*.

Filigree Panel A decorative panel having openings which have been cut to form an intricate design.

Fill Earth placed over natural ground. When fill material is brought to a site from another location, it is called IMPORT FILL; when material is removed

from a site, it is called EXPORT. See also: *Borrow*.

Filler Metal See *Arc Welding*.

Fillet Weld See *Welded Joints*.

Film A thin layer of material.

Filter See *Air Filter*.

Final Set In concrete work, the point in time at which concrete has hardened to where it can no longer be trowelled, grooved, or otherwise shaped; no longer in a plastic state. See also: *Initial Set*.

Finder's Fee Similar to a referral fee, it is a slang term for when a third party receives a fee from one of the parties to a transaction for having brought them together.

Fine Aggregate See *Aggregate*.

Fine Grading See *Grading*.

Fines See *Aggregate*.

Finger Joint See *Woodworking Joints*.

Finish Coat In plastering, the last or final coat. See also: *Plaster Finishes, Plaster Coats*.

Finish Grade In earthwork or paving, the final surface elevation (such as for graded or paved areas) as called for on the plans or in the specifications.

Finish Hardware See *Hardware*.

Finishing Concrete See *Cement Finishing*.

Finishing Nails See *Nails*.

Fink Truss See *Trusses*.

Fir (Hem-Fir) See *Softwood*.

Fire Fire is a chemical change (primarily oxidation) occurring to a material when it is exposed to a sufficiently high temperature and oxygen is present. Two forms of energy are released—heat and light. Fires are extinguished by applying materials such as water or carbon dioxide which prevent contact of the material with oxygen. Fires are divided into four classes:

CLASS A—Ordinary combustibles such as wood, paper, and cloth which, to extinguish, require cooling and quenching.

CLASS B—*Flammable Liquids* such as grease, gasoline, and lubricating oil, requiring a smothering effect.

CLASS C—Electrical, which requires a non-conducting extinguishing agent.

CLASS D—Combustible metals such as magnesium, titanium, and sodium. These require specialized extinguishers due to the danger of increasing fire intensity by use of certain chemicals.

Fire Assembly On a *Fire Door,* fire window, or *Fire Damper*, the make-up of the components (such as the door itself, its frame, hardware, etc.) so that together they meet the requirements for the time rating of the opening protection. For example, a Class A opening requires a three hour "fire assembly."

Fire Blocking Short wood pieces placed between studs, joists or rafters to prevent the spread of fire. Also called FIRE STOPS.

Fire Clay Clay which is capable of being subjected to high temperatures without fusing or softening preceptibly. It is used for laying *Fire Brick* in kilns, ovens, and similar uses.

Fire Dampers Similar to *Fire Doors* but designed to operate in *Ductwork* or to protect similar passages.

Fire Detectors Devices to indicate the presence of fire or smoke (see Figure F-2). They are of the following varieties:

FIXED TEMPERATURE DETECTOR—Operates upon the temperature reaching a fixed setting which may be 135, 180, 190, 225, 325, 450, or 600 degrees.

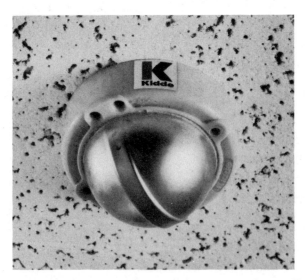

Courtesy Walter Kidde & Co.

FIGURE F-2   Fire Detector

IONIZATION DETECTOR—Operates on the products of combustion. It utilizes two chambers, one of which samples the room or unit air; the second maintains a constant separate air supply which is used for comparison purposes.

RATE OF RISE DETECTOR—Operates on a sudden temperature increase, generally exceeding 15 degrees.

SMOKE DETECTOR—Operates on a light obscuring principle using photoelectric cells.

Fire Door See *Door Types.*

Fire Extinguishers Cylindrical portable metal containers, containing chemicals which, when released and directed at a fire, will extinguish the fire by preventing oxygen from coming in contact with the material. They are effective in the initial phases of a fire but do not take the place of fire sprinkler systems or fire department equipment and techniques. The two most common types of fire extinguishers are:

DRY CHEMICAL FIRE EXTINGUISHER—The most common type of portable fire extinguisher. Basically, they contain dry chemicals stored in a cylinder having nitrogen and carbon dioxide as the pressure expellent and an extinguishing chemical. When the expellent gas is released, the dry extinguishing chemical is sprayed out with it. The most frequently used is the "ABC" type (which is effective against Class A, Class B, and Class C fires) which utilizes monoammonium phosphate as the extinguisher and is effective against any type of fire except flammable metals. Potassium bicarbonate and sodium bicarbonate are effective extinguishers for Class B and Class C fires, and sodium chloride is an effective extinguisher for Class D fires. See also: *Fire* (for "classes" of fire referred to).

CO_2 EXTINGUISHER—Carbon dioxide is stored under high pressure in liquified form within the container. When the release valve is opened, the discharge becomes a snow-like vapor which extinguishes the fire by a smothering effect, excluding the oxygen necessary for combustion. It is effective for *Flammable Liquid* and electrical fires but has little effect on paper or wood combustibles.

Fire Hose Cabinet Most *Building Codes* require fire hose cabinets to be placed at specific intervals within certain types of buildings. They usually have 75 feet of 1½" hose with a nozzle and are connected to a *Wet Stand-Pipe* (pipe system containing water under pressure). (See Figure F-3.)

Fire Hydrant A *Hydrant* used solely for fire protection pruposes.

Fire Labels Labels issued and authorized by the *Underwriters Laboratories* and placed on *Fire Doors* as proof that they meet the fire resistive time period indicated on the label.

Fire Marshal Most states have Fire Marshal Laws (having to do with fire prevention and life safety) and employ a State Fire Marshal whose duties are to enforce fire prevention laws, make investigations and coordinate fire prevention activity. Fire Marshals belong to the Fire Marshals Association of North America which is affiliated with the National Fire Protection Association. Heads of local Fire Prevention Bureaus may also be called Fire Marshals and are responsible for enforcement of State Fire Marshal laws.

Fireplace An enclosure within a building where a fire may be burned safely, with the smoke rising within a *Chimney* or *Flue* to discharge above the roof. In addition to the conventional masonry type, fireplaces are also availa-

FIGURE F-3   Fire Hose Cabinet and Portable Fire Extinguisher

ble in a variety of manufactured metal types and precast concrete.

Fire Resistive   Refers to material or assemblies of materials which prevent or slow the spread of fire.

Fire Resistive Time Periods   A system of rating building materials, walls, floors, partitions, doors, etc., in terms of their ability to resist laboratory controlled test fires for a specified period of time. Ratings vary from ¾ of an hour to 4 hours.

Fire Retardant   Describes materials or assemblies of materials (such as various roof coverings) which, by chemical treatment or composition, are more resistant to fire spread than would be the case without such treatment or special composition.

Fire-Retardant Chemical   In *Lumber* terminology, a chemical or preparation of chemicals used to reduce the flammability of wood or to retard spread of fire.

Fire Retardant Paint   See *Paint*.

Fire Shutters Similar to *Fire Doors* but designed to fire protect an opening, such as a window or vent opening, by shutter-like action.

Fire Sprinkler System See *Automatic Fire Sprinkler System*.

Fire Stops See *Fire Blocking*.

Fire Treated Wood Wood which has been pressure impregnated with a fire retardant chemical. The chemical replaces the air in the cells of the wood.

Fire Wall See *Walls*.

Fire Zone A *Building Code* subdivision of a city or jurisdiction into areas, or zones, depending upon their degree of fire hazard, and having varying construction restrictions. The spread of fire from one building to the next is of great importance, and therefore more restrictions and fire resistive requirements are usually placed on the more densely developed areas. Downtown areas—the most hazardous—are No. 1 zones; lesser degrees of commercial and industrial areas are No. 2; residential areas are No. 3 zone. A new zone (No. 4) is coming into use which includes brush or watershed areas which have the possibility of spreading fire through undeveloped property.

Fish Tape An electrician's device to aid pulling wires through *Conduit*. It is a spool of spring steel wire-like strip or *Polyethylene* cable which can be fed into the end of a conduit run and pushed around bends and corners until it emerges from the other end, where the wires to be pulled through are hooked on and the tape and wires pulled back through.

Fittings See *Pipe Fittings*.

Fixed Window See *Window Types*.

Fixture See *Plumbing Fixture*.

Fixture Unit In plumbing, a means of determining required pipe sizes for water supply and waste lines. Each type of fixture which uses water has a number value (fixture units) proportional to the amount of water it uses on an average (it is numerically equal to the gallons per minute flow divided by 7.5). By referring to charts and tables in plumbing codes a required pipe size can be determined for a specified number of fixture units.

Flagstone A type of stone which can be easily split into thin slabs and which is usually used for *Veneer* or paving.

Flame Hardening A process of hardening localized areas of *Ferrous* alloy parts by heating these areas with a high temperature flame followed by controlled cooling.

Flameproofing A chemical surface treatment of a material to impede or prevent combustion when exposed to a high heat source or to render the material self-extinguishing when the heat source is removed.

Flame Spread Classification A method of classifying wall and ceiling finish materials for their resistance to spread of flame along their surfaces. The standard test for this is the *Tunnel Test*.

Flammable *Combustible* materials or *Flammable Liquids* which readily release a gas which burns easily. The term means the same as, but is preferred to, *Inflammable*.

Flammable Liquids Any liquid having a *Flash Point* below 190°F., but this defining temperature can vary with different *Codes*. The term is used in assessing the degree of fire hazard in a building where such materials are used or stored.

Flange The widened "top" and "bottom" parts of a *Beam* or *Column* as opposed to the *Web*. (See Figure F-4.)

Flanged Joint See *Pipe Joints*.

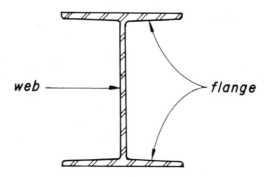

FIGURE F-4   *Web/Flange (Parts of a Beam)*

Flared Joints   See *Pipe Joints*.

Flashing

1. Sheet metal weather protection placed over a joint between different building materials or between parts of a building, in such a manner that water is prevented from entering. COUNTER-FLASHING is a second and overlapping layer of flashing where conditions are such that the first layer may not insure water tightness.

2. In painting, the irregular appearance of differences in *Color* or gloss. It is usually caused by improper sealing of porous surfaces.

Flash Point   Applies to the degree of fire hazard of *Flammable Liquids* and is the temperature of the liquid at which it gives off enough vapor such that it could ignite if a flame or spark were present. For example, the flash point of gasoline is about -35°F. and of motor oil about 350°F.

Flat   See *Apartment Nomenclature*.

Flat-Bed Truck   See *Truck Types*.

Flat Paint   See *Paint Finishes*.

Flat Slab Construction   See *Concrete Slabs*.

Flemish Bond   See *Masonry Bond Patterns*.

Flexible Conduit   See electric *Conduit*.

Flexible Pavement   A pavement which has a wearing surface of *Bituminous* material such as *Asphalt*, as opposed to *Concrete*, which is considered a "rigid" pavement.

Flexural Stress   See *Stress*.

Flexure   Bending.

Flitch   The sawn segment from the log from which *Veneers* are cut. The term also is applied to the resulting sheets of veneer which are kept together, arranged in a sequence of cutting so that adjacent sheets will have almost identical figures for matching. The term normally applies to *Hardwood Plywood* veneers.

Flitch Beam   A beam made up of one or more steel plates sandwiched between wood beams, and held in place by bolts through the assembly.

Float   A wooden trowel-like tool used by plasterers to level and smooth plaster. To provide varying textures, it can be surfaced with cork, carpet, rubber, etc. A similar tool with a very long handle is used for *Cement Finsihing* of concrete flatwork.

Float Finish   See *Plaster Finishes*.

Float Glass   See *Glass*.

Floating Foundation A raft-like concrete slab foundation which is used to distribute loads over a large area to reduce bearing pressure on the underlying soil.

Floor Diaphragm See *Diaphragm.*

Floor Drain A grate-like plumbing fixture set into and flush with a floor, into which water can be drained and carried away by an underfloor pipe to points of disposal, such as into a sewer. It has a *Trap* to prevent odors form coming up through the drain.

Floor Hardener In concrete construction, a means of improving the abrasion resistance of a slab by integrally adding, during final trowelling, material such as carborundum grits or steel filings. Other hardening processes include thin film sealing compounds such as chlorinated rubber, and chemical hardeners such as magnesium and/or magnesium zinc fluorsilicates which react with the free *Lime* to affect sealing, curing, and hydration so as to allow the concrete to develop its full potential strength. The latter penetrate the surface through chemical action with the cement and are used on existing construction. Another type of hardener is a "dust-on" type which contains an iron aggregate and is dusted on while the concrete is in a plastic state, providing a topping of about ⅛ inch. The terms *Floor Sealer* and "floor hardener" are sometimes used interchangeably when they are formulated with materials which tend to accomplish both effects.

Floor Joist See *Joists.*

Floor Plan See *Plans.*

Floor Sealer In concrete construction, a means of reducing *Dusting* of a slab by applying a liquid such as urethane, chlorinated rubber, or *Acrylic*, to the surface after the concrete has set, forming a film over the surface. Some *Curing Compounds* contain such agents.

The term "sealer" and *Floor Hardener* are sometimes used interchangeably when they are formulated with materials which tend to accomplish both effects.

Floor Sink Similar to a *Floor Drain* but having a sink recessed into the floor with or without a grate covering. It is usually used to receive a greater volume of wastes than could normally by handled by a *Floor Drain*.

Floor Tile See *Resilient Flooring, Ceramic Tile, Quarry Tile.*

Flow Line On a paved area, the line along which run-off water flows over its surface.

Flue An opening through, and within, a chimney or smokestack through which smoke passes.

Fluid Any gas, vapor, or liquid—not in a solid state.

Fluorescent Lamp A light-producing tube filled with a current conducting gas (mercury vapor) and having a fluorescent coating on the inside wall which glows as electrons flow through the tube between electrodes at opposite ends. They give more light per *Watt* of power than *Incandescent Lamps*. See also: *High Intensity Discharge Lamps.*

Flush Pull A door-opening device usually for a closet door, *Mortised* into the face of a door such that the door can be pulled or slid open or closed by inserting fingers into the pull.

Flush Valve (or Flushometer) A method of flushing, where the water used for flushing is provided directly from the water supply under pressure rather than from a storage tank on the fixture.

Flux

1. In *Welding* and *Soldering*, a material that aids in the fusion of metals by preventing oxidation.

2. In *Steel* production, a material such as limestone, sued in a *Blast Furnace* to absorb and segregate impurities.

Foamed Plastic *Plastic* material having numerous cells throughout its mass. Sponge-like, the cells may be connected with open ends or closed, depending upon the type of plastic and the method of making the foam. Foams can be flexible or rigid. Essentially, a blowing agent is added to the base *Resin* which when heated, gives off a gas expanding the resin. The most widely used foams include *Epoxy, Polystyrene, Polyvinyl Chloride,* and *Silicone.* Foamed plastic is also called CELLULAR PLASTIC or EXPANDED PLASTIC.

Fog Curing A means of *Curing a Concrete Slab* by spraying it with a mist of water to keep the surface moist during the curing period.

Foil See *Aluminum Foil.*

Folded Plate Roof A pleated type of roof structure, usually of concrete, consisting of inclined intersecting planes and used to span large areas. (See Figure F-5.)

Foot-Candle See *Illumination.*

Footings Concrete placed on or into the soil to transfer loads from a structure to the soil. Footings are further classified as ISOLATED FOOTINGS, such as a square foundation to support a column, and CONTINUOUS FOOTINGS, such as under the full length of a wall. SPREAD FOOTING, a general term including both the above, are those which are bottomed near the ground surface as opposed to *Piles* or a *Caisson.*

Foot Lambert See *Illumination.*

Foot-Pound A unit of measurement of *Torque* or *Moment*; a rotation producing force, in pounds, multiplied by its distance in feet form the point about which it tends to rotate. The term INCH-POUND has a similar meaning except the distance is in inches rather than feet.

Forced Air Furnance See *Furnace.*

Forced Air Heating See *Heating Systems.*

Forced Circulation (or Forced Draft) The movement of air by a *Fan* as opposed to *Natural Draft.*

Foreman The person in charge of a work crew. See also: *Superintendent.*

Fork Lift Truck A material handling truck having a pair of forks on the front which can be slid under a load or pallet

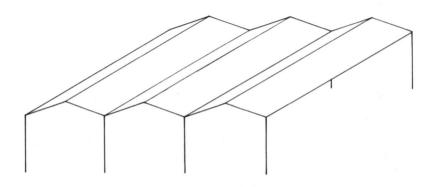

FIGURE F-5   Folded Plate Roof

and raised such that the truck can then move carrying the load. With such a truck, material can be stacked to 20' or higher. Tires are either solid rubber or pneumatic and they are powered by gasoline engines or electric (battery) motors.

Formica A trade name (American Cyanamid Corporation) for a *Plastic* laminate surfacing material, topped with an overlay of *Melamine* and used for counter tops, doors, panels, etc.

Form Ties In concrete forming, metal rods or similar devices used in *Formwork* to space two forms apart and to keep them from spreading due to the pressure of the wet concrete, and which are removed or left embedded when the forms are removed.

Formwork (or Forms) Temporary wood or metal surfaces used to contain concrete until it hardens, after which they are removed.

Foundation That part of a building which is in contact with the soil. The foundation transfers the weight of the building or structure to the soil and is almost always of concrete. Types of foundations include *Footings, Floating Foundations, Belled Piers,* and *Piles.*

Foundation Bolt See *Bolts.*

Foundation Engineer See Engineers.

Foundation Plan See *Plans.*

Fourplex See *Apartment Nomenclature*

Fracture Structural failure of a material by actual parting of the material as by breaking, cracking, or splitting.

Framing The putting together or the assembly of parts, such as *Beams, Columns, Joists, Rafters,* etc., forming a framework over which enclosing materials will be attached. See also: *Wood Framing.*

Framing Anchors Manufactured sheet metal connectors for joining wood framing members by nailing or bolting. (See Figure F-6.)

Framing Plan See *Plans.*

Freestanding Partitions Interior partitions not extending to the ceiling, which are interlocked or anchored to the floor to provide lateral support. They are manufactured in a variety of sizes, materials, configurations, and may or may not have provisions for electrical or telephone wiring. They are sometimes called BANKER PARTITIONS.

Freeze-Up Failure of a refrigerating unit to operate normally due to formation of ice at the expansion device. A valve may freeze shut or open, causing improper refrigeration in either case. On a *Coil*, it is frost formation to the extent that air flow stops or is severely restricted.

French Door See *Door Types.*

French Drain A below-ground drain consisting of a trench filled with gravel to permit movement of water through the gravel and into the ground. See also: *Drywell.*

Freon A trade name (DuPont Company) for a *Refrigerant* used in *Air Conditioning* and in Aerosol spray cans. Freon is colorless and nonflammable.

Fresco In plastering, a method of coloring the final coat of plaster by applying a water soluble paint to freshly spread *Plaster* before it dries.

Friction Bolts See *Bolts (High Strength).*

Friction Catch See *Catches.*

Friction Loss In plumbing and air

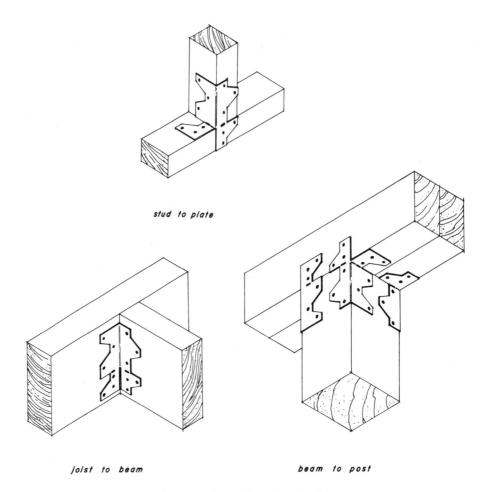

stud to plate

joist to beam   *beam to post*

FIGURE F-6   *Framing Anchors*

conditioning, the loss of pressure between the inlet and outlet of a pipe or *Duct* due to the frictional resistance, or drag, as the fluid flows through the pipe or duct.

Friction Piles   See *Piles*.

Front End Loader   See *Loader*.

Frost Action   A weathering process acting on *Soil* and rock caused by repeated cycles of freezing and thawing.

Frost Line   The maximum depth to which ground becomes frozen in winter. In the United States it ranges from about one inch in the south to 60 inches in Maine.

Frustrum (of a Cone)   The solid part of a cone or pyramidal shape which remains when a slice through it, parallel to the base, removes the upper part.

Fuel   Any substance—solid, liquid, or gaseous—which may be relatively easily ignited and burned to produce heat, light, or other usable energy. Practically all fuels consist of carbon and hydrogen in various proportions. Types of fuel include *coal* which consists chiefly of various compounds of carbon and has a B.T.U. content of between 10,500 and 14,000 B.T.U.'s per pound; *Wood,* which is rarely used as a fuel because of its scarcity and low heating value and has a B.T.U. content which varies between 5,000 and 7,200 B.T.U.'s per pound; *Natural Gas,* which consists chiefly of a mixture of *Organic* compounds of carbon and hydrogen namely methane, *Butane, Propane*, and ethane, and has a B.T.U. content of between 900 and 1,200 B.T.U.'s per cubic foot; and *fuel oil* which, broadly, is any liquid petroleum product used for the generation of heat —its B.T.U. content ranges from 130,000 to 155,000 B.T.U.'s per gallon. See also: *Combustion*.

Full-Cell Process Any process for impregnating *Wood* with preservatives or chemicals in which a vacuum is drawn to remove air from the wood before admitting the preservative. This provides large absorption and retention of the preservative in the treated portions.

Furnace That part of a boiler or warm-air heating plant in which combustion takes place. Also, a complete heating unit for transferring heat from fuel being burned to the air supplied to a heating system. A GRAVITY FURNACE de-pends upon the rising of warm (less dense) air for circulation to the spaces being heated. A FORCED AIR FUR-NACE uses a *Fan* or *Blower* to force the heated air through *Ductwork*.

Furring Wood or metal strips applied over a wall or other framing to provide a means of attaching a finished wall or ceiling.

Furring Channel A small steel *Channel* used for *Furring*.

Fuse An electrical safety device in-serted in a circuit to prevent overload. It contains a wire which will melt if exces-sive current flows through it, thereby in-terrupting current flow. Its basic differ-ence from a *Circuit Breaker* is that it de-stroys itself when an overload occurs.

Fusible Link A connecting device having a predetermined melting point, usually for the purpose of causing a *Fire Door* or hatch to close by fire melting the hold-open link. They are usually used on fire doors to close the door automatically by a system of counterweighted cables when fire melts the link.

Fusion Weld A *Welding* method whereby fusion of parts to be joined is accomplished by a high contact pressure and heat.

G

Gable An end wall of a building having a triangularly shaped upper portion formed by a sloping roof on either side of a ridge.

Gad A pointed steel bar used for prying, wedging, or breaking rock.

Gage

1. The distance between rows of bolt or rivet holes.

2. A system indicating the thickness of steel sheets or wire. See also: *American Standard Wire Gage or Brown and Sharpe Wire Gage.*

Gaging In plastering, the controlling of the setting time of a plaster mix by mixing GAGING PLASTER (which is a specially processed *Gypsum Plaster*) or *Keene's Cement*, with *Lime Putty* or *Hydrated Lime*.

Gaging Plaster See *Gaging*.

Gallon One gallon equals two quarts or eight pints. One gallon of water weighs 8.3 pounds, and there are 7.48 gallons in one cubic foot.

Galvanic Corrosion When two dissimilar metals come into contact and one of them is farther down the *Electromotive Series* than the other, and if moisture is present, current will flow from one to the other (see *Electrolysis*) causing corrosion of the one lower on the electromotive series. Example: Corrosion of *Steel* in contact with *Aluminum*.

Galvanized Steel Pipe See *Pipe*.

Galvanizing A zinc coating put on steel to prevent corrosion. The item to be galvanized is dipped into molten zinc. Generally the charge for galvanizing is based upon the amount of zinc deposited as determined by weighing the item before and after dipping.

Gambrel Roof A roof having two pitches between each eave and the ridge, the lower of which is steeper than the upper. (See Figure R-3.)

Gantry A moving framework, containing a crane, which travels on rails at ground level.

Garden Apartment See *Apartment Nomenclature*.

Gas See *Acetylene, Liquified Petroleum Gas (L.P.G.), Fuel*.

Gas Air Conditioning See *Refrigeration Equipment. (Absorption Refrigeration.)*

Gas Cutting See *Gas Welding and Cutting*.

Gasket A thin, flat seal placed between two joining parts to prevent leakage of air or water.

Gas Meter A meter used to measure and indicate the amount of *Natural Gas* flowing through a line, measured in cubic feet, and from which periodic readings are made by the utility company for billing purposes.

Gas Shielded Arc Welding See *Arc Welding*.

Gas Welding and Cutting As opposed to *Arc Welding*, the heat required to melt and fuse the metal parts together is supplied by a torch using a flammable gas burning in air or oxygen, *Propane* and *Butane* torches are used but do not produce a temperature high enough for welding some metals. Since the burning of a flammable gas in pure oxygen results in a higher temperature, the oxyacetylene torch is most often used. A mixture of oxygen and acetylene generates an intense flame which, for *Welding*, melts the metal on each side of the joint and simultaneously a *Filler Metal*—in the form of a rod—is melted into the space between the parts to be joined. For cutting, the metal is first heated where the cut is to be made and then a controllable jet of oxygen is discharged from the nozzle to produce extremely high temperatures and is directed into the area to burn through the metal.

Gate Valve See *Valves*.

Gauge Spelling variation of *Gage*.

Gazebo A small open structure, usually round, octagonal or of similar shape, set in surroundings such as gardens for the purpose of viewing them or resting in its shelter.

General Conditions (and Supplementary General Conditions) Generally occuring prior to the *Specifications*, they represent an extension of the contract agreement and contain contractual-legal requirements, as opposed to specific technical requirements contained in the 16 "Divisions" of the Specifications. They include such items as: methods of payment, responsibilities of the *Contractor* and the owner; insurance requirements; and other items of legal significance having the purpose of avoiding disagreements between the contractor and the owner. The most commonly used General Conditions are those published by the American Institute of Architects. "Supplementary General Conditions" are modifications of the General Conditions to satisfy requirements for a particular construction project. (See Figure G-1.)

General Contractor See *Contractor*.

General Services Administration An independent federal organization which primarily provides the government with an economic and efficient system of management of its properties including construction and operation of buildings, procurement and distribution of supplies, utilization and dispersal of properties. Its other functions include the transportation and stockpiling of strategic materials. Regional offices are maintained. Address: General Services Administration Building, 18th and F Streets, N.W., Washington, D.C. 20405

Generator A machine for making electricity, available in wide variety of sizes to provide, depending upon the type, either *Alternating Current* (A.C.), or *Direct Current* (D.C.).

Geodesic Dome A structurally stable, dome-shaped structure consisting entirely of straight members connected to each other to form a continuous surface of small triangles.

Geodesy The study of the size and shape of the earth.

Geologic Map A map which shows the character, distribution, and structural details of material at or near the ground surface, such as *Faults*, dips, strikes, and anticlines.

THE AMERICAN INSTITUTE OF ARCHITECTS

AIA Document A201

General Conditions of the Contract for Construction

THIS DOCUMENT HAS IMPORTANT LEGAL CONSEQUENCES; CONSULTATION WITH AN ATTORNEY IS ENCOURAGED WITH RESPECT TO ITS MODIFICATION

TABLE OF ARTICLES

This document has been approved and endorsed by The Associated General Contractors of America.

AIA DOCUMENT A201 • GENERAL CONDITIONS OF THE CONTRACT FOR CONSTRUCTION • TWELFTH EDITION • APRIL 1970 ED.
AIA® • © 1970 • THE AMERICAN INSTITUTE OF ARCHITECTS, 1735 NEW YORK AVENUE, N.W., WASHINGTON, D.C. 20006

FIGURE G-1 *General Conditions (Users should ascertain the latest edition when using AIA forms)*

INDEX

AIA DOCUMENT A201 • GENERAL CONDITIONS OF THE CONTRACT FOR CONSTRUCTION • TWELFTH EDITION • APRIL 1970 ED.
AIA® • © 1970 • THE AMERICAN INSTITUTE OF ARCHITECTS, 1735 NEW YORK AVENUE, N.W., WASHINGTON, D.C. 20006

GENERAL CONDITIONS OF THE CONTRACT FOR CONSTRUCTION

ARTICLE 1

CONTRACT DOCUMENTS

1.1 DEFINITIONS

1.1.1 THE CONTRACT DOCUMENTS

The Contract Documents consist of the Agreement, the Conditions of the Contract (General, Supplementary and other Conditions), the Drawings, the Specifications, all Addenda issued prior to execution of the Contract, and all Modifications thereto. A Modification is (1) a written amendment to the Contract signed by both parties, (2) a Change Order, (3) a written interpretation issued by the Architect pursuant to Subparagraph 1.2.5, or (4) a written order for a minor change in the Work issued by the Architect pursuant to Paragraph 12.3. A Modification may be made only after execution of the Contract.

1.1.2 THE CONTRACT

The Contract Documents form the Contract. The Contract represents the entire and integrated agreement between the parties hereto and supersedes all prior negotiations, representations, or agreements, either written or oral, including the bidding documents. The Contract may be amended or modified only by a Modification as defined in Subparagraph 1.1.1.

1.1.3 THE WORK

The term Work includes all labor necessary to produce the construction required by the Contract Documents, and all materials and equipment incorporated or to be incorporated in such construction.

1.1.4 THE PROJECT

The Project is the total construction designed by the Architect of which the Work performed under the Contract Documents may be the whole or a part.

1.2 EXECUTION, CORRELATION, INTENT AND INTERPRETATIONS

1.2.1 The Contract Documents shall be signed in not less than triplicate by the Owner and Contractor. If either the Owner or the Contractor or both do not sign the Conditions of the Contract, Drawings, Specifications, or any of the other Contract Documents, the Architect shall identify them.

1.2.2 By executing the Contract, the Contractor represents that he has visited the site, familiarized himself with the local conditions under which the Work is to be performed, and correlated his observations with the requirements of the Contract Documents.

1.2.3 The Contract Documents are complementary, and what is required by any one shall be as binding as if required by all. The intention of the Documents is to include all labor, materials, equipment and other items

as provided in Subparagraph 4.4.1 necessary for the proper execution and completion of the Work. It is not intended that Work not covered under any heading, section, branch, class or trade of the Specifications shall be supplied unless it is required elsewhere in the Contract Documents or is reasonably inferable therefrom as being necessary to produce the intended results. Words which have well-known technical or trade meanings are used herein in accordance with such recognized meanings.

1.2.4 The organization of the Specifications into divisions, sections and articles, and the arrangement of Drawings shall not control the Contractor in dividing the Work among Subcontractors or in establishing the extent of Work to be performed by any trade.

1.2.5 Written interpretations necessary for the proper execution or progress of the Work, in the form of drawings or otherwise, will be issued with reasonable promptness by the Architect and in accordance with any schedule agreed upon. Either party to the Contract may make written request to the Architect for such interpretations. Such interpretations shall be consistent with and reasonably inferable from the Contract Documents, and may be effected by Field Order.

1.3 COPIES FURNISHED AND OWNERSHIP

1.3.1 Unless otherwise provided in the Contract Documents, the Contractor will be furnished, free of charge, all copies of Drawings and Specifications reasonably necessary for the execution of the Work.

1.3.2 All Drawings, Specifications and copies thereof furnished by the Architect are and shall remain his property. They are not to be used on any other project, and, with the exception of one contract set for each party to the Contract, are to be returned to the Architect on request at the completion of the Work.

ARTICLE 2

ARCHITECT

2.1 DEFINITION

2.1.1 The Architect is the person or organization licensed to practice architecture and identified as such in the Agreement and is referred to throughout the Contract Documents as if singular in number and masculine in gender. The term Architect means the Architect or his authorized representative.

2.1.2 Nothing contained in the Contract Documents shall create any contractual relationship between the Architect and the Contractor.

2.2 ADMINISTRATION OF THE CONTRACT

2.2.1 The Architect will provide general Administration of the Construction Contract, including performance of the functions hereinafter described.

AIA DOCUMENT A201 • GENERAL CONDITIONS OF THE CONTRACT FOR CONSTRUCTION • TWELFTH EDITION • APRIL 1970 ED.
AIA® • © 1970 • THE AMERICAN INSTITUTE OF ARCHITECTS, 1735 NEW YORK AVENUE, N.W., WASHINGTON, D.C. 20006

2.2.2 The Architect will be the Owner's representative during construction and until final payment. The Architect will have authority to act on behalf of the Owner to the extent provided in the Contract Documents, unless otherwise modified by written instrument which will be shown to the Contractor. The Architect will advise and consult with the Owner, and all of the Owner's instructions to the Contractor shall be issued through the Architect.

2.2.3 The Architect shall at all times have access to the Work wherever it is in preparation and progress. The Contractor shall provide facilities for such access so the Architect may perform his functions under the Contract Documents.

2.2.4 The Architect will make periodic visits to the site to familiarize himself generally with the progress and quality of the Work and to determine in general if the Work is proceeding in accordance with the Contract Documents. On the basis of his on-site observations as an architect, he will keep the Owner informed of the progress of the Work, and will endeavor to guard the Owner against defects and deficiencies in the Work of the Contractor. The Architect will not be required to make exhaustive or continuous on-site inspections to check the quality or quantity of the Work. The Architect will not be responsible for construction means, methods, techniques, sequences or procedures, or for safety precautions and programs in connection with the Work, and he will not be responsible for the Contractor's failure to carry out the Work in accordance with the Contract Documents.

2.2.5 Based on such observations and the Contractor's Applications for Payment, the Architect will determine the amounts owing to the Contractor and will issue Certificates for Payment in such amounts, as provided in Paragraph 9.4.

2.2.6 The Architect will be, in the first instance, the interpreter of the requirements of the Contract Documents and the judge of the performance thereunder by both the Owner and Contractor. The Architect will, within a reasonable time, render such interpretations as he may deem necessary for the proper execution or progress of the Work.

2.2.7 Claims, disputes and other matters in question between the Contractor and the Owner relating to the execution or progress of the Work or the interpretation of the Contract Documents shall be referred initially to the Architect for decision which he will render in writing within a reasonable time.

2.2.8 All interpretations and decisions of the Architect shall be consistent with the intent of the Contract Documents. In his capacity as interpreter and judge, he will exercise his best efforts to insure faithful performance by both the Owner and the Contractor and will not show partiality to either.

2.2.9 The Architect's decisions in matters relating to artistic effect will be final if consistent with the intent of the Contract Documents.

2.2.10 Any claim, dispute or other matter that has been referred to the Architect, except those relating to artistic effect as provided in Subparagraph 2.2.9 and except any which have been waived by the making or acceptance of final payment as provided in Subparagraphs 9.7.5 and 9.7.6, shall be subject to arbitration upon the written demand of either party. However, no demand for arbitration of any such claim, dispute or other matter may be made until the earlier of:

2.2.10.1 The date on which the Architect has rendered his written decision, or

.2 the tenth day after the parties have presented their evidence to the Architect or have been given a reasonable opportunity to do so, if the Architect has not rendered his written decision by that date.

2.2.11 If a decision of the Architect is made in writing and states that it is final but subject to appeal, no demand for arbitration of a claim, dispute or other matter covered by such decision may be made later than thirty days after the date on which the party making the demand received the decision. The failure to demand arbitration within said thirty days' period will result in the Architect's decision becoming final and binding upon the Owner and the Contractor. If the Architect renders a decision after arbitration proceedings have been initiated, such decision may be entered as evidence but will not supersede any arbitration proceedings unless the decision is acceptable to the parties concerned.

2.2.12 The Architect will have authority to reject Work which does not conform to the Contract Documents. Whenever, in his reasonable opinion, he considers it necessary or advisable to insure the proper implementation of the intent of the Contract Documents, he will have authority to require special inspection or testing of the Work in accordance with Subparagraph 7.8.2 whether or not such Work be then fabricated, installed or completed. However, neither the Architect's authority to act under this Subparagraph 2.2.12, nor any decision made by him in good faith either to exercise or not to exercise such authority, shall give rise to any duty or responsibility of the Architect to the Contractor, any Subcontractor, any of their agents or employees, or any other person performing any of the Work.

2.2.13 The Architect will review Shop Drawings and Samples as provided in Subparagraphs 4.13.1 through 4.13.8 inclusive.

2.2.14 The Architect will prepare Change Orders in accordance with Article 12, and will have authority to order minor changes in the Work as provided in Subparagraph 12.3.1.

2.2.15 The Architect will conduct inspections to determine the dates of Substantial Completion and final completion, will receive and review written guarantees and related documents required by the Contract and assembled by the Contractor and will issue a final Certificate for Payment.

2.2.16 If the Owner and Architect agree, the Architect will provide one or more Full-Time Project Representatives to assist the Architect in carrying out his responsibilities at the site. The duties, responsibilities and limitations of authority of any such Project Representative shall be as set forth in an exhibit to be incorporated in the Contract Documents.

AIA DOCUMENT A201 • GENERAL CONDITIONS OF THE CONTRACT FOR CONSTRUCTION • TWELFTH EDITION • APRIL 1970 ED.
AIA® • © 1970 • THE AMERICAN INSTITUTE OF ARCHITECTS, 1735 NEW YORK AVENUE, N.W., WASHINGTON, D.C. 20006

2.2.17 The duties, responsibilities and limitations of authority of the Architect as the Owner's representative during construction as set forth in Articles 1 through 14 inclusive of these General Conditions will not be modified or extended without written consent of the Owner, the Contractor and the Architect.

2.2.18 The Architect will not be responsible for the acts or omissions of the Contractor, any Subcontractors, or any of their agents or employees, or any other persons performing any of the Work.

2.2.19 In case of the termination of the employment of the Architect, the Owner shall appoint an architect against whom the Contractor makes no reasonable objection whose status under the Contract Documents shall be that of the former architect. Any dispute in connection with such appointment shall be subject to arbitration.

ARTICLE 3

OWNER

3.1 DEFINITION

3.1.1 The Owner is the person or organization identified as such in the Agreement and is referred to throughout the Contract Documents as if singular in number and masculine in gender. The term Owner means the Owner or his authorized representative.

3.2 INFORMATION AND SERVICES REQUIRED OF THE OWNER

3.2.1 The Owner shall furnish all surveys describing the physical characteristics, legal limits and utility locations for the site of the Project.

3.2.2 The Owner shall secure and pay for easements for permanent structures or permanent changes in existing facilities.

3.2.3 Information or services under the Owner's control shall be furnished by the Owner with reasonable promptness to avoid delay in the orderly progress of the Work.

3.2.4 The Owner shall issue all instructions to the Contractor through the Architect.

3.2.5 The foregoing are in addition to other duties and responsibilities of the Owner enumerated herein and especially those in respect to Payment and Insurance in Articles 9 and 11 respectively.

3.3 OWNER'S RIGHT TO STOP THE WORK

3.3.1 If the Contractor fails to correct defective Work or persistently fails to supply materials or equipment in accordance with the Contract Documents, the Owner may order the Contractor to stop the Work, or any portion thereof, until the cause for such order has been eliminated.

3.4 OWNER'S RIGHT TO CARRY OUT THE WORK

3.4.1 If the Contractor defaults or neglects to carry out the Work in accordance with the Contract Documents or fails to perform any provision of the Contract, the Owner may, after seven days' written notice to the Contractor and without prejudice to any other remedy he

may have, make good such deficiencies. In such case an appropriate Change Order shall be issued deducting from the payments then or thereafter due the Contractor the cost of correcting such deficiencies, including the cost of the Architect's additional services made necessary by such default, neglect or failure. The Architect must approve both such action and the amount charged to the Contractor. If the payments then or thereafter due the Contractor are not sufficient to cover such amount, the Contractor shall pay the difference to the Owner.

ARTICLE 4

CONTRACTOR

4.1 DEFINITION

4.1.1 The Contractor is the person or organization identified as such in the Agreement and is referred to throughout the Contract Documents as if singular in number and masculine in gender. The term Contractor means the Contractor or his authorized representative.

4.2 REVIEW OF CONTRACT DOCUMENTS

4.2.1 The Contractor shall carefully study and compare the Contract Documents and shall at once report to the Architect any error, inconsistency or omission he may discover. The Contractor shall not be liable to the Owner or the Architect for any damage resulting from any such errors, inconsistencies or omissions in the Contract Documents. The Contractor shall do no Work without Drawings, Specifications or Modifications.

4.3 SUPERVISION AND CONSTRUCTION PROCEDURES

4.3.1 The Contractor shall supervise and direct the Work, using his best skill and attention. He shall be solely responsible for all construction means, methods, techniques, sequences and procedures and for coordinating all portions of the Work under the Contract.

4.4 LABOR AND MATERIALS

4.4.1 Unless otherwise specifically noted, the Contractor shall provide and pay for all labor, materials, equipment, tools, construction equipment and machinery, water, heat, utilities, transportation, and other facilities and services necessary for the proper execution and completion of the Work.

4.4.2 The Contractor shall at all times enforce strict discipline and good order among his employees and shall not employ on the Work any unfit person or anyone not skilled in the task assigned to him.

4.5 WARRANTY

4.5.1 The Contractor warrants to the Owner and the Architect that all materials and equipment furnished under this Contract will be new unless otherwise specified, and that all Work will be of good quality, free from faults and defects and in conformance with the Contract Documents. All Work not so conforming to these standards may be considered defective. If required by the Architect, the Contractor shall furnish satisfactory evidence as to the kind and quality of materials and equipment.

4.6 TAXES

4.6.1 The Contractor shall pay all sales, consumer, use and other similar taxes required by law.

AIA DOCUMENT A201 • GENERAL CONDITIONS OF THE CONTRACT FOR CONSTRUCTION • TWELFTH EDITION • APRIL 1970 ED.
AIA® • © 1970 • THE AMERICAN INSTITUTE OF ARCHITECTS, 1735 NEW YORK AVENUE, N.W., WASHINGTON, D.C. 20006

4.7 PERMITS, FEES AND NOTICES

4.7.1 The Contractor shall secure and pay for all permits, governmental fees and licenses necessary for the proper execution and completion of the Work, which are applicable at the time the bids are received. It is not the responsibility of the Contractor to make certain that the Drawings and Specifications are in accordance with applicable laws, statutes, building codes and regulations.

4.7.2 The Contractor shall give all notices and comply with all laws, ordinances, rules, regulations and orders of any public authority bearing on the performance of the Work. If the Contractor observes that any of the Contract Documents are at variance therewith in any respect, he shall promptly notify the Architect in writing, and any necessary changes shall be adjusted by appropriate Modification. If the Contractor performs any Work knowing it to be contrary to such laws, ordinances, rules and regulations, and without such notice to the Architect, he shall assume full responsibility therefor and shall bear all costs attributable thereto.

4.8 CASH ALLOWANCES

4.8.1 The Contractor shall include in the Contract Sum all allowances stated in the Contract Documents. These allowances shall cover the net cost of the materials and equipment delivered and unloaded at the site, and all applicable taxes. The Contractor's handling costs on the site, labor, installation costs, overhead, profit and other expenses contemplated for the original allowance shall be included in the Contract Sum and not in the allowance. The Contractor shall cause the Work covered by these allowances to be performed for such amounts and by such persons as the Architect may direct, but he will not be required to employ persons against whom he makes a reasonable objection. If the cost, when determined, is more than or less than the allowance, the Contract Sum shall be adjusted accordingly by Change Order which will include additional handling costs on the site, labor, installation costs, overhead, profit and other expenses resulting to the Contractor from any increase over the original allowance.

4.9 SUPERINTENDENT

4.9.1 The Contractor shall employ a competent superintendent and necessary assistants who shall be in attendance at the Project site during the progress of the Work. The superintendent shall be satisfactory to the Architect, and shall not be changed except with the consent of the Architect, unless the superintendent proves to be unsatisfactory to the Contractor and ceases to be in his employ. The superintendent shall represent the Contractor and all communications given to the superintendent shall be as binding as if given to the Contractor. Important communications will be confirmed in writing. Other communications will be so confirmed on written request in each case.

4.10 RESPONSIBILITY FOR THOSE PERFORMING THE WORK

4.10.1 The Contractor shall be responsible to the Owner for the acts and omissions of all his employees and all Subcontractors, their agents and employees, and all other persons performing any of the Work under a contract with the Contractor.

4.11 PROGRESS SCHEDULE

4.11.1 The Contractor, immediately after being awarded the Contract, shall prepare and submit for the Architect's approval an estimated progress schedule for the Work. The progress schedule shall be related to the entire Project to the extent required by the Contract Documents. This schedule shall indicate the dates for the starting and completion of the various stages of construction and shall be revised as required by the conditions of the Work, subject to the Architect's approval.

4.12 DRAWINGS AND SPECIFICATIONS AT THE SITE

4.12.1 The Contractor shall maintain at the site for the Owner one copy of all Drawings, Specifications, Addenda, approved Shop Drawings, Change Orders and other Modifications, in good order and marked to record all changes made during construction. These shall be available to the Architect. The Drawings, marked to record all changes made during construction, shall be delivered to him for the Owner upon completion of the Work.

4.13 SHOP DRAWINGS AND SAMPLES

4.13.1 Shop Drawings are drawings, diagrams, illustrations, schedules, performance charts, brochures and other data which are prepared by the Contractor or any Subcontractor, manufacturer, supplier or distributor, and which illustrate some portion of the Work.

4.13.2 Samples are physical examples furnished by the Contractor to illustrate materials, equipment or workmanship, and to establish standards by which the Work will be judged.

4.13.3 The Contractor shall review, stamp with his approval and submit, with reasonable promptness and in orderly sequence so as to cause no delay in the Work or in the work of any other contractor, all Shop Drawings and Samples required by the Contract Documents or subsequently by the Architect as covered by Modifications. Shop Drawings and Samples shall be properly identified as specified, or as the Architect may require. At the time of submission the Contractor shall inform the Architect in writing of any deviation in the Shop Drawings or Samples from the requirements of the Contract Documents.

4.13.4 By approving and submitting Shop Drawings and Samples, the Contractor thereby represents that he has determined and verified all field measurements, field construction criteria, materials, catalog numbers and similar data, or will do so, and that he has checked and coordinated each Shop Drawing and Sample with the requirements of the Work and of the Contract Documents.

4.13.5 The Architect will review and approve Shop Drawings and Samples with reasonable promptness so as to cause no delay, but only for conformance with the design concept of the Project and with the information given in the Contract Documents. The Architect's approval of a separate item shall not indicate approval of an assembly in which the item functions.

4.13.6 The Contractor shall make any corrections required by the Architect and shall resubmit the required number of corrected copies of Shop ·Drawings or new Samples until approved. The Contractor shall direct spe-

AIA DOCUMENT A201 • GENERAL CONDITIONS OF THE CONTRACT FOR CONSTRUCTION • TWELFTH EDITION • APRIL 1970 ED.
AIA® • © 1970 • THE AMERICAN INSTITUTE OF ARCHITECTS, 1735 NEW YORK AVENUE, N.W., WASHINGTON, D.C. 20006

cific attention in writing or on resubmitted Shop Drawings to revisions other than the corrections requested by the Architect on previous submissions.

4.13.7 The Architect's approval of Shop Drawings or Samples shall not relieve the Contractor of responsibility for any deviation from the requirements of the Contract Documents unless the Contractor has informed the Architect in writing of such deviation at the time of submission and the Architect has given written approval to the specific deviation, nor shall the Architect's approval relieve the Contractor from responsibility for errors or omissions in the Shop Drawings or Samples.

4.13.8 No portion of the Work requiring a Shop Drawing or Sample submission shall be commenced until the submission has been approved by the Architect. All such portions of the Work shall be in accordance with approved Shop Drawings and Samples.

4.14 USE OF SITE

4.14.1 The Contractor shall confine operations at the site to areas permitted by law, ordinances, permits and the Contract Documents and shall not unreasonably encumber the site with any materials or equipment.

4.15 CUTTING AND PATCHING OF WORK

4.15.1 The Contractor shall do all cutting, fitting or patching of his Work that may be required to make its several parts fit together properly, and shall not endanger any Work by cutting, excavating or otherwise altering the Work or any part of it.

4.16 CLEANING UP

4.16.1 The Contractor at all times shall keep the premises free from accumulation of waste materials or rubbish caused by his operations. At the completion of the Work he shall remove all his waste materials and rubbish from and about the Project as well as all his tools, construction equipment, machinery and surplus materials, and shall clean all glass surfaces and leave the Work "broom-clean" or its equivalent, except as otherwise specified.

4.16.2 If the Contractor fails to clean up, the Owner may do so and the cost thereof shall be charged to the Contractor as provided in Paragraph 3.4.

4.17 COMMUNICATIONS

4.17.1 The Contractor shall forward all communications to the Owner through the Architect.

4.18 INDEMNIFICATION

4.18.1 The Contractor shall indemnify and hold harmless the Owner and the Architect and their agents and employees from and against all claims, damages, losses and expenses including attorneys' fees arising out of or resulting from the performance of the Work, provided that any such claim, damage, loss or expense (1) is attributable to bodily injury, sickness, disease or death, or to injury to or destruction of tangible property (other than the Work itself) including the loss of use resulting therefrom, and (2) is caused in whole or in part by any negligent act or omission of the Contractor, any Subcontractor, anyone directly or indirectly employed by any of them or anyone for whose acts any of them may be liable,

regardless of whether or not it is caused in part by a party indemnified hereunder.

4.18.2 In any and all claims against the Owner or the Architect or any of their agents or employees by any employee of the Contractor, any Subcontractor, anyone directly or indirectly employed by any of them or anyone for whose acts any of them may be liable, the indemnification obligation under this Paragraph 4.18 shall not be limited in any way by any limitation on the amount or type of damages, compensation or benefits payable by or for the Contractor or any Subcontractor under workmen's compensation acts, disability benefit acts or other employee benefit acts.

4.18.3 The obligations of the Contractor under this Paragraph 4.18 shall not extend to the liability of the Architect, his agents or employees arising out of (1) the preparation or approval of maps, drawings, opinions, reports, surveys, Change Orders, designs or specifications, or (2) the giving of or the failure to give directions or instructions by the Architect, his agents or employees provided such giving or failure to give is the primary cause of the injury or damage.

ARTICLE 5

SUBCONTRACTORS

5.1 DEFINITION

5.1.1 A Subcontractor is a person or organization who has a direct contract with the Contractor to perform any of the Work at the site. The term Subcontractor is referred to throughout the Contract Documents as if singular in number and masculine in gender and means a Subcontractor or his authorized representative.

5.1.2 A Sub-subcontractor is a person or organization who has a direct or indirect contract with a Subcontractor to perform any of the Work at the site. The term Sub-subcontractor is referred to throughout the Contract Documents as if singular in number and masculine in gender and means a Sub-subcontractor or an authorized representative thereof.

5.1.3 Nothing contained in the Contract Documents shall create any contractual relation between the Owner or the Architect and any Subcontractor or Sub-subcontractor.

**5.2 AWARD OF SUBCONTRACTS AND OTHER
CONTRACTS FOR PORTIONS OF THE WORK**

5.2.1 Unless otherwise specified in the Contract Documents or in the Instructions to Bidders, the Contractor, as soon as practicable after the award of the Contract, shall furnish to the Architect in writing for acceptance by the Owner and the Architect a list of the names of the Subcontractors proposed for the principal portions of the Work. The Architect shall promptly notify the Contractor in writing if either the Owner or the Architect, after due investigation, has reasonable objection to any Subcontractor on such list and does not accept him. Failure of the Owner or Architect to make objection promptly to any Subcontractor on the list shall constitute acceptance of such Subcontractor.

5.2.2 The Contractor shall not contract with any Subcontractor or any person or organization (including those who are to furnish materials or equipment fabricated to a special design) proposed for portions of the Work designated in the Contract Documents or in the Instructions to Bidders or, if none is so designated, with any Subcontractor proposed for the principal portions of the Work who has been rejected by the Owner and the Architect. The Contractor will not be required to contract with any Subcontractor or person or organization against whom he has a reasonable objection.

5.2.3 If the Owner or Architect refuses to accept any Subcontractor or person or organization on a list submitted by the Contractor in response to the requirements of the Contract Documents or the Instructions to Bidders, the Contractor shall submit an acceptable substitute and the Contract Sum shall be increased or decreased by the difference in cost occasioned by such substitution and an appropriate Change Order shall be issued; however, no increase in the Contract Sum shall be allowed for any such substitution unless the Contractor has acted promptly and responsively in submitting for acceptance any list or lists of names as required by the Contract Documents or the Instructions to Bidders.

5.2.4 If the Owner or the Architect requires a change of any proposed Subcontractor or person or organization previously accepted by them, the Contract Sum shall be increased or decreased by the difference in cost occasioned by such change and an appropriate Change Order shall be issued.

5.2.5 The Contractor shall not make any substitution for any Subcontractor or person or organization who has been accepted by the Owner and the Architect, unless the substitution is acceptable to the Owner and the Architect.

5.3   SUBCONTRACTUAL RELATIONS

5.3.1 All work performed for the Contractor by a Subcontractor shall be pursuant to an appropriate agreement between the Contractor and the Subcontractor (and where appropriate between Subcontractors and Subsubcontractors) which shall contain provisions that:

.1 preserve and protect the rights of the Owner and the Architect under the Contract with respect to the Work to be performed under the subcontract so that the subcontracting thereof will not prejudice such rights;

.2 require that such Work be performed in accordance with the requirements of the Contract Documents;

.3 require submission to the Contractor of applications for payment under each subcontract to which the Contractor is a party, in reasonable time to enable the Contractor to apply for payment in accordance with Article 9;

.4 require that all claims for additional costs, extensions of time, damages for delays or otherwise with respect to subcontracted portions of the Work shall be submitted to the Contractor (via any Subcontractor or Sub-subcontractor where appropriate) in sufficient time so that the Con-

tractor may comply in the manner provided in the Contract Documents for like claims by the Contractor upon the Owner;

.5 waive all rights the contracting parties may have against one another for damages caused by fire or other perils covered by the property insurance described in Paragraph 11.3, except such rights as they may have to the proceeds of such insurance held by the Owner as trustee under Paragraph 11.3; and

.6 obligate each Subcontractor specifically to consent to the provisions of this Paragraph 5.3.

5.4   PAYMENTS TO SUBCONTRACTORS

5.4.1 The Contractor shall pay each Subcontractor, upon receipt of payment from the Owner, an amount equal to the percentage of completion allowed to the Contractor on account of such Subcontractor's Work, less the percentage retained from payments to the Contractor. The Contractor shall also require each Subcontractor to make similar payments to his subcontractors.

5.4.2 If the Architect fails to issue a Certificate for Payment for any cause which is the fault of the Contractor and not the fault of a particular Subcontractor, the Contractor shall pay that Subcontractor on demand, made at any time after the Certificate for Payment should otherwise have been issued, for his Work to the extent completed, less the retained percentage.

5.4.3 The Contractor shall pay each Subcontractor a just share of any insurance moneys received by the Contractor under Article 11, and he shall require each Subcontractor to make similar payments to his subcontractors.

5.4.4 The Architect may, on request and at his discretion, furnish to any Subcontractor, if practicable, information regarding percentages of completion certified to the Contractor on account of Work done by such Subcontractors.

5.4.5 Neither the Owner nor the Architect shall have any obligation to pay or to see to the payment of any moneys to any Subcontractor except as may otherwise be required by law.

ARTICLE 6

SEPARATE CONTRACTS

6.1   OWNER'S RIGHT TO AWARD SEPARATE CONTRACTS

6.1.1 The Owner reserves the right to award other contracts in connection with other portions of the Project under these or similar Conditions of the Contract.

6.1.2 When separate contracts are awarded for different portions of the Project, "the Contractor" in the contract documents in each case shall be the contractor who signs each separate contract.

6.2   MUTUAL RESPONSIBILITY OF CONTRACTORS

6.2.1 The Contractor shall afford other contractors reasonable opportunity for the introduction and storage of their materials and equipment and the execution of their work, and shall properly connect and coordinate his Work with theirs.

6.2.2 If any part of the Contractor's Work depends for proper execution or results upon the work of any other separate contractor, the Contractor shall inspect and promptly report to the Architect any apparent discrepancies or defects in such work that render it unsuitable for such proper execution and results. Failure of the Contractor so to inspect and report shall constitute an acceptance of the other contractor's work as fit and proper to receive his Work, except as to defects which may develop in the other separate contractor's work after the execution of the Contractor's Work.

6.2.3 Should the Contractor cause damage to the work or property of any separate contractor on the Project, the Contractor shall, upon due notice, settle with such other contractor by agreement or arbitration, if he will so settle. If such separate contractor sues the Owner or initiates an arbitration proceeding on account of any damage alleged to have been so sustained, the Owner shall notify the Contractor who shall defend such proceedings at the Owner's expense, and if any judgment or award against the Owner arises therefrom the Contractor shall pay or satisfy it and shall reimburse the Owner for all attorneys' fees and court or arbitration costs which the Owner has incurred.

**6.3   CUTTING AND PATCHING
          UNDER SEPARATE CONTRACTS**

6.3.1 The Contractor shall be responsible for any cutting, fitting and patching that may be required to complete his Work except as otherwise specifically provided in the Contract Documents. The Contractor shall not endanger any work of any other contractors by cutting, excavating or otherwise altering any work and shall not cut or alter the work of any other contractor except with the written consent of the Architect.

6.3.2 Any costs caused by defective or ill-timed work shall be borne by the party responsible therefor.

6.4   OWNER'S RIGHT TO CLEAN UP

6.4.1 If a dispute arises between the separate contractors as to their responsibility for cleaning up as required by Paragraph 4.16, the Owner may clean up and charge the cost thereof to the several contractors as the Architect shall determine to be just.

ARTICLE 7

MISCELLANEOUS PROVISIONS

7.1   GOVERNING LAW

7.1.1 The Contract shall be governed by the law of the place where the Project is located.

7.2   SUCCESSORS AND ASSIGNS

7.2.1 The Owner and the Contractor each binds himself, his partners, successors, assigns and legal representatives to the other party hereto and to the partners, successors, assigns and legal representatives of such other party in respect to all covenants, agreements and obligations contained in the Contract Documents. Neither

party to the Contract shall assign the Contract or sublet it as a whole without the written consent of the other, nor shall the Contractor assign any moneys due or to become due to him hereunder, without the previous written consent of the Owner.

7.3   WRITTEN NOTICE

7.3.1 Written notice shall be deemed to have been duly served if delivered in person to the individual or member of the firm or to an officer of the corporation for whom it was intended, or if delivered at or sent by registered or certified mail to the last business address known to him who gives the notice.

7.4   CLAIMS FOR DAMAGES

7.4.1 Should either party to the Contract suffer injury or damage to person or property because of any act or omission of the other party or of any of his employees, agents or others for whose acts he is legally liable, claim shall be made in writing to such other party within a reasonable time after the first observance of such injury or damage.

**7.5   PERFORMANCE BOND AND
          LABOR AND MATERIAL PAYMENT BOND**

7.5.1 The Owner shall have the right to require the Contractor to furnish bonds covering the faithful performance of the Contract and the payment of all obligations arising thereunder if and as required in the Instructions to Bidders or elsewhere in the Contract Documents.

7.6   RIGHTS AND REMEDIES

7.6.1 The duties and obligations imposed by the Contract Documents and the rights and remedies available thereunder shall be in addition to and not a limitation of any duties, obligations, rights and remedies otherwise imposed or available by law.

7.7   ROYALTIES AND PATENTS

7.7.1 The Contractor shall pay all royalties and license fees. He shall defend all suits or claims for infringement of any patent rights and shall save the Owner harmless from loss on account thereof, except that the Owner shall be responsible for all such loss when a particular design, process or the product of a particular manufacturer or manufacturers is specified, but if the Contractor has reason to believe that the design, process or product specified is an infringement of a patent, he shall be responsible for such loss unless he promptly gives such information to the Architect.

7.8   TESTS

7.8.1 If the Contract Documents, laws, ordinances, rules, regulations or orders of any public authority having jurisdiction require any Work to be inspected, tested or approved, the Contractor shall give the Architect timely notice of its readiness and of the date arranged so the Architect may observe such inspection, testing or approval. The Contractor shall bear all costs of such inspections, tests and approvals unless otherwise provided.

7.8.2 If after the commencement of the Work the Architect determines that any Work requires special inspection, testing, or approval which Subparagraph 7.8.1

does not include, he will, upon written authorization from the Owner, instruct the Contractor to order such special inspection, testing or approval, and the Contractor shall give notice as in Subparagraph 7.8.1. If such special inspection or testing reveals a failure of the Work to comply (1) with the requirements of the Contract Documents or (2), with respect to the performance of the Work, with laws, ordinances, rules, regulations or orders of any public authority having jurisdiction, the Contractor shall bear all costs thereof, including the Architect's additional services made necessary by such failure; otherwise the Owner shall bear such costs, and an appropriate Change Order shall be issued.

7.8.3 Required certificates of inspection, testing or approval shall be secured by the Contractor and promptly delivered by him to the Architect.

7.8.4 If the Architect wishes to observe the inspections, tests or approvals required by this Paragraph 7.8, he will do so promptly and, where practicable, at the source of supply.

7.8.5 Neither the observations of the Architect in his Administration of the Construction Contract, nor inspections, tests or approvals by persons other than the Contractor shall relieve the Contractor from his obligations to perform the Work in accordance with the Contract Documents.

7.9 INTEREST

7.9.1 Any moneys not paid when due to either party under this Contract shall bear interest at the legal rate in force at the place of the Project.

7.10 ARBITRATION

7.10.1 All claims, disputes and other matters in question arising out of, or relating to, this Contract or the breach thereof, except as set forth in Subparagraph 2.2.9 with respect to the Architect's decisions on matters relating to artistic effect, and except for claims which have been waived by the making or acceptance of final payment as provided by Subparagraphs 9.7.5 and 9.7.6, shall be decided by arbitration in accordance with the Construction Industry Arbitration Rules of the American Arbitration Association then obtaining unless the parties mutually agree otherwise. This agreement to arbitrate shall be specifically enforceable under the prevailing arbitration law. The award rendered by the arbitrators shall be final, and judgment may be entered upon it in accordance with applicable law in any court having jurisdiction thereof.

7.10.2 Notice of the demand for arbitration shall be filed in writing with the other party to the Contract and with the American Arbitration Association, and a copy shall be filed with the Architect. The demand for arbitration shall be made within the time limits specified in Subparagraphs 2.2.10 and 2.2.11 where applicable, and in all other cases within a reasonable time after the claim, dispute or other matter in question has arisen, and in no event shall it be made after the date when institution of legal or equitable proceedings based on such claim, dispute or other matter in question would be barred by the applicable statute of limitations.

7.10.3 The Contractor shall carry on the Work and maintain the progress schedule during any arbitration proceedings, unless otherwise agreed by him and the Owner in writing.

ARTICLE 8

TIME

8.1 DEFINITIONS

8.1.1 The Contract Time is the period of time alloted in the Contract Documents for completion of the Work.

8.1.2 The date of commencement of the Work is the date established in a notice to proceed. If there is no notice to proceed, it shall be the date of the Agreement or such other date as may be established therein.

8.1.3 The Date of Substantial Completion of the Work or designated portion thereof is the Date certified by the Architect when construction is sufficiently complete, in accordance with the Contract Documents, so the Owner may occupy the Work or designated portion thereof for the use for which it is intended.

8.1.4 The term day as used in the Contract Documents shall mean calendar day.

8.2 PROGRESS AND COMPLETION

8.2.1 All time limits stated in the Contract Documents are of the essence of the Contract.

8.2.2 The Contractor shall begin the Work on the date of commencement as defined in Subparagraph 8.1.2. He shall carry the Work forward expeditiously with adequate forces and shall complete it within the Contract Time.

8.2.3 If a date or time of completion is included in the Contract, it shall be the Date of Substantial Completion as defined in Subparagraph 8.1.3, including authorized extensions thereto, unless otherwise provided.

8.3 DELAYS AND EXTENSIONS OF TIME

8.3.1 If the Contractor is delayed at any time in the progress of the Work by any act or neglect of the Owner or the Architect, or by any employee of either, or by any separate contractor employed by the Owner, or by changes ordered in the Work, or by labor disputes, fire, unusual delay in transportation, unavoidable casualties or any causes beyond the Contractor's control, or by delay authorized by the Owner pending arbitration, or by any cause which the Architect determines may justify the delay, then the Contract Time shall be extended by Change Order for such reasonable time as the Architect may determine.

8.3.2 All claims for extension of time shall be made in writing to the Architect no more than twenty days after the occurrence of the delay; otherwise they shall be waived. In the case of a continuing cause of delay only one claim is necessary.

8.3.3 If no schedule or agreement is made stating the dates upon which written interpretations as set forth in Subparagraph 1.2.5 shall be furnished, then no claim for delay shall be allowed on account of failure to furnish such interpretations until fifteen days after demand is made for them, and not then unless such claim is reasonable.

8.3.4 This Paragraph 8.3 does not exclude the recovery of damages for delay by either party under other provisions of the Contract Documents.

ARTICLE 9

PAYMENTS AND COMPLETION

9.1 CONTRACT SUM

9.1.1 The Contract Sum is stated in the Agreement and is the total amount payable by the Owner to the Contractor for the performance of the Work under the Contract Documents.

9.2 SCHEDULE OF VALUES

9.2.1 Before the first Application for Payment, the Contractor shall submit to the Architect a schedule of values of the various portions of the Work, including quantities if required by the Architect, aggregating the total Contract Sum, divided so as to facilitate payments to Subcontractors in accordance with Paragraph 5.4, prepared in such form as specified or as the Architect and the Contractor may agree upon, and supported by such data to substantiate its correctness as the Architect may require. Each item in the schedule of values shall include its proper share of overhead and profit. This schedule, when approved by the Architect, shall be used only as a basis for the Contractor's Applications for Payment.

9.3 PROGRESS PAYMENTS

9.3.1 At least ten days before each progress payment falls due, the Contractor shall submit to the Architect an itemized Application for Payment, supported by such data substantiating the Contractor's right to payment as the Owner or the Architect may require.

9.3.2 If payments are to be made on account of materials or equipment not incorporated in the Work but delivered and suitably stored at the site, or at some other location agreed upon in writing, such payments shall be conditioned upon submission by the Contractor of bills of sale or such other procedures satisfactory to the Owner to establish the Owner's title to such materials or equipment or otherwise protect the Owner's interest including applicable insurance and transportation to the site.

9.3.3 The Contractor warrants and guarantees that title to all Work, materials and equipment covered by an Application for Payment, whether incorporated in the Project or not, will pass to the Owner upon the receipt of such payment by the Contractor, free and clear of all liens, claims, security interests or encumbrances, hereinafter referred to in this Article 9 as "liens"; and that no Work, materials or equipment covered by an Application for Payment will have been acquired by the Contractor, or by any other person performing the Work at the site or furnishing materials and equipment for the Project, subject to an agreement under which an interest therein or an encumbrance thereon is retained by the seller or otherwise imposed by the Contractor or such other person.

9.4 CERTIFICATES FOR PAYMENT

9.4.1 If the Contractor has made Application for Payment as above, the Architect will, with reasonable promptness but not more than seven days after the receipt of the Application, issue a Certificate for Payment to the Owner, with a copy to the Contractor, for such amount as he determines to be properly due, or state in writing his reasons for withholding a Certificate as provided in Subparagraph 9.5.1.

9.4.2 The issuance of a Certificate for Payment will constitute a representation by the Architect to the Owner, based on his observations at the site as provided in Subparagraph 2.2.4 and the data comprising the Application for Payment, that the Work has progressed to the point indicated; that, to the best of his knowledge, information and belief, the quality of the Work is in accordance with the Contract Documents (subject to an evaluation of the Work for conformance with the Contract Documents upon Substantial Completion, to the results of any subsequent tests required by the Contract Documents, to minor deviations from the Contract Documents correctable prior to completion, and to any specific qualifications stated in his Certificate); and that the Contractor is entitled to payment in the amount certified. In addition, the Architect's final Certificate for Payment will constitute a further representation that the conditions precedent to the Contractor's being entitled to final payment as set forth in Subparagraph 9.7.2 have been fulfilled. However, by issuing a Certificate for Payment, the Architect shall not thereby be deemed to represent that he has made exhaustive or continuous on-site inspections to check the quality or quantity of the Work or that he has reviewed the construction means, methods, techniques, sequences or procedures, or that he has made any examination to ascertain how or for what purpose the Contractor has used the moneys previously paid on account of the Contract Sum.

9.4.3 After the Architect has issued a Certificate for Payment, the Owner shall make payment in the manner provided in the Agreement.

9.4.4 No certificate for a progress payment, nor any progress payment, nor any partial or entire use or occupancy of the Project by the Owner, shall constitute an acceptance of any Work not in accordance with the Contract Documents.

9.5 PAYMENTS WITHHELD

9.5.1 The Architect may decline to approve an Application for Payment and may withhold his Certificate in whole or in part, to the extent necessary reasonably to protect the Owner, if in his opinion he is unable to make representations to the Owner as provided in Subparagraph 9.4.2. The Architect may also decline to approve any Applications for Payment or, because of subsequently discovered evidence or subsequent inspections, he may nullify the whole or any part of any Certificate for Payment previously issued, to such extent as may be necessary in his opinion to protect the Owner from loss because of:

    .1 defective work not remedied,

    .2 third party claims filed or reasonable evidence indicating probable filing of such claims,

.3 failure of the Contractor to make payments properly to Subcontractors or for labor, materials or equipment,

.4 reasonable doubt that the Work can be completed for the unpaid balance of the Contract Sum,

.5 damage to another contractor,

.6 reasonable indication that the Work will not be completed within the Contract Time, or

.7 unsatisfactory prosecution of the Work by the Contractor.

9.5.2 When the above grounds in Subparagraph 9.5.1 are removed, payment shall be made for amounts withheld because of them.

9.6   FAILURE OF PAYMENT

9.6.1 If the Architect should fail to issue any Certificate for Payment, through no fault of the Contractor, within seven days after receipt of the Contractor's Application for Payment, or if the Owner should fail to pay the Contractor within seven days after the date of payment established in the Agreement any amount certified by the Architect or awarded by arbitration, then the Contractor may, upon seven additional days' written notice to the Owner and the Architect, stop the Work until payment of the amount owing has been received.

9.7   SUBSTANTIAL COMPLETION AND FINAL PAYMENT

9.7.1 When the Contractor determines that the Work or a designated portion thereof acceptable to the Owner is substantially complete, the Contractor shall prepare for submission to the Architect a list of items to be completed or corrected. The failure to include any items on such list does not alter the responsibility of the Contractor to complete all Work in accordance with the Contract Documents. When the Architect on the basis of an inspection determines that the Work is substantially complete, he will then prepare a Certificate of Substantial Completion which shall establish the Date of Substantial Completion, shall state the responsibilities of the Owner and the Contractor for maintenance, heat, utilities, and insurance, and shall fix the time within which the Contractor shall complete the items listed therein. The Certificate of Substantial Completion shall be submitted to the Owner and the Contractor for their written acceptance of the responsibilities assigned to them in such Certificate.

9.7.2 Upon receipt of written notice that the Work is ready for final inspection and acceptance and upon receipt of a final Application for Payment, the Architect will promptly make such inspection and, when he finds the Work acceptable under the Contract Documents and the Contract fully performed, he will promptly issue a final Certificate for Payment stating that to the best of his knowledge, information and belief, and on the basis of his observations and inspections, the Work has been completed in accordance with the terms and conditions of the Contract Documents and that the entire balance found to be due the Contractor, and noted in said final Certificate, is due and payable.

9.7.3 Neither the final payment nor the remaining retained percentage shall become due until the Contractor submits to the Architect (1) an Affidavit that all payrolls, bills for materials and equipment, and other indebtedness connected with the Work for which the Owner or his property might in any way be responsible, have been paid or otherwise satisfied, (2) consent of surety, if any, to final payment and (3), if required by the Owner, other data establishing payment or satisfaction of all such obligations, such as receipts, releases and waivers of liens arising out of the Contract, to the extent and in such form as may be designated by the Owner. If any Subcontractor refuses to furnish a release or waiver required by the Owner, the Contractor may furnish a bond satisfactory to the Owner to indemnify him against any such lien. If any such lien remains unsatisfied after all payments are made, the Contractor shall refund to the Owner all moneys that the latter may be compelled to pay in discharging such lien, including all costs and reasonable attorneys' fees.

9.7.4 If after Substantial Completion of the Work final completion thereof is materially delayed through no fault of the Contractor, and the Architect so confirms, the Owner shall, upon certification by the Architect, and without terminating the Contract, make payment of the balance due for that portion of the Work fully completed and accepted. If the remaining balance for Work not fully completed or corrected is less than the retainage stipulated in the Agreement, and if bonds have been furnished as required in Subparagraph 7.5.1, the written consent of the surety to the payment of the balance due for that portion of the Work fully completed and accepted shall be submitted by the Contractor to the Architect prior to certification of such payment. Such payment shall be made under the terms and conditions governing final payment, except that it shall not constitute a waiver of claims.

9.7.5 The making of final payment shall constitute a waiver of all claims by the Owner except those arising from:

.1 unsettled liens,

.2 faulty or defective Work appearing after Substantial Completion,

.3 failure of the Work to comply with the requirements of the Contract Documents, or

.4 terms of any special guarantees required by the Contract Documents.

9.7.6 The acceptance of final payment shall constitute a waiver of all claims by the Contractor except those previously made in writing and still unsettled.

ARTICLE 10

PROTECTION OF PERSONS AND PROPERTY

10.1   SAFETY PRECAUTIONS AND PROGRAMS

10.1.1 The Contractor shall be responsible for initiating, maintaining and supervising all safety precautions and programs in connection with the Work.

10.2   SAFETY OF PERSONS AND PROPERTY

10.2.1 The Contractor shall take all reasonable precautions for the safety of, and shall provide all reasonable protection to prevent damage, injury or loss to:

AIA DOCUMENT A201 • GENERAL CONDITIONS OF THE CONTRACT FOR CONSTRUCTION • TWELFTH EDITION • APRIL 1970 ED.
AIA® • © 1970 • THE AMERICAN INSTITUTE OF ARCHITECTS, 1735 NEW YORK AVENUE, N.W., WASHINGTON, D.C. 20006

.1 all employees on the Work and all other persons who may be affected thereby;

.2 all the Work and all materials and equipment to be incorporated therein, whether in storage on or off the site, under the care, custody or control of the Contractor or any of his Subcontractors or Sub-subcontractors; and

.3 other property at the site or adjacent thereto, including trees, shrubs, lawns, walks, pavements, roadways, structures and utilities not designated for removal, relocation or replacement in the course of construction.

10.2.2 The Contractor shall comply with all applicable laws, ordinances, rules, regulations and lawful orders of any public authority having jurisdiction for the safety of persons or property or to protect them from damage, injury or loss. He shall erect and maintain, as required by existing conditions and progress of the Work, all reasonable safeguards for safety and protection, including posting danger signs and other warnings against hazards, promulgating safety regulations and notifying owners and users of adjacent utilities.

10.2.3 When the use or storage of explosives or other hazardous materials or equipment is necessary for the execution of the Work, the Contractor shall exercise the utmost care and shall carry on such activities under the supervision of properly qualified personnel.

10.2.4 All damage or loss to any property referred to in Clauses 10.2.1.2 and 10.2.1.3 caused in whole or in part by the Contractor, any Subcontractor, any Sub-subcontractor, or anyone directly or indirectly employed by any of them, or by anyone for whose acts any of them may be liable, shall be remedied by the Contractor, except damage or loss attributable to faulty Drawings or Specifications or to the acts or omissions of the Owner or Architect or anyone employed by either of them or for whose acts either of them may be liable, and not attributable to the fault or negligence of the Contractor.

10.2.5 The Contractor shall designate a responsible member of his organization at the site whose duty shall be the prevention of accidents. This person shall be the Contractor's superintendent unless otherwise designated in writing by the Contractor to the Owner and the Architect.

10.2.6 The Contractor shall not load or permit any part of the Work to be loaded so as to endanger its safety.

10.3 EMERGENCIES

10.3.1 In any emergency affecting the safety of persons or property, the Contractor shall act, at his discretion, to prevent threatened damage, injury or loss. Any additional compensation or extension of time claimed by the Contractor on account of emergency work shall be determined as provided in Article 12 for Changes in the Work.

ARTICLE 11

INSURANCE

11.1 CONTRACTOR'S LIABILITY INSURANCE

11.1.1 The Contractor shall purchase and maintain such insurance as will protect him from claims set forth below which may arise out of or result from the Contractor's operations under the Contract, whether such operations be by himself or by any Subcontractor or by anyone directly or indirectly employed by any of them, or by anyone for whose acts any of them may be liable:

.1 claims under workmen's compensation, disability benefit and other similar employee benefit acts;

.2 claims for damages because of bodily injury, occupational sickness or disease, or death of his employees;

.3 claims for damages because of bodily injury, sickness or disease, or death of any person other than his employees;

.4 claims for damages insured by usual personal injury liability coverage which are sustained (1) by any person as a result of an offense directly or indirectly related to the employment of such person by the Contractor, or (2) by any other person; and

.5 claims for damages because of injury to or destruction of tangible property, including loss of use resulting therefrom.

11.1.2 The insurance required by Subparagraph 11.1.1 shall be written for not less than any limits of liability specified in the Contract Documents, or required by law, whichever is greater, and shall include contractual liability insurance as applicable to the Contractor's obligations under Paragraph 4.18.

11.1.3 Certificates of Insurance acceptable to the Owner shall be filed with the Owner prior to commencement of the Work. These Certificates shall contain a provision that coverages afforded under the policies will not be cancelled until at least fifteen days' prior written notice has been given to the Owner.

11.2 OWNER'S LIABILITY INSURANCE

11.2.1 The Owner shall be responsible for purchasing and maintaining his own liability insurance and, at his option, may purchase and maintain such insurance as will protect him against claims which may arise from operations under the Contract.

11.3 PROPERTY INSURANCE

11.3.1 Unless otherwise provided, the Owner shall purchase and maintain property insurance upon the entire Work at the site to the full insurable value thereof. This insurance shall include the interests of the Owner, the Contractor, Subcontractors and Sub-subcontractors in the Work and shall insure against the perils of Fire, Extended Coverage, Vandalism and Malicious Mischief.

11.3.2 The Owner shall purchase and maintain such steam boiler and machinery insurance as may be required by the Contract Documents or by law. This insurance shall include the interests of the Owner, the Contractor, Subcontractors and Sub-subcontractors in the Work.

11.3.3 Any insured loss is to be adjusted with the Owner and made payable to the Owner as trustee for the insureds, as their interests may appear, subject to the requirements of any applicable mortgagee clause and of Subparagraph 11.3.8.

AIA DOCUMENT A201 • GENERAL CONDITIONS OF THE CONTRACT FOR CONSTRUCTION • TWELFTH EDITION • APRIL 1970 ED.
AIA® • © 1970 • THE AMERICAN INSTITUTE OF ARCHITECTS, 1735 NEW YORK AVENUE, N.W., WASHINGTON, D.C. 20006

11.3.4 The Owner shall file a copy of all policies with the Contractor before an exposure to loss may occur. If the Owner does not intend to purchase such insurance, he shall inform the Contractor in writing prior to commencement of the Work. The Contractor may then effect insurance which will protect the interests of himself, his Subcontractors and the Sub-subcontractors in the Work, and by appropriate Change Order the cost thereof shall be charged to the Owner. If the Contractor is damaged by failure of the Owner to purchase or maintain such insurance and so to notify the Contractor, then the Owner shall bear all reasonable costs properly attributable thereto.

11.3.5 If the Contractor requests in writing that insurance for special hazards be included in the property insurance policy, the Owner shall, if possible, include such insurance, and the cost thereof shall be charged to the Contractor by appropriate Change Order.

11.3.6 The Owner and Contractor waive all rights against each other for damages caused by fire or other perils to the extent covered by insurance provided under this Paragraph 11.3, except such rights as they may have to the proceeds of such insurance held by the Owner as trustee. The Contractor shall require similar waivers by Subcontractors and Sub-subcontractors in accordance with Clause 5.3.1.5.

11.3.7 If required in writing by any party in interest, the Owner as trustee shall, upon the occurrence of an insured loss, give bond for the proper performance of his duties. He shall deposit in a separate account any money so received, and he shall distribute it in accordance with such agreement as the parties in interest may reach, or in accordance with an award by arbitration in which case the procedure shall be as provided in Paragraph 7.10. If after such loss no other special agreement is made, replacement of damaged work shall be covered by an appropriate Change Order.

11.3.8 The Owner as trustee shall have power to adjust and settle any loss with the insurers unless one of the parties in interest shall object in writing within five days after the occurrence of loss to the Owner's exercise of this power, and if such objection be made, arbitrators shall be chosen as provided in Paragraph 7.10. The Owner as trustee shall, in that case, make settlement with the insurers in accordance with the directions of such arbitrators. If distribution of the insurance proceeds by arbitration is required, the arbitrators will direct such distribution.

11.4 LOSS OF USE INSURANCE

11.4.1 The Owner, at his option, may purchase and maintain such insurance as will insure him against loss of use of his property due to fire or other hazards, however caused.

ARTICLE 12

CHANGES IN THE WORK

12.1 CHANGE ORDERS

12.1.1 The Owner, without invalidating the Contract, may order Changes in the Work within the general scope of the Contract consisting of additions, deletions or other revisions, the Contract Sum and the Contract Time being adjusted accordingly. All such Changes in the Work shall be authorized by Change Order, and shall be executed under the applicable conditions of the Contract Documents.

12.1.2 A Change Order is a written order to the Contractor signed by the Owner and the Architect, issued after the execution of the Contract, authorizing a Change in the Work or an adjustment in the Contract Sum or the Contract Time. Alternatively, the Change Order may be signed by the Architect alone, provided he has written authority from the Owner for such procedure and that a copy of such written authority is furnished to the Contractor upon request. A Change Order may also be signed by the Contractor if he agrees to the adjustment in the Contract Sum or the Contract Time. The Contract Sum and the Contract Time may be changed only by Change Order.

12.1.3 The cost or credit to the Owner resulting from a Change in the Work shall be determined in one or more of the following ways:

    .1 by mutual acceptance of a lump sum properly itemized;

    .2 by unit prices stated in the Contract Documents or subsequently agreed upon; or

    .3 by cost and a mutually acceptable fixed or percentage fee.

12.1.4 If none of the methods set forth in Subparagraph 12.1.3 is agreed upon, the Contractor, provided he receives a Change Order, shall promptly proceed with the Work involved. The cost of such Work shall then be determined by the Architect on the basis of the Contractor's reasonable expenditures and savings, including, in the case of an increase in the Contract Sum, a reasonable allowance for overhead and profit. In such case, and also under Clause 12.1.3.3 above, the Contractor shall keep and present, in such form as the Architect may prescribe, an itemized accounting together with appropriate supporting data. Pending final determination of cost to the Owner, payments on account shall be made on the Architect's Certificate for Payment. The amount of credit to be allowed by the Contractor to the Owner for any deletion or change which results in a net decrease in cost will be the amount of the actual net decrease as confirmed by the Architect. When both additions and credits are involved in any one change, the allowance for overhead and profit shall be figured on the basis of net increase, if any.

12.1.5 If unit prices are stated in the Contract Documents or subsequently agreed upon, and if the quantities originally contemplated are so changed in a proposed Change Order that application of the agreed unit prices to the quantities of Work proposed will create a hardship on the Owner or the Contractor, the applicable unit prices shall be equitably adjusted to prevent such hardship.

12.1.6 Should concealed conditions encountered in the performance of the Work below the surface of the ground be at variance with the conditions indicated by the Contract Documents or should unknown physical conditions below the surface of the ground of an unusual nature,

AIA DOCUMENT A201 • GENERAL CONDITIONS OF THE CONTRACT FOR CONSTRUCTION • TWELFTH EDITION • APRIL 1970 ED.
AIA® • © 1970 • THE AMERICAN INSTITUTE OF ARCHITECTS, 1735 NEW YORK AVENUE, N.W., WASHINGTON, D.C. 20006

differing materially from those ordinarily encountered and generally recognized as inherent in work of the character provided for in this Contract, be encountered, the Contract Sum shall be equitably adjusted by Change Order upon claim by either party made within twenty days after the first observance of the conditions.

12.1.7 If the Contractor claims that additional cost is involved because of (1) any written interpretation issued pursuant to Subparagraph 1.2.5, (2) any order by the Owner to stop the Work pursuant to Paragraph 3.3 where the Contractor was not at fault, or (3) any written order for a minor change in the Work issued pursuant to Paragraph 12.3, the Contractor shall make such claim as provided in Paragraph 12.2.

12.2 CLAIMS FOR ADDITIONAL COST

12.2.1 If the Contractor wishes to make a claim for an increase in the Contract Sum, he shall give the Architect written notice thereof within twenty days after the occurrence of the event giving rise to such claim. This notice shall be given by the Contractor before proceeding to execute the Work, except in an emergency endangering life or property in which case the Contractor shall proceed in accordance with Subparagraph 10.3.1. No such claim shall be valid unless so made. If the Owner and the Contractor cannot agree on the amount of the adjustment in the Contract Sum, it shall be determined by the Architect. Any change in the Contract Sum resulting from such claim shall be authorized by Change Order.

12.3 MINOR CHANGES IN THE WORK

12.3.1 The Architect shall have authority to order minor changes in the Work not involving an adjustment in the Contract Sum or an extension of the Contract Time and not inconsistent with the intent of the Contract Documents. Such changes may be effected by Field Order or by other written order. Such changes shall be binding on the Owner and the Contractor.

12.4 FIELD ORDERS

12.4.1 The Architect may issue written Field Orders which interpret the Contract Documents in accordance with Subparagraph 1.2.5 or which order minor changes in the Work in accordance with Paragraph 12.3 without change in Contract Sum or Contract Time. The Contractor shall carry out such Field Orders promptly.

ARTICLE 13

UNCOVERING AND CORRECTION OF WORK

13.1 UNCOVERING OF WORK

13.1.1 If any Work should be covered contrary to the request of the Architect, it must, if required by the Architect, be uncovered for his observation and replaced, at the Contractor's expense.

13.1.2 If any other Work has been covered which the Architect has not specifically requested to observe prior to being covered, the Architect may request to see such Work and it shall be uncovered by the Contractor. If such Work be found in accordance with the Contract Documents, the cost of uncovering and replacement

shall, by appropriate Change Order, be charged to the Owner. If such Work be found not in accordance with the Contract Documents, the Contractor shall pay such costs unless it be found that this condition was caused by a separate contractor employed as provided in Article 6, and in that event the Owner shall be responsible for the payment of such costs.

13.2 CORRECTION OF WORK

13.2.1 The Contractor shall promptly correct all Work rejected by the Architect as defective or as failing to conform to the Contract Documents whether observed before or after Substantial Completion and whether or not fabricated, installed or completed. The Contractor shall bear all cost of correcting such rejected Work, including the cost of the Architect's additional services thereby made necessary.

13.2.2 If, within one year after the Date of Substantial Completion or within such longer period of time as may be prescribed by law or by the terms of any applicable special guarantee required by the Contract Documents, any of the Work is found to be defective or not in accordance with the Contract Documents, the Contractor shall correct it promptly after receipt of a written notice from the Owner to do so unless the Owner has previously given the Contractor a written acceptance of such condition. The Owner shall give such notice promptly after discovery of the condition.

13.2.3 All such defective or non-conforming Work under Subparagraphs 13.2.1 and 13.2.2 shall be removed from the site if necessary, and the Work shall be corrected to comply with the Contract Documents without cost to the Owner.

13.2.4 The Contractor shall bear the cost of making good all work of separate contractors destroyed or damaged by such removal or correction.

13.2.5 If the Contractor does not remove such defective or non-conforming Work within a reasonable time fixed by written notice from the Architect, the Owner may remove it and may store the materials or equipment at the expense of the Contractor. If the Contractor does not pay the cost of such removal and storage within ten days thereafter, the Owner may upon ten additional days' written notice sell such Work at auction or at private sale and shall account for the net proceeds thereof, after deducting all the costs that should have been borne by the Contractor including compensation for additional architectural services. If such proceeds of sale do not cover all costs which the Contractor should have borne, the difference shall be charged to the Contractor and an appropriate Change Order shall be issued. If the payments then or thereafter due the Contractor are not sufficient to cover such amount, the Contractor shall pay the difference to the Owner.

13.2.6 If the Contractor fails to correct such defective or non-conforming Work, the Owner may correct it in accordance with Paragraph 3.4.

13.3 ACCEPTANCE OF DEFECTIVE OR NON-CONFORMING WORK

13.3.1 If the Owner prefers to accept defective or non-conforming Work, he may do so instead of requiring its

AIA DOCUMENT A201 • GENERAL CONDITIONS OF THE CONTRACT FOR CONSTRUCTION • TWELFTH EDITION • APRIL 1970 ED.
AIA® • © 1970 • THE AMERICAN INSTITUTE OF ARCHITECTS, 1735 NEW YORK AVENUE, N.W., WASHINGTON, D.C. 20006

removal and correction, in which case a Change Order will be issued to reflect an appropriate reduction in the Contract Sum, or, if the amount is determined after final payment, it shall be paid by the Contractor.

ARTICLE 14

TERMINATION OF THE CONTRACT

14.1 TERMINATION BY THE CONTRACTOR

14.1.1 If the Work is stopped for a period of thirty days under an order of any court or other public authority having jurisdiction, or as a result of an act of government, such as a declaration of a national emergency making materials unavailable, through no act or fault of the Contractor or a Subcontractor or their agents or employees or any other persons performing any of the Work under a contract with the Contractor, or if the Work should be stopped for a period of thirty days by the Contractor for the Architect's failure to issue a Certificate for Payment as provided in Paragraph 9.6 or for the Owner's failure to make payment thereon as provided in Paragraph 9.6, then the Contractor may, upon seven days' written notice to the Owner and the Architect, terminate the Contract and recover from the Owner payment for all Work executed and for any proven loss sustained upon any materials, equipment, tools, construction equipment and machinery, including reasonable profit and damages.

14.2 TERMINATION BY THE OWNER

14.2.1 If the Contractor is adjudged a bankrupt, or if he makes a general assignment for the benefit of his creditors, or if a receiver is appointed on account of his insolvency, or if he persistently or repeatedly refuses or fails, except in cases for which extension of time is provided, to supply enough properly skilled workmen or proper materials, or if he fails to make prompt payment to Subcontractors or for materials or labor, or persistently disregards laws, ordinances, rules, regulations or orders of any public authority having jurisdiction, or otherwise is guilty of a substantial violation of a provision of the Contract Documents, then the Owner, upon certification by the Architect that sufficient cause exists to justify such action, may, without prejudice to any right or remedy and after giving the Contractor and his surety, if any, seven days' written notice, terminate the employment of the Contractor and take possession of the site and of all materials, equipment, tools, construction equipment and machinery thereon owned by the Contractor and may finish the Work by whatever method he may deem expedient. In such case the Contractor shall not be entitled to receive any further payment until the Work is finished.

14.2.2 If the unpaid balance of the Contract Sum exceeds the costs of finishing the Work, including compensation for the Architect's additional services, such excess shall be paid to the Contractor. If such costs exceed such unpaid balance, the Contractor shall pay the difference to the Owner. The costs incurred by the Owner as herein provided shall be certified by the Architect.

AIA DOCUMENT A201 • GENERAL CONDITIONS OF THE CONTRACT FOR CONSTRUCTION • TWELFTH EDITION • APRIL 1970 ED.
AIA® • © 1970 • THE AMERICAN INSTITUTE OF ARCHITECTS, 1735 NEW YORK AVENUE, N.W., WASHINGTON, D.C. 20006

Geophysics The application of procedures used in physics to study the properties of *Soil* and *Rocks*.

Gin Pole A hoisting assembly consisting of a boom-like guide pole usually having a block and tackle mechanism at the top.

Girder A large *Beam*, generally one into which other beams frame.

Girt In an industrial type building, a horizontal side member spanning between columns and to which the siding is attached.

Glass Glass is manufactured from a molten mixture of sand, lime, soda, and other materials. Briefly described, the process by which the molten material is formed into sheets is as follows: from the glass melting tank the glass flows horizontally with the aid of forming rolls or across molten metal, or is drawn vertically, virtually untouched except by side edge-guides, until the semi-solid continuous ribbon cools and solidifies. It next enters an annealing oven to control stresses and strains by gradual, controlled cooling. Finally, it may be ground and polished, or further processed prior to cutting, inspecting, and packaging. Types of glass include:

ANNEALED GLASS—Used synonymously with *Heat Strengthened glass*.

BULLET RESISTANT GLASS—Three to five sheets of polished *Plate Glass* laminated together to a total thickness of from ¾″ to 3″ in a maximum sheet size of 66″ x 120″.

DOUBLE STRENGTH SHEET GLASS—See *Sheet Glass*.

FIGURED GLASS—Flat glass which has a pattern on one or both sides.

FLOAT GLASS—Glass which has its bottom surfaces formed by floating on molten metal, the top surface being gravity formed, producing a high optical quality of *Plate Glass* with parallel surfaces and, without polishing and grinding, the fire-finished brilliance of the finest *Sheet Glass*. Float glass is replacing *Plate Glass*.

GLARE REDUCING GLASS—See *Tinted Glass*.

HEAT ABSORBING GLASS—Refers to *Tinted Glass*.

HEAT STRENGTHENED GLASS—Glass which is reheated, after forming, just below melting point and then cooled. A compressed surface is formed which increases its strength.

INSULATING GLASS—Insulating glass refers to two pieces of glass spaced apart and hermetically sealed to form a single-glazed unit with an air space between. The normal air space is approximately ½″ thick. Heat transmission through this type of glass may be as low as half that without such an air space. It is also called Double Glazing.

LAMINATED GLASS—Two or more sheets with an inner layer of transparent plastic to which the glass adheres if broken.

OBSCURE GLASS—See *Pattern Glass*.

PATTERN GLASS—Mainly used for decoration, diffusion, or privacy. The design is pressed into the glass during the rolling process. There are over thirty patterns available in both *Plate Glass* and *Wire Glass*.

PLATE GLASS—Polished plate glass is a rolled, ground, and polished product with true flat parallel plane surfaces affording excellent vision. It has less surface polish than *Sheet Glass* and is available in thickness varying from ⅛″ to 1¼″.

SAFETY GLASS—A glazing material

predominantly ceramic in character including *Laminated Glass, Tempered Glass* and *Wire Glass*.

SHATTER RESISTANT GLASS— Two or more layers of glass with a layer, or layers, of transparent vinyl plastic sandwiched between layers under heat and pressure. When broken, the glass cracks with the resulting pieces tending to adhere to the plastic. Maximum size is 66″ x 120″.

SHEET GLASS—A transparent, flat glass whose surface has a characteristic waviness. There are three basic classifications of sheet glass:

1. SINGLE STRENGTH (S.S.): 3/32″ thick.

2. DOUBLE STRENGTH (D.S.): ⅛″ thick.

3. Heavy Sheet which has three available thicknesses: 3/16″, 7/32″, and ¼″.

SINGLE STRENGTH GLASS—See *Sheet Glass*.

SOLAR GLASS—See *Tinted Glass*.

TEMPERED GLASS—As with *Heat Strengthened Glass*, it is reheated to just below the melting point but suddenly cooled. When shattered it breaks into small pieces.

THERMOPANE GLASS—Trade name (Libby-Owens Ford); same as *Insulating Glass*.

TINTED GLASS—A mineral *Admixture* is incorporated in the glass, resulting in a degree of tinting. Any tinting reduces both visual and radiant transmittance.

TRANSPARENT MIRRORS—Often incorrectly referred to as "one-way" glass; however, no glass is truly "one-way." Such mirrors are glass coated with a thin transparent reflective coating producing a one-way vision effect when the balance of the light on either side is in an effective ratio.

WINDOW GLASS—See *Sheet Glass*.

WIRE GLASS—Polished or clear glass, ¼″ thick. Wire mesh is embedded within the glass such that the glass will not shatter when broken. The wire pattern is available in many types. It is frequently used in skylights, overhead glazing, and locations where a fire-retardant glass is required.

Glass Blocks Manufactured hollow blocks of glass which are laid up with *Mortar Joints* to form light-transmitting walls. They are made by sealing together two square, pan-shaped glass castings into a hollow, glass-faced unit.

Glazed Tile See *Ceramic Tile*.

Glazing

1. The installation of *Glass*.

2. In painting, the application of a transparent or translucent coat over a painted surface.

Glitter Finish See *Plaster Finishes*.

Globe Valve See *Valves*.

Gloss Finish See *Paint Finishes*.

Glue A gelatinous, sticky material used as an *Adhesive*. Although there are numerous glues available, such as animal and vegetable glues, the most commonly used in the construction field are as follows:

CASEIN—A glue derived from a constituent of milk (*Casein*), mixed with other ingredients. It is a "dry use" glue (such as for interior doors) since water dissolves its adhesive properties.

LIQUID GLUE—A glue sold in prepared form and generally used for small, lightweight items, usually wood joining.

RESIN GLUES—These are synthetically made glues which are replacing the older, more conventional glues (i.e., casein) used in construction. They include:

LIQUID PVA—(polyvinyl acetate) also referred to as "white glue." A water-resistant glue which has good flexibility and is fast setting at ambient temperatures. It is used for furniture and other small items.

MELAMINE—A colorless glue suitable for coatings and castings. It sets at high temperatures and is waterproof.

PHENOLIC RESIN—A waterproof glue mixed with an aldehyde and sold in prepared form. It is used in the manufacture of exterior *Plywood*, and sets at high temperatures.

PHENOL RESORCINOL—A waterproof glue dries at room temperature. It is used in plywood *Stressed-Skin Panels, Glued Laminated Beams,* and plywood *Box Beams.*

UREA RESIN—A type of glue which hardens slowly, is water resistant and is used for *Hardwood Plywood* and *Particleboard.*

Glued Laminated Lumber Built-up *Beams, Girders,* and *Columns* made from small boards (laminations) glued together. They are most often made in a shop specializing in this type of work and are made in any depth a customer's design calls for, and in widths corresponding to standard widths of dimension lumber, usually finish widths of 5-1/4″ (using 2 x 6's or 6-3/4″ (using 2 x 8's). *Scarf Joints* or *Finger Joints* are used where ends of boards meet to give greater *Glue* contact area. (See Figure G-2.)

Glue Gun A hand-held machine for applying strips (*Beads*) of glue, such as to glue plywood sheets onto supporting joists or rafters.

Glue Line In *Glued Laminated Lumber,* the area occupied by the glue between pieces of wood which are joined by *Adhesive,* which is called a "line,"

Courtesy of Beven-Herron, Inc.

FIGURE G-2   Installation of Glued Laminated Beam Roof Structure

since only the edge is visible in a finished member.

Gooseneck In plumbing, a flexible connection used in sinks and lavatories allowing greater flexibility for connections.

Government Survey See *Land Description*.

G.P.M. Abbreviation for gallons per minute, a standard measurement of the quantity of a liquid flow.

Gradation The particle size distribution of *Aggregate* as determined by separation with standard screens (sieves). See also: *Grain Size Distribution Curve*.

Grade The surface of the ground; also, the act of *Grading*.

Grade Beam A *Continuous Footing* reinforced to act as a beam. It is used to bridge over localized weak soil or spread concentrated loads over a larger soil bearing area.

Grademark See *Lumber Grading*.

Grader In earthwork, a self-propelled machine used for grading and final levelling. It has a transverse blade between the front and rear wheels which can be adjusted hydraulically to be raised, lowered, angled, or tilted such that the earth is levelled or "scraped" to the desired shape. It is also called a MOTOR GRADER or in slang, a *Blade*.

Grading In earthwork, the levelling or otherwise shaping of the ground surface, usually by means of earthmoving and excavating equipment. ROUGH GRADING is the shaping of the ground to approximately the desired configuration, usually to within plus or minus one-tenth of a foot; FINE GRADING is the final levelling, such as immediately before paving.

Grading (of Lumber) See *Lumber Grading*.

Grading Plan See *Plans*

Grain A small particle, as of *Sand*.

Grain (of Wood) The direction, size, arrangement, appearance, or quality of the fibers in wood. Descriptions of grain types include:

CLOSE-GRAIN WOOD—Wood with narrow, inconspicuous *Annual Rings*. The term is sometimes used to designate wood having small and closely spaced pores, but in this sense the term "fine textured" is more often used.

COARSE-GRAIN WOOD—Wood with wide, conspicuous annual rings in which there is considerable difference between *Springwood* and *Summerwood*. The term is sometimes used to designate wood with large pores, such as oak, ash, chestnut, and walnut, but in this sense the term "coarse textured" is more often used.

CROSS-GRAIN WOOD—Wood in which the fibers deviate from a line parallel to the sides of the piece. Cross grains may be either diagonal or spiral grain or a combination of the two.

CURLY-GRAIN WOOD—Wood in which the fibers are distorted so that they have a curled appearance. The areas showing curly grain may vary up to several inches in diameter.

DIAGONAL-GRAIN WOOD—Wood in which the annual rings are at an angle with the axis of a piece as a result of sawing at an angle with the *Bark* of the tree or log. A form of *Cross-Grain*.

EDGE-GRAIN WOOD—See *Close-Grain Wood*.

FLAT-GRAIN WOOD—Lumber that has been sawed in a plane approximately perpendicular to a radius of the log.

Lumber is considered flat grained when the annual growth rings make an angle of less than 45° with the surface of the piece.

INTERLOCKED-GRAIN WOOD —Wood in which the fibers are inclined in one direction in a number of rings of annual growth, then gradually reverse and are inclined in an opposite direction in succeeding growth rings, then reverse again.

OPEN-GRAIN WOOD—Common classification by painters for woods with large pores, such as oak, ash, chestnut, and walnut. Also known as "coarse textured."

RAISED GRAIN—A roughened condition of the surface of "dressed size" *Lumber* in which the hard *Summerwood* is raised above the shorter *Springwood* but not torn loose from it.

SPIRAL-GRAIN WOOD—Wood in which the fibers take a spiral course about the trunk of the tree instead of the normal vertical course. The spiral may extend in a right-handed or left-handed direction around the tree trunk. Spiral grain is a form of *Cross-Grain*.

STRAIGHT-GRAIN WOOD—Wood in which the fibers run parallel to the axis of a piece.

VERTICAL-GRAIN WOOD—Another term for *Straight-Grain Wood*.

WAVY-GRAIN WOOD—Wood in which the fibers collectively take the form of waves or undulations.

Grain Raising A condition whereby wood fibers swell as a result of the absorption of stains or other materials.

Grain Size Distribution Curve A plot (graph) of the percentage of each size of grain in a *Soil* or *Aggregate* specimen. (See Figure G-3)

Gram Unit of weight in the *Metric System*. It is the weight of one cubic cen-timeter of water. Twenty-eight grams equals one ounce.

Grand Fir See *Softwood*.

Grand Master Key See *Keying*.

Granite A type of *Igneous Rock* consisting of *Quartz*, feldspar and mica. It is used in building construction due to its hardness and ability to take polish.

Graph A two-dimensional chart-like visual representation of the quantity relationship between several things.

Grapple A type of *Clamshell* bucket having three or more leaf-like sides; also called an *Orangepeel* bucket.

Grating A closely spaced framework of bars which are all connected together—usually welded—and in one plane. Gratings are used to cover openings in floors, such as over pits and trenches, and as decking around elevated industrial equipment.

Gravel Rock particles which, by convention, range from ¼ to 3 inches in diamter. See also: *Cobbles, Sand*.

Gravel Stop An angle-shaped sheet metal trim member at the edge of a roof, having a slightly raised lip at the roof edge to retain gravel surfacing material.

Gravity The force acting on all masses and attracting them towards the center of the earth; the weight of a mass is the magnitude of the force of gravity.

In structural design, the term (abbreviated "g") is used to mean weight or force, as applied in a horizontal direction from an earthquake. For example, 0.15 g =.15 times the weight of the structure acting horizontally. See also: *Center of Gravity, Specific Gravity*.

Gravity Ventilator A roof ventilator depending for its operation, on the natural upward draft of the less dense air from within the building.

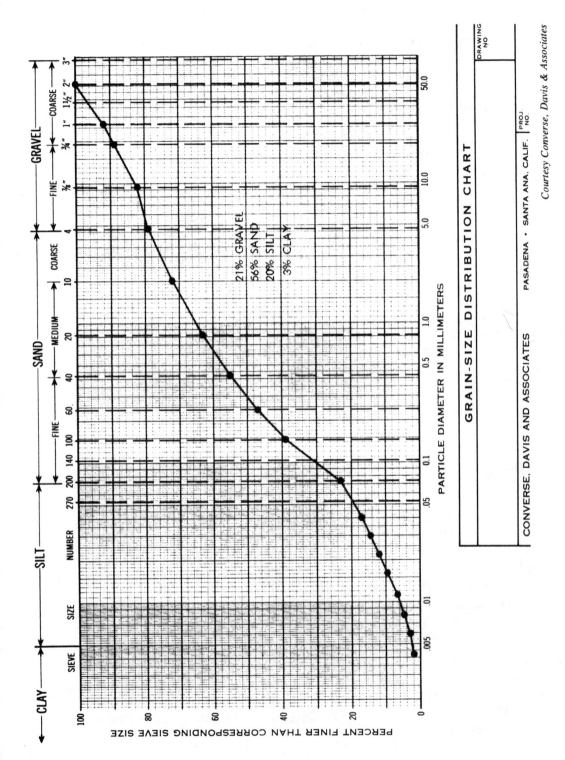

FIGURE G-3 Sample Grain Size Distribution Chart

Grease Trap See *Interceptors*.

Green Lumber Lumber which has been recently cut from a tree or has not had sufficient time to dry out. The high shrinkage (during drying) of green lumber makes it undesirable for light framing or house construction but acceptable for rough work such as concrete *Formwork*. Lumber cut from a tree usually has in excess of 30% moisture.

Green Plaster Wet or damp plaster, which has not yet completely hardened.

Grid Any network of evenly spaced parallel lines which intersect at right angles. Example—graph paper.

Grillage Tiers of short beams laid in alternating layers to form a box-like support which is usually temporary—as under a house which has been jacked up, ready to be moved.

Grille A louvered or perforated covering for air passage through a wall, ceiling, or floor. See also: *Diffuser, Register*.

Groin For shoreline protection, a low wall built out into the water perpendicular to the shoreline. Its purpose is to prevent beach erosion by causing water-borne sand to be trapped and deposited as it is carried along the shoreline by wave action and tides.

Gross Area The full area without deducting for holes, cut-outs, etc., as opposed to the *Net Area*.

Grotto A cave-like shelter which is either naturally formed or artificially contrived.

Ground

1. An electrical *Conductor* in contact with the earth such as by attachment to a *Ground Rod*, a steel plate buried in the earth, or a steel pipe which runs underground, such as a cold water line.

2. In a building, the continuing of an electric *Circuit* back to the supply source and which, at some point in its run, is grounded (as per the above description).

Ground Conductor A *Conductor* connected to a *Ground* to dissipate unwanted electrical charges. It is usually color coded green and is also called an EQUIPMENT GROUND.

Ground Rod A steel rod, usually *Copper* clad, driven into the earth to form a *Ground*.

Grounds In plastering, wood or metal strips attached to the framing or plaster base with the exposed surface acting as a depth guide to control the thickness and plane of the plaster.

Ground Slab Same as *Floating Foundation*.

Ground Water Water standing or flowing between soil particles below the ground surface. It is caused primarily by rainfall percolating into the soil but stopped by dense soil or rock masses. The top surface is called the WATER TABLE. If the ground water intersects the surface it forms a spring (if on a hillside) or swamp (if on level ground).

Grout A fluid mixture of *Portland Cement, Aggregate*, and water, with or without the inclusion of an *Admixture*. Its chief use is in masonry work where it is poured into the hollow cells or joints of a wall to encase reinforcing steel and bond the units together.

Grouted Masonry Masonry in which spaces within the wall are filled with *Grout*. When such spaces contain *Reinforcing Bars*, it is called REINFORCED GROUTED MASONRY.

Grouting The act of placing *Grout*.

Growth Ring (Wood) See *Annual Rings*.

Grub See *Clear and Grub*.

Guaranty A written assurance by a material or equipment supplier or subcontractor that such supplier or subcontractor will be liable for non-performance of such material, equipment, or installation, for a stated length of time. It differs from a WARRANTY in that the latter provides additional assurance that, at the risk of voiding the original *Contract*, all statements and claims made in writing by the seller to the buyers are, in fact, as stated.

Guard Rail A protective railing at the edge of a raised area, such as a *Mezzanine,* or *Loading Dock. Building Codes* usually specify their required height. See also: *Handrail.*

Guillotine Door See *Door Types.*

Gumbo A highly plastic *Clay,* sticky when wet and which develops large shrinkage cracks when dry.

Gunite *Pneumatic* placing of concrete whereby a mixture of cement and sand is discharged under pressure from a hose and combined with water at a nozzle, and "shot" (sprayed) onto a backing surface (such as against earth, in the case of a swimming pool). See also: *Wet Mix Shotcrete.*

Gusset Plate A triangular *Plate,* usually of steel, which stiffens the corner of a structural joint. Also, a steel plate projecting from a *Beam* or *Column* to provide a means for connecting another member by bolting or welding.

Gutter

1. A horizontal sheet metal trough for conveying roof run-off water to a *Downspout* or other point of discharge.

2. In electrical work, a rectangular metal enclosure used to distribute *Conductors* to motors, switches, etc. See also: *Wireway.*

Guy An inclined wire or cable used to brace a pole, tower or similar structure.

Gypsum A whitish mineral which occurs naturally in sedimentary rocks and which, after special processing, hardens when mixed with water. It is used as the main ingredient in *Plaster*, *Gypsum Board*, and similar products. It is not suitable for exterior use since moisture causes deterioration of its cementing capability. Chemically, gypsum is hydrous calcium sulphate having the approximate chemical formula $CaSO_4$, $2H_2O$, having a water content of about 20%. In manufacture, about three-fourths of this chemically combined water is removed from the gypsum rock by means of a controlled calcining (burning to a powder) process. In the case of plaster, water is added at the job, causing the material to crystallize (set) and thereby revert to its original chemical composition.

Gypsum Board A panel consisting of a core of *Gypsum* faced with paper on each side. It is also called WALLBOARD and is used as a wall and/or ceiling covering material. It is most commonly available in ½" and ⅝" thickness and in sheet sizes of 4' x 8' and 4' x 12'. For *Fire Wall* rating requirements the panels are avilable with a core of greater fire resistive material designated "Type X" gypsum board. See also: *Gypsum Lath.*

Gypsum Board Nail See *Nails.*

Gypsum Lath A rigid plaster base which is 16" and 24" × 48" and ⅜" or ½" thick, having a gypsum core with a paper surfacing on each side to receive a *Gypsum Plaster* application, the plaster bonding by suction to the paper. *Perforated Gypsum Lath* is also available, the holes being for mechanical bonding of the plaster.

Gypsum Plaster See *Plaster*.

Gypsum Roof Deck A roof deck made by pouring a concrete-like mixture of gypsum, *Aggregate* and water onto fiber or gypsum form boards, which are in turn usually supported by bulb tee purlins which span between steel roof framing members. It hardens to form a solid deck over which a weather protective built-up roof is applied. (See Figure G-4.)

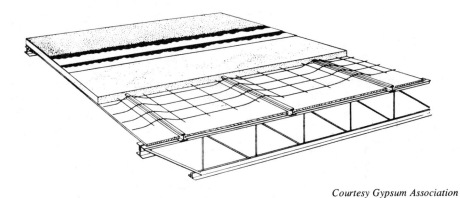

Courtesy Gypsum Association

FIGURE G-4   Gypsum Roof Deck

H

Hack Saw See *Saws*.

Hand (of a Door) A term used to describe which way a door swings with respect to its hinged side when viewed from the outside. A LEFT HAND DOOR has hinges on the left side and swings away from the viewer; a RIGHT HAND DOOR has hinges on the right side when swinging away from the viewer. A LEFT HAND REVERSE DOOR has hinges on the left side but swings toward the viewer; a RIGHT HAND REVERSE DOOR has hinges on the right side and also swings toward the viewer.

Hand Level In surveying, a small hand-held sighting tube used as a *Level* for determining points of equal *Elevation*. It has cross hairs on the outer lens and a built-in bubble type level which, by means of a mirror, can be viewed when sighting through the tube to determine when it is being held level. Some models have a small amount of magnification.

Handrail A protective railing, usually along the open side of a stairway. Also, a railing mounted on a wall alongside a stairway for support while using the stairway. See also: *Guard Rail*.

Hand Saws See *Saws*.

Hardboard A dense building panel consisting of fibrous or laminated material formed by compressing the material under controlled conditions of moisture, pressure and heat, the resulting panel having a density between 50 and 80 p.s.f. An example is "Masonite," a trade name of Masonite Corporation.

Hardened Steel Steel which has been hardened by *Heat Treating*.

Hardener See *Floor Hardener*.

Hardpan A layer of *Soil* which has been naturally cemented by the leaching of salts down from the soil strata above.

Hard Soldering See *Brazing*.

Hardwall Plaster A term usually used to mean *Gypsum Plaster*.

Hardware A general term for all metal or plastic items used for fasteners or connectors on a building, and for the fastening and/or operation of doors, windows, cabinets and similar movable items, also known collectively as BUILDER'S HARDWARE. Hardware is further subdivided into ROUGH HARDWARE which includes connection type items such as *Bolts*, *Nails*, timber connectors, etc., and FINISH HARDWARE which includes such items as *Hinges*, *Locksets*, *Latchsets*, *Closers*, etc., which are generally for movable items and are visible after installation.

Hardware Finishes To simplify and standardize a wide range and combination of metal finishes for hardware, and the base material used, a numbering system was established by the National Bureau of Standards of the U.S. Department of Commerce. This system uses the letters US followed by numbers and/or letters indicating the finish and base material. Another system has been established by the Builder's Hardware Manufacturers Association which uses a three digit number starting with the number "6." For example, the designation US 2 C means a cadmium plated finish on steel. Its equivalent in the B.H.M.A.'s system is "602." Specifications for hardware usually include these designations to indicate the type of finish desired.

Hard Water A term used to describe water with poor sudsing ability due to impurities in the calcium and magnesium ions of the water. *Water Softeners* are used to remove the impurities and provide easier sudsing.

Hardwood One of the two botanical categories of *Wood*, the other being *Softwood*. Hardwoods are from the usually deciduous variety of trees, generally broad-leaved. The chief uses of hardwood are for *Cabinetwork*, furniture, *Millwork*, *Pallets*, and *Veneer*, are opposed to structural framing lumber which is chiefly softwood. The agency responsible for most hardwood *Lumber Grading* is the National Hardwood Lumber Association. The principal species of hardwood are:

ALDER (Red Alder)—Grown along the Pacific Coast between Alaska and California. It varies from white to pinkish-brown. It is suitable for furniture and is coming into use for construction.

ASH—A light colored, open grained wood. Different varieties are grown on the east and west coasts. It is used for tool and implement handles and as a paneling veneer.

ASPEN—Grown principally in Canada, and in the Northeastern and Lake states, it is grayish-white to light grayish-brown and is usually straight grained. Its uses include pulp, excelsior, matches, carting, and some construction.

BALSA—Grown in tropical South America with the principal sources in Ecuador. It is white to pale gray and coarse textured. Because of its very low density it is used for speciality items as model airplanes, flotation gear, and *Insulation*. This is an example of a hardwood being "softer" than a softwood.

BIRCH—A light colored, close-grained wood grown in Northeastern and Lake states. It is widely used for furniture and as plywood veneer.

CHERRY—A light to dark reddish-brown wood produced chiefly in the Middle Atlantic states. It is used primarily for furniture due to the beauty of its grain.

HICKORY—Includes a number of species growing throughout the eastern half of the United State. Subgrouped into true hickories and pecan hickories, this very tough and strong wood is mainly used for tool handles, flooring (pecan), pallets and specialty items.

LAUAN—Also known as Philippine mahogany, it comprises a large group of species from the Philippines ranging in color from grayish, or very light red, to dark reddish-brown. It is widely used for paneling, plywood, and cabinetwork.

MAHOGANY—True mahogany is grown in Central and South America and is called "Honduras" mahogany. It has a pleasing grain configuration, and ranges from light pink to brown-red. "African" mahogany is similar in appearance and properties and may be loosely considered as also being a true mahogany. "Philippine" mahogany is not a true mahogany and should be properly referred to as

Lauan. Mahogany is easy to work and dimensionally stable. Its uses include fine furniture, paneling, models, and patterns.

MAPLE—Grown principally in the Middle Atlantic and Great Lake states but various species grow in almost all parts of the United States. Maple is generally straight grained and reddish-brown in color. It makes excellent flooring and furniture, veneer (sometimes with curly, wavy, and *Birdseye* patterns) and is the standard wood for bowling pins.

OAK—The most widely used hardwood, comprising two specie subgroups, "Red Oak" and "White Oak." Both are heavy and strong. The former is grown principally in the southern states and has a color ranging from reddish-tan to brown. The latter is grown in the Eastern states, is the more decay resistant of the two, and has a color range from light grayish-tan to brown. It finds a wide variety of uses ranging from railroad ties (red oak must be treated) to veneer for paneling, furniture and flooring.

RED OAK—See *Oak*.

ROSEWOOD—Grown in Central and South America, it has a hard coarse, even grain. Colors are dark brown and purple. It is principally used as a decorative veneer.

TEAK—Grown principally in southern Asia, it contains an oil that gives it a pleasant odor and repels insects. The color is golden yellow with a coarse open grain. It is an expensive wood chiefly used in quality furniture.

WALNUT—Most production of walnut is in the Central states. It is a heavy, strong wood normally straight grained with *Heartwood* varying from light to dark brown. A very fine wood used for paneling, furniture, and gun stocks.

WHITE OAK—See *Oak*.

YELLOW POPLAR—Grown from Connecticut and New York southward to Florida and westward to Missouri, with the greatest production in the south. It is generally straight grained and the *Heartwood* is sometimes streaked purple, green, black, blue, or red. It is used as veneer, plywood, for furniture and *Millwork*.

Hasp A door or gate locking device consisting of a hinged strap having a slotted hole which engages, when closed, a semicircular loop through which a padlock can be placed.

Hatch A small door or covering over an opening which provides access to a roof or *Attic* area.

Haunch A flared-out, shoulder-like support at an end of an *Arch*, or a similar support for a *Beam* or *Girder*.

Hawk In plastering, a flat wood or metal tray-like tool 10″ to 14″ square with a handle, used by plasterers to carry *Plaster* from the mixer to the place of application.

"H" Beam See *Structural Shapes*.

"H" Block See *Concrete Blocks*.

Head The top part of a window or door opening.

Header

1. In carpentry, the framing member at the end of a floor or roof opening into which a *Joist* or *Rafter* is framed.

2. In *Asphaltic Concrete Paving*, a wood member around the perimeter of a paved area.

3. In *Masonry* construction, a unit laid flat with its greatest dimension perpendicular to the face of the wall. It is generally used to tie two *Tiers (Wythes)* of masonry together.

Head Joint In *Masonry*, the vertical *Mortar* joint between abutting ends of masonry units.

Hearth That part of the floor of a *Fireplace* which is in front of the fireplace opening.

Heartwood See *Wood*.

Heat Absorbing Glass See *Glass*.

Heat Exchanger A device specifically designed to transfer heat between two physically separated fluids (air or liquid) at different temperatures. Where a coil containing hot water is surrounded by cooler air, the coil wall is the exchanger for heat transfer from the water to the air.

Heat Flow See *Heat Transmission*.

Heat Gain (or Loss) Gain or loss of heat from a room or area caused by *Heat Transmission* through the enclosing surfaces (or openings). Design of heating and *Air Conditioning* systems must take into account heat gains or losses so that desired temperatures can be maintained by introducing the correct amount of air at the right temperature and humidity.

Heating Systems The following are the principal types of heating systems:

FORCED AIR HEATING—A *Furnace* having a *Fan* such that the warm air is "forced" through a *Duct* distribution system to the various rooms. The furnace contains a *Heat Exchanger* which uses gas, oil, or electrical energy to heat the air.

INFRA-RED HEATERS—A principle of heating using a heat producing wavelength just below visible light ("infra" meaning below). Its major heating function is by "line-of-sight" contact with objects; it does not heat the air through which it passes.

PERIMETER WARM AIR HEATING—A warm air heating system combining panel and *Convection* types. Warm air ducts are embedded in a concrete floor slab or ceiling, around their perimeter.

The ducts receive heated air from a furnace and deliver it, through registers, to the space to be heated. Air is returned to the furnace through a series of return air ducts from registers centrally located.

RADIANT HEATING—Heating a room by means of coils embedded within a floor, ceiling, or wall, through which hot water or air is passed.

RADIATOR—An antiquated means of heating, consisting of visually exposed units, generally mounted near the floor or on it. The radiator transfers heat by *Radiation* to the surrounding air, which in turn is circulated by natural *Convection*. The term "radiator" has been established by long usage.

SPACE HEATERS—Unit heaters used to heat specific areas or spaces. They are gas or electric fired and contain a *Fan* to direct the discharged, heated air. Space heaters are generally hung from a roof or ceiling.

STEAM HEATING—A system in which heat is transferred from a boiler or other heat source through a heat exchanger by means of steam at, above, or below, atmospheric pressure.

WARM AIR HEATING—A system consisting of a heating unit, such as a fuel burning furnace enclosed in a casing, from which the heated air is distributed to various rooms of the building through ducts.

Heat of Hydration The heat given off by the chemical reaction between *Cement* (generally referring to *Portland Cement*) and water during the hardening process.

Heat Pump An electrically operated *Air Conditioning* unit used for both heating and cooling. It can be thought of as an air conditioner which, for heating, has the air flow reversed such that it cools the outside air and discharges the heat thus absorbed into the room; a reversed cycling *Air Conditioning* unit.

Heat Strengthened Glass   See *Glass*.

Heat Transmission   The rate of flow of heat through a material. For example, the heat transmission through a particular wall might be expressed by the number of *B.T.U.*'s per hour per degree of temperature difference (from one side to the other). See also: *Coefficient of Heat Transmission*.

Heat Treating   A combination of heating and cooling operations, timed and applied to a metal or alloy in its solid state in a way that will produce specific properties. These may include strength, hardness, *Ductility*, wear resistance (such as machinability and weldability), or a change in electrical or magnetic characteristics. Varieties of heat treating include: *Annealing, Carburizing, Case Hardening, Cyaniding, Flame Hardening, Normalizing, Quenching, Stress Relieving,* and *Tempering*.

Heavy Timber Construction   Construction in which fire resistance is attained by placing limitations on the minimum size, thickness, or composition of all load carrying wood members; by avoiding concealed spaces under floors and roofs; by using approved fastening construction details and adhesives; and by providing the required degree of fire resistance on exterior and interior walls.

Heel

1. The end of a *Truss*.

2. The bottom inside edge of a retaining wall footing—the edge under the side retaining the earth.

Heliarc Welding   A trade name (Linde Division of Union Carbide Company) for TIG (Tungsten-Inert Gas) *Welding*.

Helix   A spiral-like coil. The threads of a screw form a helix.

Hem-Fir   See *Softwood*.

Hemp   A natural fiber used in one type of *Rope*; its use is declining.

Hermetic Compressor   See *Compressor*.

Herringbone Bridging (or Blocking)   Short wood blocks between *Studs* or *Joists* which are alternately offset so they can be end-nailed into place.

Hertz   Cycles per second, abbreviated "Hz." This term is replacing the term "cycles per second."

Hexagon   A six-sided geometric figure.

Hickey   A device, or tool, for bending *Reinforcing Bars* and electrical *Conduit*.

High Early Strength Cement   See *Portland Cement*.

High Intensity Discharge (HID) Lamps   A type of lamp using an arc discharge through a metallic vapor. Ballasts are required to start the lamp and control the arc. Common types are:

MERCURY VAPOR LAMPS—These lamps consist of a quartz or glass arc tube and produce light by passing an electric arc through a high pressure mercury vapor contained in the tube.

METAL-HALIDE LAMPS—This is basically the same as the "mercury vapor lamp" but salts such as thalium, indium, or sodium metals are added to the arc tube. This type of lamp produces warmer colors.

SODIUM VAPOR LAMPS—This lamp utilizes a ceramic arc tube and contains xenon, mercury, and sodium, and operates similarly to the other discharge lamps. It produces a warm yellow light.

High Lift Grouting   In masonry construction, a method of *Grouting* a brick wall whereby a considerable height of wall is constructed before grouting the space between the inner and outer faces *(Wythes)*. To keep the faces from pushing outward from the pressure of the fluid grout, metal *Ties* are placed periodically between the faces as they are constructed.

High Piled Storage A term used in evaluating fire risk for combustible goods stacked in high piles, usually defined as more than 15 feet for closely packed materials or 12 feet if stored on pallets, but the definition varies with the risk evaluating agency involved and the hazard classification of the product stored. Where such areas are protected by an *Automatic Fire Sprinkling System*, special precautions are required, such as closer *Sprinkler Head* spacing. Such areas are often required to have *Smoke Vents* and *Draft Curtains*.

High-Rise Building A building which is considerably higher than either plan dimension.

High-Strength Bolt See *Bolts*.

Hinge Types A device to permit rotation of one part with respect to another. Basic types of hinges include:

BUTT HINGE—Hinges having equal and rectangular leaves intended to be mounted against the edges of a door and a *Jamb*. They are normally used to hang doors and are commonly called BUTTS. They have removable pins to facilitate door removal without removing either *Leaf* of the hinge and are not visible when the door is closed.

CABINET HINGE—A decorative hinge used on cabinet work.

INVISIBLE HINGE—A type of hinge installed into the edge of a door and frame so it is not visible when viewing the door.

PIVOT HINGE—A type of hinge used to attach a gate or door to a frame where

the movable part needs to swing freely in both directions. It is available in both gravity and spring types.

STRAP HINGE—Consisting of elongated, strap-like leaves (flat parts) mounted on the face of two abutting parts.

"T" HINGE—Having the shape of a sidewise letter "T."

Hip On a roof, the line where two adjacent upward slopes meet. It is the opposite of a *Valley*.

Hipped Roof A roof having *Hips*. For illustration see *Roof Shapes*.

Hod A wooden device used by a plaster tender or brick tender to carry *Plaster* or *Mortar* from the mixer to the place of application.

Hog Ring A heavy galvanized wire staple applied with a *Pneumatic* gun which forms a closed, clinched ring to connect *Furring Channels*, *Metal Studs*, *Rods*, etc.

Hoist A self-contained lifting unit with or without a motor and having a drum or similar device to receive the lifting *Cable* or chain. See also: *Crane*.

Hoisting Equipment See *Bridge Crane, Climbing Crane, Derrick, Gantry, Jib Crane, Monorail Crane, Tower Crane, Truck Crane, Winch*.

Hold Harmless Clause An agreement by an insurance carrier that they will assume a contractural obligation made by their named insured and be responsible for certain acts which otherwise might be the obligation of the other party to the *Contract*.

Hollow Core Door See *Door Types*.

Hollow Metal Door See *Door Types*.

Hollow Unit Masonry See *Masonry*.

Honduras Mahogany See *Hardwood*.

Honeycomb

1. A pitted appearance of a concrete surface caused by incomplete penetration of the concrete against the *Formwork*.

2. In *Sandwich Panel* construction, a means of holding the face panels apart by an inter-connected network of thin, web-like material, usually *Resin* impregnated paper, (resembling a layer of a "honeycomb"), and glued at each edge to a face panel.

Hooke's Law The characteristic of a material to deform in exact proportion to the intensity of the *Stress* within it, and to resume its original shape after the removal of stress. The ratio of stress to strain is a constant for each material and is called its *Modulus of Elasticity*.

Hoop See *Tie*.

Hopper A bin in the shape of an inverted cone or pyramid which, due to its sloping sides, will empty its stored material when a gate at the bottom is opened.

Horizontal Pivoted Window See *Window Types*.

Horizontal Shear *Shear* forces within a horizontal plane of a *Beam* which have the effect of causing imaginary "layers" of the beam to slide past one another such as would cause slippage, during bending, between two beams placed one atop the other.(See Figure H-1.)

Horizontal Sheathing See *Sheathing*.

Horizontal Sliding Window See *Window Types*.

Horsepower A unit of measurement of *Power*. One horsepower = 550 foot-pounds (one foot-pound = one pound lifted one foot) per second or 33,000 foot-pounds per minute. One horsepower = 746 watts of electrical energy or 20,000 B.T.U.'s of heat energy.

Hose Bib A water faucet with a threaded male fitting intended for the attachment of a hose. See also: *Valves*.

Hose Clamp An ajustable, strap-like, circular clamp tightened by a special fastening device.

Hot Dip Galvanizing See *Galvanizing*.

Hot-Poured Joint See *Pipe Joints*.

Hot Rolling Forming of structural steel shapes by forcing a molten bar

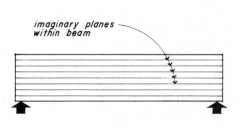

Unloaded Beam

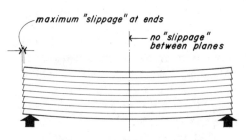

"Horizontal Shear" is the tendency of imaginary horizontal planes of a beam to slip with respect to one another when the beam is bent.

FIGURE H-1 Horizontal Shear

through a succession of rollers, each one of which changes the shape slightly toward the shape desired, such as to an *Angle, Bar, Channel,* or *I-Beam*.

House Drain See *Drainage and Vent Piping*.

House Sewer See *Drainage and Vent Piping*.

Housing A general term meaning dwelling units, but the term most often means low-income "housing" which can include apartments as well as individual houses.

Howe Truss See *Trusses*.

"H" Pile See *Piles*.

Hub An enlarged end of a pipe into which an adjoining pipe is fitted and into which *Caulking* is installed to form a seal. It is the "bell" in *Bell and Spigot Pipe Joints*.

Hubless Joint See *Pipe Joints*.

Hue See *Color*.

Humidifier A device to add moisture to the air, usually by forcing air through a spray of water or steam.

Humidistat A regulatory device, actuated by changes in *Humidity*, used for the automatic control of *Relative Humidity*.

Humidity The amount of moisture in the air. See also: *Relative Humidity*.

Hybrid Consisting of two or more different materials, such as a *Plate Girder* made from one type of steel for the *Flanges* and another type for the *Web*.

Hydrant Similar in use to a water faucet, but referring to a *Gate Valved* outlet connected directly to a water main. It is usually used to mean a *Fire Hydrant*.

Hydrated Lime A processed type of

Quicklime which is added to *Mortar* and *Plaster* mixes to increase workability. It is obtained by adding just enough water (called *Slaking*) to quick-lime such that a lumpy, hydrated, dry material results, which chemically is calcium hydroxide, and which is ground to a powder and marketed in bags for mixing with *Cement*. It is also factory mixed with *Portland Cement* which is marketed as *Masonry Cement*.

Hydration The chemical action resulting from combining a material with water, but usually referring to the hardening process when *Cement* combines with water. See also: *Heat of Hydration*.

Hydraulic Cement See *Cement*.

Hydraulic Elevator See *Elevators*.

Hydraulic Fill In earthwork, a *Fill* made by depositing soil suspended in water, such as from a dredge. After placement, the water drains away leaving the soil deposit.

Hydraulics The science of the behavior of fluids.

Hydrometer A calibrated float designed to determine the *Specific Gravity* of liquids.

Hydrometer Analysis In *Soil Mechanics*, a test to determine the percentage of *Clay*-sized particles in a soil sample by determining the *Specific Gravity* of a solution with the particles in colloidal suspension.

Hydronics A coined word meaning the art and practice of heating and cooling with water.

Hydrostatic Pressure Pressure exerted by water.

Hygrometer An instrument for measuring *Relative Humidity*.

Hyperbolic Paraboloid A saddle-

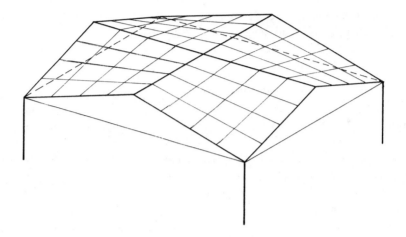

FIGURE H-2   *Hyperbolic Paraboloid*

shaped structural surface usually of *Thin Shell Concrete* and usually used for architecturally dramatic roofs. Its mathematical formulation is based upon the laws of the *Parabola* and the hyperbola. One significant feature is that its surface can be generated by a series of straight lines whose ends move along prescribed lines, hence *Formwork* for such surfaces is simplified. (See Figure H-2.)

Hypotenuse The side of a right triangle opposite the 90° angle.

Hz Abbreviation for *Hertz*, which means the same as, but is preferred to, cycles per second.

I

"I" Beam A structural shape resembling the letter "I" in cross section. Its *Flanges* are narrower than *Wide Flange Beams*. For illustration, see *Structural Shapes*.

Ice Ice weighs approximately 56 pounds per cubic foot. Water, when frozen, expands by about 10%.

Identification Index See *Plywood*.

Igneous Rock One of the three classifications of *Rock* (the other two being *Metamorphic Rock* and *Sedimentary Rock*) which is formed by solidification of a molten mixture of minerals. Examples would include *Granite*, basalt, and *Pumice*.

Illumination Common terms used to describe illumination include:

CANDLE—The standard unit of measurement of luminous (light) intensity.

CANDLEPOWER—The luminous intensity of one light source in a specific direction as measured in *Candles*.

FOOT-CANDLE—A unit of measurement for the average illumination on a surface. One foot-candle = one lumen per square foot.

FOOT LAMBERT—The unit of brightness of a light source or of an illuminated surface.

LUMEN—A unit of light output from one source. It is equal to the light falling on one square foot of surface on an imaginary sphere having one foot radius around one candle. A light source of one candle will give out 12.56 lumens (surface area of a sphere one foot in radius). Standard lamp specifications include lumens as well as *Watts, Volts,* and lamp life.

Immiscible The inability of one liquid to mix with another thereby forming separate layers or appearing cloudy.

Impact Load See *Loads*.

Impact Wrench An electrically or pneumatically operated wrench which causes a rapid hammer-like rotation of the socket for fast turning and tightening of bolts.

Impervious Incapable of being penetrated.

Import Fill In earthwork, *Fill* material which is brought to a site from another location, also called BORROW. Fill removed from a site is called EXPORT.

Inactive Door That leaf of a pair of doors that is bolted when closed and to which the lock strike is fastened to receive the *Latch* of the *Active Door* and to which the *Astragal* is attached.

Incandescent Lamps A household type of lamp having a fine tungsten wire (filament) which "burns" white hot when current passes through it. The bulb is filled with an inert gas so the filament will not oxidize (burn). See also: *Fluorescent Lamp, High Intensity Discharge Lamps*.

Incense Cedar See *Softwood*.

Inch-Pound See *Foot-Pound*.

Incombustible Construction A construction system using materials that are not *Combustible* under normal fire conditions, such as metal or concrete.

Indeterminate Structure See *Statically Indeterminate Structure*.

Indicator Post See *Automatic Fire Sprinkler System*.

Indirect Lighting Lighting in which the source of the light is not visible to persons in a room and illumination is by light reflected from a wall or ceiling.

Indoor-Outdoor Carpeting See *Carpet*.

Industrialized Building A system of building where a complete building unit, or most of the major parts thereof, are manufactured in a plant so that labor at the job site is minimized, as opposed to conventional construction where nearly all the assembly of materials is done at the job site. See also: *Factory Built Housing, Modular Construction, Sectional House, Pre-Engineered Steel Building*.

Industrial Park A planned grouping of industrial buildings whereby one developer subdivides the land, installs streets and utilities, and either sells the lots or constructs speculative industrial buildings for sale or lease.

Industrial Wastes In plumbing, liquid wastes from an industrial process which may contain chemicals or other materials as opposed to human waste or rain water.

Inert Gas A gas which will not chemically combine with another material; for example, argon and nitrogen. See also: *Welding*.

Infiltration In *Air Conditioning*, outdoor air flowing into a conditioned space as through a wall, crack, or other opening.

Inflammable Means the same as *Flammable*; the latter term is preferred.

Infra-Red Rays felt as heat which are longer in wavelength than visible light rays. One use of infra-red heat is in the painting industry for baked finishes; another is in *Heating Systems*.

Infra-Red Heaters See *Heating Systems*.

Ingot A first-cast bar of metal such as of *Aluminum, Iron,* or *Steel* for melting, storing, transporting to a manufacturer, or otherwise processing into a finished product.

Initial Set In concrete work, that point in time after mixing with water, when the concrete hardens sufficiently to hold its shape, but can still be worked, as for trowelling, as opposed to *Final Set*.

Inlaid A piece, or pieces, set into a surface so as to form a smooth surface.

Inorganic Generally meaning an absence of carbon, or referring to matter that is not animal or vegetable.

Insert An item of hardware embedded in *Concrete* or *Masonry* to provide a means of attaching something.

In Situ Test A test made at the site of origin, for example, testing in-place soil or rock.

Inspection Card A card issued with a *Building Permit* which is posted at the job site and which is signed by the building inspector as various portions of work are approved by him. The use of such cards varies with applicable *Building Codes* and locality.

Inspector See *Building Inspector, Deputy Inspector*.

Instructions to Bidders Printed instructions issued to bidders on larger projects indicating the time and date set for opening of *Bids* and other items of information pertaining to the legality of the bids.

Insulating Glass See *Glass*.

Insulation Any material having the ability to reduce *Heat Transmission*. The term *"thermal* insulation" is often used to distinguish this use from the use of these materials for "sound insulation," as when they are placed within walls or over ceilings for sound reduction. Four general classifications are:

BATT INSULATION and BLANKET INSULATION (also referred to as FLEXIBLE INSULATION—Batt insultaion is made of loosely matted mineral or vegetable fibers in thickness of from 2 to 6 inches. It is generally available in 15 and 23 inch widths and in 24 and 48 inch lengths. Batt insulation may have one or both sides faced with paper or *Aluminum Foil*. Blanket insulation is made of the same fibrous material as batt insulation but is furnished in rolls or strips in thickness of 1 to 3 inches and in widths suited to standard *Stud* and *Joist* spacing. This type of insulation is usually covered with paper with side tabs for attachment and in some cases the covering sheet is surfaced with aluminum foil.

LOOSE FILL INSULATION—This type of insulation is made from a variety of materials in the form of loose, fluffy pieces. It is used in bulk form and is supplied in bags or bales. Each bag generally covers a 25 square foot area to a depth of 4 inches. It can be "poured" over a ceiling area or can be pneumatically pumped or blown into place. It is also called POURING WOOL INSULATION or BLOWING WOOL INSULATION.

REFLECTIVE INSULATION— Usually a metal foil or foil surfaced material. Reflective insulation is effective against the transmission of radiant heat energy by reflecting the heat waves off the shiny metallic surface. It differs from other types of insulation in that the number of reflecting surfaces (not the thickness of the material) determines its insulating value. The metal foil must be exposed to an air space, perferably ¾ inches or more in depth. The foil is available in sheets or corrugations supported on paper, or in accordion shaped multiple spaced sheets.

RIGID INSULATION—Insulation made of pulp (wood, cane, or other fiber) and assembled into lightweight boards. It is available in ½ to 1 inch thicknesses and in a wide range of width and length sizes. Foamed glass, cork board, and foamed plastic are also used for rigid insulation purposes.

Insulation Board Boards manufactured chiefly for *Insulation* value and over which finish materials are applied, although some types have a finished surface on one side. They are lightweight and usually have a low structural value.

Insurance A method of protection against loss. There are a variety of types of insurance; however, the most common in the building industry are:

BUILDER'S RISK INSURANCE

—Insurance taken out for the duration of construction by an owner or builder *(Contractor)* to provide protection during construction against fire, wind, theft, vandalism, etc.

ERRORS AND OMISSION—Same as *Professional Liability Insurance*.

PROFESSIONAL LIABILITY INSURANCE—Insurance carried by an architect or engineer to protect him from claims based on malpractice.

TITLE INSURANCE An insurance policy issued to the buyer and seller of a property describing the state of title at a given point in time. It is issued by a title insurance company guaranteeing that the title to the property has been searched and is free from ENCUMBERANCES (obstructions), defects, etc., not stated in the *Title Report*. If a property is sold the title insurance is no longer valid and a new policy must be issued.

WORKMEN'S COMPENSATION INSURANCE—Insurance required of employers to provide payments to employees for job related accidents. Legal requirements vary with different states.

Intangible Value Something that represents value but cannot be directly converted to money. For example, a lake view lot as opposed to one not having such a view, or the "good will" established by a business.

Interceptors In plumbing, box-like devices in a building *Drainage and Vent Piping* system to trap and retain material not suitable for discharge into a public sewer, such as grease, oil, or sand. They consist of an inlet and outlet pipe and a baffle or baffles to prevent through flow of the deleterious material. They must be cleaned out periodically. According to their intended use they are called GREASE TRAPS and SAND TRAPS. Interceptors are also called SEPARATORS.

Interim Financing Financing of a construction project from the start of construction through its completion after which time "long term," "permanent," or other similarly termed financing takes over. Interim financing is also called "temporary financing" or a CONSTRUCTION LOAN.

Internal Friction In *Soil Mechanics*, the resistance to sliding between grains of soil.

Interpolation In mathematics, determining a value, or quantity, between two known values, by its proportional relationship between the two.

Invert The lowest point, usually referring to a pipe or ditch.

Iron Iron, from which *Steel* is made, is produced from a *Blast Furnace* with the resulting metal called *Pig Iron* which is cast into *Ingots*. It is obtained from the ore Hematite, *Limestone*, and Magnetite and, as it comes from the blast furnace, is about 95% iron, 4% *Carbon*, and 1% other elements. Varieties of iron include:

CAST IRON—Castings made from iron with 2-1/2 - 4% carbon content.

MALLEABLE IRON—Cast iron which has been heat treated to reduce brittleness. See also: *Steel*.

WROUGHT IRON—Cast iron or low carbon steel which has been worked by shaping and pounding for functional purposes.

Iron Work A comprehensive term for ornamental type iron and steel, such as railings and decorative grillework, as opposed to structural steel framing.

Iron Worker Trade classification for workmen fabricating and erecting iron work and structural steel. They belong to the International Association of Bridge, Sturctural, and Ornamental Iron Workers.

Isometric Drawing A drawing of an object which, in one view, shows three sides of the object. It differs from a *Perspective Drawing* (which illustrates depth through the use of lines converging to a "vanishing point").

Isotropic Descriptive of a material having the same physical properties in all directions. Steel is isotropic; wood is not.

J

Jack Hammer A *Pneumatic* Drill-like hammer operated by one man, usually used for breaking rock or concrete paving.

Jack Pine See *Softwood*.

Jack Rafter Short *Rafters* framing into a *Hip* or *Valley*.

Jalousie Window See *Window Types*.

Jamb The side of a window or door opening.

Jamb Block See *Concrete Blocks*.

Japanning Application of a *Lacquer* coating, which is air dried, to a hard glossy finish. The method reached its peak of technique in Japan, hence its name.

"J" Bolt See *Bolts*.

Jerry-Built A slang term meaning built in an unprofessional and slipshod manner.

Jetting

1. Driving *Piles* into the ground by jetting water under high pressure ahead of the pile tip as it is being driven.

2. A method of consolidating granular soil by injecting a high velocity stream of water into the soil mass to agitate the grains into closer contact.

Jib Crane A stationary, rotating crane having a vertical "mast" on a fixed base and a horizontal "boom" which can pivot and from which the load is lifted.

Joinery In woodworking, the means by which boards are joined together. See also: *Woodworking Joints*.

Joint See *Construction Joint, Contraction Joint, Expansion Joint, Milled Joint, Pipe Joints, Woodworking Joints*.

Jointer A power driven woodworking tool used to straighten edges of boards for joining. By the use of attachments it can also perform some *Router* applications.

Jointing See *Masonry Joints*.

Joint Venture Two or more companies jointly entering into a contract for a specfic venture. It is usually done to increase financial capability or "know-how" beyond that which either could provide alone.

Joist One of a series of parallel beams used to support a floor, called FLOOR JOISTS, or a ceiling only called CEILING JOISTS. See also: *Rafter, Tail Joist*.

Joist Anchor A metal strap having one end embedded in a masonry or concrete wall and the other end nailed or bolted to a *Joist* or *Rafter* for the purpose of providing a lateral tie for earthquake or wind resistance, between the wall and a floor or roof.

Joist Hanger Any of a variety of strap-like metal hangers used to support the ends of wood *Joists* or *Rafters*. (For illustration, see Figure L-3)

Journeyman A workman who is thoroughly qualified in a trade, having passed through an *Apprenticeship,* and who earns the prevailing standard wage for the trade.

Junction Box See *Electric Box.*

Jute Fibers obtained from the jute plant, which is a tall shrub. Most jute fibers come from India and are used in *Burlap, Rope,* and *Oakum,* and as a backing for *Carpet.*

K

Kalamein Door See *Door Types*.

Kalsomine See *Paint (Calcimine)*.

Kaolin A *Clay*, mainly hydrous *Aluminum* silicate, from which porcelain is made. It is also known as china clay.

Keene's Cement In plastering, a specially processed type of *Gypsum Plaster* used principally to obtain a very hard, smooth *Finish Coat*.

Keeper See *Strike (of a Lock)*.

Kerf A slot made by the cut of a saw.

Keyhole Saw See *Saws*.

Keying Refers to the number of different keys used for the locks on a building. A MASTER KEY will open several locks, each of which is individually keyed, and a GRAND MASTER KEY will open several groups of master keyed locks.

Keystone In masonry construction, a wedge-shaped piece at the top of an *Arch* which binds, or locks, adjacent pieces together.

Kick Plate A metal plate near the bottom of a door used to protect the door against kicking and marring. It is usually used on doors having heavy traffic.

Kiln A high temperature *Furnace*. The building industry's use of a kiln includes "burning" clay to make bricks, *Vitrification* of clay products, and drying of lumber to reduce moisture content.

Kiln Dried Lumber The artificial drying of wood, in opposition to air drying, to reduce moisture content and thereby reduce further shrinkage.

Kilowatt One thousand *Watts*.

Kilowatt-Hour A unit of electrical power consumption equal to 1,000 *Watts* acting for one hour.

Kinetic Energy See *Energy*.

King Post Truss See *Trusses*.

Kiosk A small, free-standing structure used for the purpose of sales, display, or advertisement.

Kip One thousand pounds. An abbreviation frequently used in engineering calculations, such as 12k = 12,000 lbs.

Knee Brace A short diagonal brace between a *Beam* and a *Column* to make a rigid, braced connection.

Knot That portion of a branch or limb which has been surrounded by subsequent growth of the wood of the tree. As a knot appears on the cut surface it is actually only a section of the entire knot. Its shape depends upon the direction of the cut.

Knotholes In *Lumber* terminology,

voids produced by the dropping of *Knots* from the wood in which they were originally embedded.

Kraft Paper A heavy, utilitarian, moisture resistant paper or paperboard made from wood pulp. Kraft paper comes in a variety of grades, can be *Asphalt* impregnated or bleached to appear white.

L

Labeled Door See *Door Types*.

Labor (Skilled, Unskilled) "Un-skilled labor" (also called COMMON LABOR) includes, as opposed to SKILLED LABOR, those jobs and tasks which require very little judgment and which can be learned by nearly anyone in a short period of time with little instruction. "Skilled labor" involves training and experience and varying degrees of judgment depending upon the complexity of the job.

Labor Unions See *Trade Unions*.

Lacquer See *Paint*.

Lag Bolt See *Bolts*.

Lagging Horizontal supports spanning between *Soldier Piles* to retain a vertical excavation.

Lally Column Generally, a round steel *Column* filled with *Concrete*. The term "Lally Column" is a trade name of the Lally Column Co., which also manufactures other concrete filled or encased structural columns.

Lamella Roof A roof frame consisting of a series of intersecting *Skewed* arches made up of relatively short members called "lamellas," fastened together at an angle so that each is intersected by two similar adjacent members at its mid-point, thus forming an interlocking diamond-patterned network. The framework is then covered with *Sheathing* and a roofing material.

Laminar Flow Fluid or air flow in which each particle moves in a smooth path substantially parallel to the paths followed by all other particles.

Laminated Fabricated in layers which are glued or otherwise bonded together. See also *Glued Laminated Lumber, Laminated Plastic*.

Laminated Glass See *Glass*.

Laminated Plastic Layers of paper, cloth, or wood impregnated with a *Resin* and pressed together under high temperature to form finish surfacing boards or sheets, such as are used on counter tops in which case they have a *Melamine* overlay. Formica (trade name of American Cyanimide Corp.) is an example.

Laminated Wood See *Glued Laminated Lumber*.

Lamps The three most common types of lamps used for *Illumination* are: *Fluorescent Lamps, High Intensity Discharge Lamps,* and *Incandescent Lamps*.

Land Contract Similar to a "conditional sales contract" whereby the seller retains title to the property until the buyer has fulfilled all payment obligations —usually installment payments over a period of time.

Land Description The location and boundaries of any piece of land can be described by one of three methods detailed below:

GOVERNMENT SURVEY—A reference system established by the U.S. General Land Office to locate any parcel of land within the public domain within the continental United States, except the original 13 colonies (states). The system is based upon using the nearest reference point, which points are strategically located throughout the public domain, and which are identified by the intersection of a principal *Meridian* and *Base* line. From each of these points, imaginary lines running north and south are spaced six miles apart east and west of the point (called RANGE LINES). Corresponding lines running east and west are spaced six miles apart north and south of the point (called TOWNSHIP LINES). The land included within a square between such lines is called a TOWNSHIP, and consists nominally of 36 square miles (six miles on a side). A township is further subdivided into 36 SECTIONS, each of which is nominally one mile square (numbered as shown in Figure L-1) and contains about 640 *Acres*. A sample land description based upon the government survey method might be: The N.W. ¼ of the S.W. ¼ of Section 21, T 3 N, R 5 W, San Bernardino Meridian, in the county of San Bernardino, State of California, as shown on the official *Plat* of said land.

METES AND BOUNDS DESCRIPTION—Land description whereby distances and angular bearings are described starting at one point and continuing around the property and back to the beginning point. (See Figure L-2.)

SUBDIVISION MAP—A map showing controlling survey *Monuments*, dimensions, and configuration of the lots of *Parcels*, and appropriate signatures and dedications of the owners of the subdivision. A sample subdivision map description would read: Lot 3, Block 40, Allen Tract.

Lap Joint See *Woodworking Joints*.

Laser A thin, concentrated beam of light. The word ''Laser'' (rhymes with ''razor'') is an acronym for Light Amplification by Stimulated Emission of Radiation. Presently its chief use in construction is as a surveying tool whereby the beam of light can serve as a reference line—like a stretched string—to guide, for example, the excavation of a tunnel. Although the beam is invisible (and harmless at the low energy level used for this purpose) it can be located at any point by interrupting it with an opaque object and measuring from the illuminated point of light. In actual practice a considerably more complex sensing device is used. At a very high energy level an intensity can be produced which will cut through rock, but at present not as economically as by mechanical means.

Latch A general term meaning a device that holds a door in the closed position when it is pushed shut.

Latch Set See *Lockset*.

Lateral Buckling A structural condition applicable to *Beams* wherein the top *Flange* (in compression) has a tendency to bow out sidewise (laterally) causing a localized buckling of the flange. The compression in the flange, which causes it to *Buckle*, is a function of the amount of bending in the beam.

Lateral Loads (or Forces) See *Load*.

Latewood See *Annual Rings*.

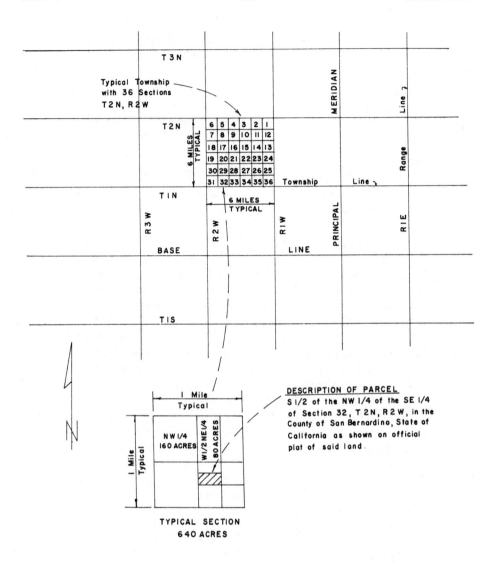

FIGURE L-1   Land Description by the Government Survey Method

Latex   Very fine natural or synthetic rubber, or plastic particles suspended in a water solution and used in rubber goods, *Adhesive, Paint*, etc. It also refers to a milky liquid found in certain plants and trees, containing *Resins*, proteins, etc., and used as the basis of rubber.

Latex Paint   See *Paint*.

Lath   A base against which *Plaster* is applied. See also: *Gypsum Lath, Metal Lath, Wood Lath*.

Latitude   Imaginary lines on the earth's surface which are parallel to the equator and measured in degrees either north or south from the equator. For example, Los Angeles is 34° north latitude. See also: *Longitude*.

Lattice Bars   Short diagonal strap-

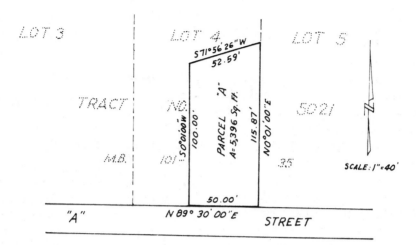

PARCEL "A":

That portion of Lot 4, Tract No. 5021 in the City of Los Angeles, County of Los Angeles, State of California, as shown on map recorded in Book 101, Page 35 of Maps in the office of the County Recorder of said County, described as follows:

Beginning at the southeasterly corner of said Lot; thence N 0°01'00"E, along the easterly line of said Lot, a distance of 115.87 feet; thence S 71°56'26"W 52.59 feet; thence S 0°01'00"W, parallel with said easterly line, 100.00 feet to the northerly line of "A" Street; thence N 89°30'00"E, along "A" Street, 50.00 feet to the point of beginning.

FIGURE L-2   Land Description by the Metes and Bounds Method

like bars used in a zig-zag pattern to connect two separated structural members.

Lauan   See *Hardwood*.

Lavatory   A fixture designed for the washing of the hands and face. It is also called a WASH BASIN.

Leach Field   A means of disposing of the liquid wastes from a *Septic Tank* whereby the liquid from the tank flows into perforated clay pipe in a gravel-filled trench, thus allowing the liquid to be spread over the length of the trench and percolate into the soil. It is also termed a DISPOSAL FIELD.

Leaching   Cesspool   Same   as *Cesspool*.

Lead   A very heavy but relatively soft

metal. It is chiefly mined as the mineral galena (lead sulfide). Lead is used with *Tin* to make *Solder*, as radiation shielding, for caulking *Pipe Joints*, and in some types of *Paint* (such as red lead).

Lead   and   Tack   (L   &   T)   In *Surveying*, a lead plug driven into a hole in concrete such as through a curb, pavement, wall etc., with a tack driven into the lead to denote a precise survey point.

Lead and Sulfur   See *Sulfur Cement*.

Lead Capped Nail   See *Nails*.

Leaded Joint   See *Pipe Joints*.

Leader   Same as *Downspout*.

Lead Pipe   See *Pipe*.

Lean-To   A building or covered area

having a roof sloped in one plane and attached to another building,—"leaning" against such building.

Lease An agreement by one party (the *Lessee*) to rent, for a fixed period of time, property owned by the other party (the *Lessor*). If the lessee in turn leases to a third party the agreement is called a SUB-LEASE. Types of leases in terms of expenses related to the lease are sometimes classified as follows:

NET LEASE—A lease whose terms require that the lessee, in addition to paying the rental amount, also pays his own utility bills (except in offices).

NET-NET LEASE—In addition to *Net Lease*, the lessee pays for insurance.

NET-NET-NET LEASE—In addition to *Net-Net Lease*, the lessee pays the real estate taxes on the building he leases. With this type of lease there are virtually no operating expenses for the lessor. It is also called an Absolute lease, or TRIPLE-NET LEASE.

PERCENTAGE LEASE—A lease where the rent is based upon a percentage of the lessee's income or sales. It applies mainly to retail stores. In some cases the lessee may be subject to a minimum monthly rental fee.

Leaseback A *Lease* arrangement whereby one party buys a property or builds (creates) it, then sells it to a second party from whom he then leases it.

Leasehold Property held under the terms of a *Lease*.

Leasehold Improvements Improvements to a property done and paid for by either the owner *(Lessor)* or tenant *(Lessee)* which is considered as having value just for that tenant for the period of the lease (example: offices to suit tenant's specific needs) but which improvements could, without other arrangements, be forfeited to the owner upon termination of the lease.

Ledger A horizontal wood or steel member fastened against the face of a masonry or concrete wall to support the ends of *Joists* or *Rafters*. (See Figure L-3.)

Left Hand Door See *Hand* (of a *Door*).

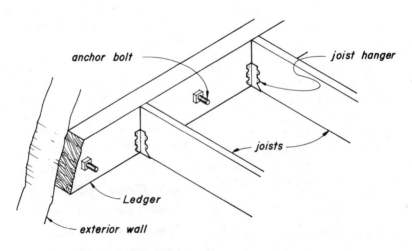

FIGURE L-3   Wood Ledger Showing Joists and Joist Hanger

Legal Description (of Land) See *Land Description*.

Lessee The party to a lease who pays rent.

Lessor The party to a lease who receives the rent and owns the facilities being leased.

Let-in Brace A diagonal brace on the side of a *Stud* wall which is notched in at each stud crossing so as to be flush with the outside face of the studs.

Level In *Surveying*, an instrument to determine relative heights by sighting through a small telescope mounted on a tripod. The magnifying sighting tube is leveled on its mount and, when sighting, has cross hairs which can be focused for precise reading on a *Rod*. A level differs from a *Transit* in that the former is not used to measure angles. Types of levels include:

AUTOMATIC LEVEL—A level utilizing a system of prisms hanging in pendulum fashion to compensate for any tilt that may remain after the head is roughly leveled; a bubble is used only for approximate leveling of the instrument.

DUMPY LEVEL—A type of level having four set screws such that it can be leveled on its base by successively leveling it in each opposing direction by knobs on each pair of the screws. Dumpy levels are now largely replaced to *Automatic Levels* because of ease of leveling the latter.

ENGINEER'S LEVEL—Leveled by means of a bubble.

WYE LEVEL—A kind of *Engineer's Level* where the telescope may be removed from the wye shaped support and reversed.

Leveling Plate A thin steel plate used under the *Bearing Plate* of a *Beam* or *Column*. Its purpose is to provide a preset level surface, at the proper height, such that the beam or column set upon it will be at the correct height.

Leveling Rod See *Rod*.

Liability Responsibility, or obligation, under the terms of a *Contract*.

Liability Insurance See *Insurance*.

Lien A claim against *Property* for non-payment of a debt relating to the property. See also: *Mechanic's Lien*.

Lift-Slab A method for constructing multi-floor buildings whereby *Concrete Slabs* are cast one above the other (stacked) at ground level directly below their final position and then lifted into place by lifting devices attached to previously erected columns.

Light

1. Light is visible electromagnetic wave energy in the spectrum range from 3,800 to 7,200 *Angstrom Units*. Light of longer wavelength (lower frequency) is termed INFRA-RED and that of shorter wavelength (higher frequency), ULTRA-VIOLET. Light is measured by *Candle, foot-candle,* and *lumen*. See *Illumination*.

2. An individual piece of glass (pane) in a window.

Light Fixtures Light fixtures within a building are of three basic types: RECESSED LIGHT FIXTURES are those where the lamp is within a housing above the ceiling surface and the lens is flush, or nearly so, with the ceiling; the SURFACE MOUNTED LIGHT FIXTURE, which is a unit attached to, but projecting down from, the ceiling (or out from a wall) surface; and the PENDANT LIGHT FIXTURE which is suspended from the ceiling by a stem, rod, or chain, such as a decorative chandelier.

Light Gage Steel Usually refers to structural members such as *Metal Decking* fabricated from sheets less than ¼″ thick. The design and specification standard for such members is the "Light Gage Cold-Formed Steel Design Manual" published by the American Iron and Steel Institute.

Light Hazard (Fire Sprinklers) See *Automatic Fire Sprinkler System*.

Light Panel See *Panelboard*.

Lightning Protection A system of metal rods above the highest part of a building which are connected to a *Conductor* leading to a *Ground* such that damage caused by lightning striking the building is minimized.

Lightweight Blocks See *Concrete Blocks*.

Lightweight Concrete Concrete made with lightweight *Aggregate* such as *Vermiculite, Perlite,* or *Pumice*. The resulting weight is less than normal *Concrete*—usually about 110 lbs. per cubic foot versus 150 lbs. per cubic foot.

Lightweight Concrete Fill A term describing concrete light enough to float, usually made with *Vermiculite* or *Pumice* as *Aggregate,* and offering negligible structural value, hence the term "fill."

Lignin In *Wood*, the thin material acting as a binder between the cells, adding strength and stiffness to cell walls.

Lime One of the oldest materials of construction due to its cementing properties. It is made by "burning" *Limestone* with a resulting loss of water content changing limestone from calcium carbonite to calcium oxide. The resulting grayish-white powder is also called QUICKLIME. It has a high attraction for water and with the controlled addition of

water it changes to calcium hydroxide, called HYDRATED LIME. The addition of quicklime and hydrated lime to *Mortar* and *Plaster* mixes increases PLASTICITY (workability).

Lime Putty *Quicklime* (which is processed *Limestone*) mixed with enough water to form a putty-like mixture which is added to a *Mortar* mix to increase *Plasticity* (workability).

Limestone A white to grey *Sedimentary Rock* which is chiefly composed of *Calcium Carbonate*. It is used as a source for *Lime*.

Limited Partnership A legal means whereby a number of persons can invest in a particular real estate venture with liability limited to the extent of their investment. It usually operates as follows: one or more General Partners conceive of, plan, and promote the venture. They solicit investments from persons who would then become limited Partners (sometimes called SPECIAL PARTNERS) whose liability is limited to the amount they invest. The General Partners who manage the property are liable for any losses and, if profitable, take a pre-established fee for having organized the venture. The remainder of the profits are divided among the Limtied Partners in proportion to their investment.

Limit Switch An electrical cut-off switch, usually for safety purposes, installed on *Craneways* and similar moving equipment to cut off power at the end of a run in the event the operator does not do so. They are usually mechanically activated by an arm on the moving part.

Lineal A term meaning in a straight line.

Linear Foot A length measurement one foot long along a line, as distinguished from a square foot or a cubic foot. It is synonymous with, but preferred to, *Lineal* foot.

Linoleum See *Resilient Flooring.*

Linseed Oil An oil made from the seed of the flax plant, having hard, weather resistant qualities when dried. Linseed oil is used in *Paints, Varnishes,* and *Linoleum.*

Lintel A *Beam* over an opening.

Lintel Block See *Concrete Blocks.*

Liquidated Damages An amount of money agreed to by both parties to a *Contract* as a settlement in the event the contract is not fulfilled.

Liquid Limit In *Soil Mechanics,* the water content at which soil passes from a plastic to a fluid state.

Liquid PVA See *Glue.*

Liquifaction A temporary fluid condition in saturated, loose, sandy soil caused by shock, such as an earthquake. It can cause serious settlement or failure of foundations.

Liquified Petroleum Gas (L.P.G.) Gas stored under sufficient pressure so that it is in liquid form. Examples are *Propane* and *Butane.* It changes to a gas upon release from its cylinder.

Liter A unit of volume measurement in the *Metric System.* One liter = 61.024 cubic inches; also 1.057 quarts.

Lithium Bromide An *Inorganic* salt with a high affinity for water, which is used in *Absorption Refrigeration* systems.

Live Load See *Load.*

Load A force or weight. In *Structural Design,* various loads are taken into consideration to assure safety. They include the following:

CONCENTRATED LOAD—A load (force) acting at one point only, as opposed to a *Uniformly Distributed Load.*

DEAD LOAD—A load which is fixed and permanent, such as the weight of the parts of a structure.

IMPACT LOAD—The sudden application of a load. It results in larger stresses than if the load were applied gradually and is most often applicable to the design of Craneways.

LATERAL LOAD—Forces acting in a horizontal (sidewise) direction, usually referring to wind or *Seismic* forces.

LIVE LOAD—Refers to loads which are subject to change such as furniture, people, etc. *Building Codes* specify minimum design live loads for various uses—for example, 50 lbs. per square foot for offices; 125 lbs. per square foot for light storage.

SEISMIC LOAD—Earthquake loading, acting both horizontally and vertically. Its Code-established magnitude is a function of ground acceleration during an earthquake and its design value varies with locality, type of building, and other factors.

UNIFORMLY DISTRIBUTED LOAD —A load spread evenly along a *Beam* or other structural member, as opposed to a *Concentrated Load.*

WIND LOAD—Pressure exerted by wind, measured in pounds per square foot. Building codes specify wind loads which vary with locality, and increase with height above the ground. Except for hurricane areas, wind loads generally vary from 15 to about 30 pounds per square foot. One generally accepted formula which relates wind velocity to its pressure is: $P = .004V^2$, where P is the wind pressure in pounds per square foot and V is the wind velocity in miles per hour.

Load Bearing Wall See *Bearing Walls*.

Loader A *Tractor*, either *Crawler* mounted or wheel mounted, having a *Bucket* on the front for excavating material primarily for loading onto trucks. It is also called a FRONT END LOADER or TRACTOR LOADER (See Figure L-4). Bucket capacities can range upward from ½ cubic yard. A SKIP LOADER is a smaller, rubber tired loader of less than one cubic yard capacity, usually having a *Backhoe* attachment.

Loblolly Pine See *Softwood*.

Lock A means of securing a door, usually operated by either a key or thumb turn. The term usually refers to either a *Deadlock*, or *Lockset*.

Lock Nut A specially designed nut used with *Screws* or *Bolts* to prevent their working loose from the pieces joined.

Lockset A standard type of door locking assembly, having a knob on one or both sides (see Figure L-5). Many locking arrangements are used depending upon

Courtesy Caterpillar Tractor Co.

FIGURE L-4   Front-End Loader

Loading Dock A platform-like area on the exterior of a building at the building floor level but elevated from the adjacent grade such that a truck can back against the dock with its bed at the same level as the dock. It can similarly be used for rail car loading and unloading. See also: *Dock Levelers*.

Loam A soil classification meaning a mixture of *Sand, Clay,* and *Organic* matter.

the use of the door. For example, a front door lockset is often operated by a key in the outside knob and a thumb turn on the inside; a bathroom lockset, usually called a PRIVACY SET, has a locking button or thumb turn only on the inside. If there are no means for locking—the assembly consisting only of knobs—it is called a LATCH SET or PASSAGEWAY SET. See also: *Deadlock*.

Lock Strike See *Strike* (of a *Lock*).

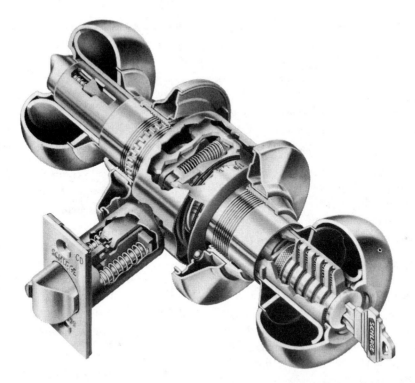

Courtesy Schlage Lock Co.

FIGURE L-5   Lockset

Lock Washer   See *Washer*.

Lodgepole Pine   See *Softwood*.

Loess   A homogeneneous, non-layered loose *Soil* deposited by wind action.

Loft   The top floor in a warehouse-type building; a storage floor directly beneath the roof.

Log   A cut length of the trunk of a tree suitable for further processing into commercial products such as *Lumber*. Terminology for the means by which lumber is sawed from a log is shown in Figure S-2. See also: *Boring Log* (as applied to *Soil* testing).

Logarithmic   A non-linear proportionality between two things, usually being a ten times increase of one number for a unit change in another.

Longitude   Imaginary lines on the earth's surface which encircle the earth and pass through both the north and south poles. They are measured as degrees of arc along the equator either east or west from Greenwich, England, the 0° reference point. For example: New York is (74°) west longitude. See also: *Latitude*.

Longleaf Pine   See *Softwood*.

Lookouts   Short wood framing members nailed to the sides of *Joists* or *Rafters* to extend a roof or floor beyond an exterior wall line.

Loose Fill Insulation   See *Insulation*.

Lot Split   A legal division of a large

Property into two or more smaller *Parcels*.

Louver An opening covered by a series of slanted vanes; usually of *Sheet Metal* such that air can pass through but rain cannot enter. Such openings usually have a screen backing to exclude insects.

Low-Boy Trailer See *Truck Types*.

Low Voltage Lighting Lighting operated at a lower voltage (usually 12 volts) for safety reasons. It is most often used in conjunction with landscaping.

LPG See *Liquified Petroleum Gas*.

Lucite Trade name (DuPont Co.) for a form of *Acrylic* plastic used in acrylic paints and for molding materials.

Lumber The term "lumber" refers to the boards, timbers, etc., produced from the sawmill whereas "wood" refers to the material itself. As indicated under *Wood*, lumber is divided botanically into *Softwood* and *Hardwood* (the latter not necessarily being harder than the former). Construction lumber is available in a variety of sizes. Excluding sheathing boards, framing lumber is available in the following nominal sizes: 2″ x 2″ through 2″ x 16″, 3″ x 3″ through 3″ x 14″ (3″ x are not commonly available), 4″ x 4″ through 4″ x 16″, 6″ x 6″ through 6″ x 16″, and 8″ x 8″ through 8″ x 18″, as well as larger timbers. Maximum lengths readily available are most often in the 16 to 20 foot length, but are obtainable to about 32 feet. Unless lumber is delivered ROUGH (as it comes from the sawmill), it is surfaced, or DRESSED (after surfacing with a planing machine), so that its actual dimensions are less than the "nominal dimension" by which it is ordered. Generally, all grading agencies use five basic size classifications as follows:

BOARDS—Up to 1-1/2″ thick and 2″ and wider.

LIGHT FRAMING (L.F.)—Consisting of members 2″ to 4″ thick and 2″ to 4″ wide.

JOISTS AND PLANKS (J & P) —Consisting of members 2″ to 4″ thick and 6″ or wider.

BEAMS AND STRINGERS (B & S)—Consisting of members 5″ and thicker by 8″ and wider.

POSTS AND TIMBERS (P & T) —Consisting of members 5″ x 5″ and greater, approximately square.

For each species (*Commercial Group Name*) and size classification, the grading agencies establish several stress-quality grades, for example: Select Structural, No. 1, No. 2, No. 3, etc., and assign allowable stresses for each.

See also: *Hardwood, Lumber Grading, Softwood*.

Lumber Defects The principal defects in lumber, affecting *Lumber Grading*; they include: *Checks, Knots, Shakes,* and *Splits*.

Lumber Grading The inspection, at the mill, of each individual piece of *Lumber* to determine its quality classification in accordance with standards developed by the industry organization representing the species being graded. For example, the West Coast Lumber Inspection Bureau, supported by the lumber producers, publishes "Standard Grading Rules for West Coast Lumber, effective 6-1-72." Grading rules for all species are under the review of the *American Softwood Lumber Standards* committee of the U.S. Department of Commerce. Most lumber is stamped with a "grademark," identifying its classification (see Figure L-6). VISUAL GRADING, used for most lumber, notes grain characteristics and defects (such as *Checks, Knots, Shakes,* and *Splits*), whereas MACHINE GRAD-

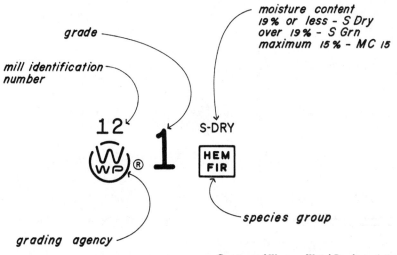

Courtesy of Western Wood Products Association

FIGURE L-6   Lumber Grade Mark

ING, which is coming into greater use, involves the *Stress* required to produce a certain *Deflection* under a test load in addition to visually determining defects.

Lumber Species   See *Wood*.

Lumber Standards   See *American Softwood Lumber Standard*.

Lumen   See *Illumination*.

Luminous Paints   See *Paint*.

Lump Sum   A fixed price for a specified amount of work, as opposed to a *Unit Cost*.

M

Machine Bolt See *Bolts*.

Machine Grading (of Lumber) See *Lumber Grading*.

Magnaflux A magnetic means of locating minute cracks in metal which are usually caused by *Heat Treating* or *Fatigue*.

Magnesia Cement See *Cement*.

Magnesite See *Cement (Magnesium Oxychloride)*.

Magnesium A lightweight metal, lighter than *Aluminum*, but not as corrosion resistant, and more costly. Chemically, it is used as an alloying element to increase *Ductility*.

Magnesium Oxychloride Cement See *Cement*.

Magnetic Catch See *Catches*.

Magnetic Declination The difference in degrees by which "north," as indicated by a magnetic compass, differs —either east or west—from true north. It varies from place to place and year to year. For example, in Los Angeles, in 1974 it was 14° 50′ east of true north.

Magnetic North "North" as indicated on a magnetic type compass. It differs from true north by the *Magnetic Declination*.

Magnetic Particle Testing See *Weld Testing*.

Mahogany See *Hardwood*.

Main That part of a distribution system, such as electrical or piping, that connects directly to a primary supply source and into which *Branch* lines connect.

Make-Up Air See *Air, Make-Up*.

Mall A wide walkway between rows of shops or store buildings.

Malleable Iron See Iron.

Man Door See *Door Types*.

Manganese An alloying metal used in the production of *Steel* to increase hardness and strength, such as for rails and earthmoving equipment, and also used in the manufacture of steel to absorb oxygen and sulfur by combining with them in the manufacturing process.

Manhole A shaft-like access for workmen to enter a *Sewer* or similar underground area from the surface.

Manilla Rope See *Rope*.

Mansard Roof A roof having two pitches on all four sides, the lower of which is steeper than the upper. For illustration see *Roof Shapes*.

Maple See *Hardwood*.

Marble A whitish rock of metamorphosized *Limestone*. It is used as a building facing material due to its hardness and its ability to take a polish. Its varied colors are due to impurities. See also: *Travertine*.

Marblecrete See *Plaster Finishes*.

Marquee A roof overhang on the front of a building which is entirely supported by the building.

Masking Tape See *Tape*.

Mason One who works with and sets up *Masonry*.

Masonite Trade name (Masonite Corporation) for a *Hardboard*.

Masonry Any assemblage of individual concrete, clay, or stone building units which are usually bonded together with *Mortar*. Masonry units having interior *Cores* or holes into which *Reinforcing Bars* can be placed and grouted (usually *Concrete Blocks*) are referred to as *Hollow Unit Masonry*. See also: *Bricks*.

Masonry Bond Patterns The pattern or arrangement, by which units of masonry are laid up in a wall. For illustration see Figure M-1.

Masonry Cement See *Portland Cement*.

Masonry Joints *Joints* between *Bricks* or *Concrete Blocks*, which are shaped to serve various decorative or weather-resistant features as shown in Figure M-2.

Masonry Nail See *Nails*.

Masonry Saw See *Saws*.

Master Key See *Keying*.

Mastic A term used to describe compounds which remain elastic and pliable, rather than hardening with age.

Mat Foundation See *Floating Foundation*.

Matrix In mathematics, a group of numbers (or things) arranged in a rectangular fashion and subject to algebraic rules.

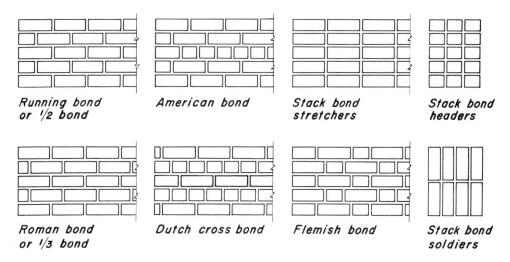

Running bond or ½ bond

American bond

Stack bond stretchers

Stack bond headers

Roman bond or ⅓ bond

Dutch cross bond

Flemish bond

Stack bond soldiers

FIGURE M-1 Masonry Bond Patterns

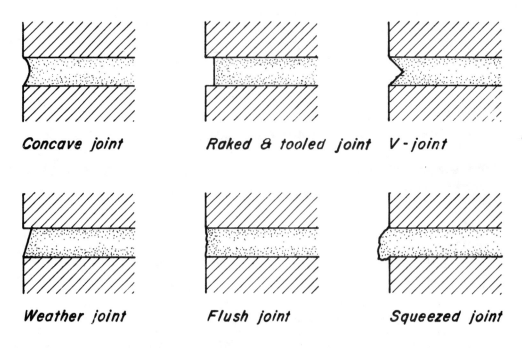

Concave joint   Raked & tooled joint   V - joint

Weather joint   Flush joint   Squeezed joint

FIGURE M-2 *Masonry Joints*

Mathematical Model A mathematical formula that corresponds to a real situation such that the real situation can be studied by studying the formula it represents.

Maximum Density In *Soil Mechanics*, the maximum amount of *Compaction* that a soil can be given under optimum conditions of moisture content. For example, the term "90% compaction" means the soil is compacted to 90% of its maximum possible density.

Mechanical Analysis A determination of *Soil* or *Aggregate* gradation (see *Grain Size Distribution Curve*) by screening through a series of progressively smaller sieves.

Mechanical Engineer See *Engineers*.

Mechanical Joint See *Pipe Joints*.

Mechanical Refrigeration See *Refrigeration Equipment*.

Mechanics Lien A legally enforceable lien (claim), made against a partially or wholly completed building, for nonpayment of an obligation due a *Contractor* or sub-contractor.

Median A mathematical term for the middle number in a series of numbers, with just as many numbers above as below it. The "median" of the numbers 3, 9, 14, 16, 24 is 14 (two numbers above and two below). See also: *Average*.

Meeting Rail A horizontal member in a window separating an upper and lower pane of *Glass*.

Melamine A *Thermosetting* plastic *Resin* made by reaction to the addition of formaldehyde. In powder form, dissolved in water or ethyl alcohol, it is used for laminating (such as for counter tops). It is colorless, odorless, and resistant to organic solvents. Melamine is also used for molding and, when combined with other resins, can be used as a coating material.

Membrane Waterproofing Waterproofing a surface such as a concrete or masonry wall below ground, by applying a thin layer (membrane) of an impervious material such as *Polyethylene* or *Asphalt* to the surface, or by the use of a penetrating surface sealer.

Mensuration The determination of lengths, areas, and volumes; measurement.

Mercury Vapor Lamp See *High Intensity Discharge Lamps*.

Meridian A north-south line of longitude. In *Land Description* by the Government Survey method, it is a north-south government established reference line which, at its intersection with a *Base Line*, forms a point of reference for locating and describing property in the surrounding area. There are a number of such reference points in the United States.

Mesh A network of wires woven, twisted, or welded together to form a variety of opening sizes and patterns. Uses include fencing (see *Chain Link Fencing*, *Woven Wire Fencing*); *Concrete* reinforcement (see *Welded Wire Fabric*); as plaster *Lath* (see *Stucco Mesh*, also called *Chicken Wire, and Metal Lath*); and for insect screening, where it is called *Wire Cloth*. For the latter, a mesh number is sometimes used to indicate the number of openings per linear inch.

Metal Clad Door See *Door Types*.

Metal Decking Formed light-gage metal sheets available in a wide variety of configurations for varying span conditions. (see Figure M-3). Such decking is

Courtesy Republic Steel Corp.

FIGURE M-3 Metal Decking Installation

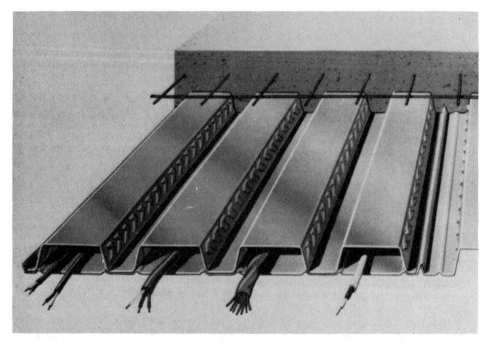

Courtesy of Elwin G. Smith
Division of Cyclops Corporation

FIGURE M-4   Cellular Metal Floor

generally available in depths from 1-1/2″ to 8″ or more, and gages from 14 to 22 gage. CELLULAR METAL FLOOR (see Figure M-4) is a type of decking used for floors, over which *Concrete* is placed, having a configuration such that spaces (cells) are created within the deck which can be used for installing electrical *Conduits* (wiring). See also: *Corrugated Metal Decking*.

Metal Foil   See *Aluminum Foil*.

Metal Halide Lamps   See *High Intensity Discharge Lamps*.

Metal Lath   Also called EXPANDED METAL LATH, sheets of metal having diamond shaped openings, which are formed by making numerous slits in a sheet of metal and stretching the sheets

such that the slits open into the diamond shape. Metal lath is usually used under *Cement Plaster* and is available in the following types:

DIAMOND MESH—Flat sheets formed as described above.

FLAT EXPANDED METAL LATH —Same as *Diamond Mesh*.

FLAT RIB METAL LATH—Same as the general description above having approximately ⅛″ deep ribs to stiffen the sheets.

PAPER-BACKED METAL LATH —Any metal lath affixed to a paper backing.

RIB METAL LATH—Same as the general description above but having rib depths of ⅜″ or ¾″.

SELF-FURRING METAL LATH—A type of *Flat Expanded Metal Lath* having ¼" protrusions such that it can be placed with the protrusions having the effect of spacing the lath out from the supporting surface.

Metallic Paint See *Paint*.

Metal Primers See *Paint*.

Metal Studs *Studs* formed of *Light-Gage Steel* and used as framing for a wall in a manner similar to wood studs. They are channel-shaped and have either solid or punched *Webs*; they are available in depths from 2-1/2" to 6" or more, and gages from 16 to 20 gage. Some types called NAILABLE METAL STUDS have a crease in each edge for inserting a nail.

Metallurgy The science of metals; their extraction, refining and preparation for use.

Metamorphic Rock One of the three classifications of *Rock* (the other two being *Igneous Rock* and *Sedimentary Rock*). These are rocks which have changed from some previous structure due to pressure, heat, water or other agency. For example, the change of *Limestone* (a sedimentary rock) into *Marble*.

Meter In the *Metric System* of measurement, one meter = 39.37 inches or 3.28 feet. One meter = 100 centimeters.

Metes and Bounds Description See *Land Description*.

Metric System A standard system of weight and measures having the *Meter* as the unit of length, the *Liter* as the measure of volume, and the *Gram* as the unit of weight.

Mezzanine A partial floor between two floors, occupying a lesser area than the floor below.

Micarta A trade name (Westinghouse Corporation) for a *Laminated Plastic*.

Micron A measurement of particle size. One micron = one millionth of a *Meter*. For visualization, 800 microns end to end would equal approximately 1/32 of an inch.

MIG Welding See *Arc Welding (Gas Shielded)*.

Mil A unit of measurement of the thickness of coatings. One mil = one-thousandth of an inch.

Mil (Circular) See *Circular Mil*.

Milcor Trade name (Inland-Ryerson Construction Products) for a number of *Metal Lath* and *Drywall* accessories, and related products.

Mild Steel See *Steel*.

Mill Building A general term, which is declining in use but has meant an industrial type building framed of *Steel* or *Wood*—usually with a *Truss* roof—and having wood or corrugated metal siding and roof covering. See also: *Pre-Engineered Steel building*.

Milled Joint A joint between two abutting members, usually metal, whose ends have been ground or planed (milled) to provide full contact.

Millwork A general term applying to wood items manufactured in planing mills and woodworking shops and including such items as door and window frames and molding. The term can also include *Cabinetwork*.

Millwright A trade classification originating with workmen in grain mills who installed spouting and machinery and now generally referring to the installation of machinery. Depending upon locality, union millwrights usually belong to the United Brotherhood of Carpenters and

Joiners of America or the International Association of Bridge, Structural, and Ornamental Iron Workers.

Mineral Spirits See *Solvents*.

Mineral Wool A fluffy insulating material composed of thread-like mineral fibers made by jetting steam through molten *Slag*. It is also called ROCK WOOL.

Minute In angular measure, one-sixtieth (1/60) of one *Degree*. Each minute is further subdivided into 60 seconds.

Mirrors, Transparent See *Glass (Transparent Mirrors)*.

Mission Tile Semi-circular Spanish-style clay tile used for roofing.

Miter To cut at an angle.

Miter Box A frame to guide *Handsaws* while making *Miter* cuts.

Mitered Joint See *Woodworking Joints*.

Mixing Box In *Central Air Conditioning*, a device in a *Double Duct System* which is located at or near a room being conditioned and which mixes hot and cold air from separate *Ducts*, in thermostatically controlled proportions, for discharge into the room through a *Diffuser*.

Mobile Home A factory built house on a towable frame with wheels, which generally does not conform to *Building Code* standards for houses. See also: *Factory Built Housing, Industrialized Building, Modular Construction*.

Model Codes See *Building Code*.

Modular Construction Construction of a building (house, apartments, etc.) using factory built *Modules*. A module can be a complete unit in itself, such as a single-family house, or separate function units which can be placed side by side or stacked to form a complete building.

Modular House A house constructed from *Modules* which are transported to the job site and erected on a prepared foundation. It differs from a SECTIONAL HOUSE in that the latter is generally considered as constructed from "parts" (such as wall and roof sections), rather than complete modules. Also, a modular house differs from a MOBILE HOME in that the latter is potentially movable and generally does not conform to *Building Code* standards for houses. See also: *Factory Built Housing, Industrialized Housing, Modular Construction*.

Modular System A method of planning a building on a grid system such that all materials are sized and spaced so they are in multiples of the module dimension (for example: 48″ on center in each direction).

Module A factory built complete unit of a house, apartment, etc. (such as one room or a group of rooms), or a three-dimensional section of a building, and which is usually constructed to *Building Code* standards. See also: *Modular Construction, Modular House*.

Modulus of Elasticity The ratio of *Stress* to *Strain* for an elastic material. By knowing the modulus of elasticity of a material, the amount of strain (deformation or movement) can be computed for a given amount of stress. Its symbol is "E". For steel: E=30,000,000 lbs. per sq. in. per inch; for wood: E = approximately 1,600,000 lbs. per sq. in. per inch.

Modulus of Rupture The unit *Stress* at which a particular *Beam* fails in bending when tested to destruction, such stress being determined by dividing the *Bending Moment* at failure by the calculated *Section Modulus* at the cross section where failure occurs.

Modulus of Subgrade Reaction In foundation design, a measure of the supporting capacity of a pavement *Subgrade*, determined by dividing the wheel load by the amount by which the pavement deflects under the load.

Moisture Content The amount of water contained in a particular sample of material (wood, for example) expressed as a percentage of its weight compared to its weight after having been oven dried.

Moisture Content (of Soils) See *Optimum Moisture Content*.

Molding A decoratively cut or shaped strip of material, usually wood, placed into a corner to form a finish trim, such as where a wall meets a ceiling.

Molly Bolt See *Bolts*.

Molybdenum An alloying element used in the manufacture of *Steel* to increase hardness and shock resistance, particularly at high temperatures, such as for boilers and jet engines.

Moment See *Bending Moment*.

Moment of Inertia In structural design, a mathematical property of the cross section of a *Beam* or *Column* relating to stiffness, which is used chiefly to compute *Deflection*.

Moment of Stability Resistance of a structure, or portion thereof, to being overturned, as by wind or earthquake forces. See also: *Overturning Moment*.

Monel Metal A trade name (U.S. Steel) for a highly corrosion resistant alloy, chiefly of *Nickel* and *Copper*.

Monitor Roof A raised, sawtooth-like portion of a roof, generally on an industrial building, whereby the steeper sloped side has windows and/or ventilators to provide additional light and/or ventilation within the building. For illustration see *Roof Shapes*.

Monolithic *Concrete* placed at one time without construction joints; placed in one "pour."

Monorail Hoist A hoist suspended from, and traveling along, a single rail as opposed to a *Bridge Crane*.

Monotube Piles See *Piles*.

Monument In *Surveying*, a permanent reference marker. It often consists of a brass plate on concrete with an inscribed "x" and such applicable data as to its elevation and/or what reference point it represents.

Mortar A thick, soupy mixture of *Cement*, sand, and water which is used to bond *Masonry* units together, but differing from *Grout* in that the latter is more fluid and is used to fill cavities within the wall. *Hydrated Lime* or *Lime Putty* is often used in the mortar mix to improve its workability.

Mortise To cut a rectangular slot-like hole in the side of a wood member into which something is to fit.

Mortise Lock A door lock recessed ("mortised") into a slot cut into a door as opposed to a surface mounted *Rim Lock* or a *Cylinder Lock*, the latter being the most commonly used.

Mortise and Tenon Joint See *Woodworking Joints*.

Mosaic Tile See *Ceramic Tile*.

Motor Grader See *Grader*.

Muck Semi-fluid mud. Also, the material excavated during construction of a tunnel.

Mud A slang term for any *Cement* mixture while it is in the plastic state. It is used to mean *Concrete*, *Mortar*, or *Plaster*.

Mud Jacking See *Pressure Grouting*.

Mudsill A horizontal wood support *Sill* piece set onto a concrete foundation or slab and onto which the wood *Studs* are set; the bottom most horizontal piece of a wood stud wall, which is in contact with the foundation.

Mullion A framing, or separating, member between adjacent door or window sections.

Multi-Zone Air Conditioning See *Central Air conditioning*.

Muntin Framing members between panes of glass in a window.

N

Nailable Metal Studs See *Metal Studs*.

Nail Driving Tool A hand operated, gun-like tool for automatically driving nails. They are pneumatically operated by trigger action and can drive nails, depending upon the size of the gun, from 6d to 16d; certain models can drive up to a rate of 7,000 per hour. (See Figure N-1.)

Nailer A wood piece bolted to a steel *Beam* to provide a means for nailing on *Sheathing, Joists,* or *Rafters*.

Nailing Machines See *Nail Driving Tool*.

Nailing Strip Any wood strip to provide a nailable backing for an overlain material.

Nail Pop The protrusion of a nail head above the surface resulting from swelling of the framing member into which the nail is driven.

Nails Nails are usually made from *Steel* wire, but are also available in *Stainless Steel, Aluminum, Copper,* and *Bronze,* the latter three being rust-proof. The shank (meaning the shaft part of the nail) may have surface protrusions, as described under types of nails listed below, to increase their holding power. Nail heads are of three basic shapes: flat, which is the type most commonly used;

Brad head, which is used in finish work and which usually has a depression in the top requiring the use of a *Nail Set* to drive it below the surface: and oval and/or rounded head used specifically for appearance. Lengths of nails are usually designated by their *Penny* size which varies from 2d (about 1″ long) to 60d (about 6″ long). Nails longer than 6″ are called SPIKES. The lengths of some types of nails, such as *Brads, Concrete Nails,* and *Roofing Nails* are indicated in inches. Nail types include:

BARBED NAIL—Has indentations the full length of the shank for greater gripping power.

BOAT NAIL—Light or heavy duty round wire nails with oval, countersunk heads. They may be annularly or helically threaded and are used for improved holding power.

BOX NAIL—Similar to a *Common Nail* but having a smaller shank diameter and used primarily for light framing, boxes, etc.

BRAD—Similar to a *Finishing Nail* but obtainable in lighter gages.

CASING NAIL—Similar to a *Finishing Nail* but thicker and having a cone-shaped head.

CEMENT COATED NAIL—Nails coated with *Cement* which soaks up the

Courtesy of Paslode Company, Division of Signode

FIGURE N-1   Nail Driving Tool

water in wood to resist corrosion and improve holding power.

CLINCH NAIL (or Clout Nail) —Intended to have the pointed end bent over after going through a second material.

COMMON NAIL—The standard nail used in wood framing—it is supplied routinely if no other specific nail is specified.

CONCRETE NAIL—A specially hardened steel nail such that it can be driven into *Concrete* or *Masonry*.

COOLER NAIL—Usually cement coated, slender nails with flat heads, generally used for nailing *Drywall*.

CUT NAIL—Nails cut from flat stock rather than wire, having a square shank. They are infrequently used.

DOUBLE HEADED NAIL—A nail primarily used for temporary work which must be taken apart (such as concrete *Formwork*), having two flat heads spaced one above the other, so it will seat against the lower head and be easily withdrawn by a *Claw Hammer* under the upper, protruding head.

DOWEL PINS—A type of nail used in *Mortise and Tenon Joints* (*Woodworking Joints*). They have a barbed shank for greater holding power and no head but a recess at the top of the shank for use with a nail set.

DRIVE SCREW—A nail having steeply inclined threads (screw-like) for improved holding power but which is hammered into place.

DRYWALL NAIL—Similar to a

Roofing Nail but having a smaller head.

DUPLEX NAIL—Same as *Double Headed Nail*.

FINISHING NAIL—A thin nail with a small, barrel shaped head with a slight recess which is driven below the surface (with a *Nail Set*) and the remaining depression filled with *Putty*. It is primarily used for *Cabinetwork* and interior paneling.

GYPSUM BOARD NAILS—Used for applying *Gypsum Board*. They are blue in color and sterilized since carpenters frequently hold them in their mouth.

LEAD CAPPED NAIL—Having a lead head which, due to its degree of softness, seals the nail hole; used primarily for exterior work.

MASONRY NAIL—Similar to a *Concrete Nail*, but with a larger shank diameter and usually zinc coated.

PLYWOOD NAIL—Like a *Common Nail* but shorter and usually with a barbed shank and countersunk heads.

PURLIN NAIL—See *Straw Nail*.

RATCHET NAIL—A nail having heightened annular rings and a flat head for nailing into metal studs to attach *Drywall or Gypsum Lath*.

RING SHANK NAIL—Having annular rings along the shank for improved holding power.

ROOFING NAIL—A nail with a large flat head for greater "hold down" of roofing felt and *Fiberboard*. Most roofing nails are galvanized; some have neoprene *Washers* to give weatherproof seal when used on aluminum and fiberglass roofs and siding. They are available with either plain or screw shanks and usually have a barbed shank.

SCREW NAIL—A nail with a spirally grooved shank used principally as a

Roofing Nail but with a smaller head. It drives more easily and rotates as it enters the wood, and provides greater holding power.

SHINGLE NAIL—These nails have thin shanks and small heads and are used for *Cedar Shingles*—two for each shingle.

SPIRAL SHANK NAIL—Same as *Ring Shank Nail* but having spiral protrusions.

STAPLES—A form of nail made from plain and/or galvanized wire; they come in a variety of sizes and are generally in the shape of a "U."

STRAW NAIL (or *Purlin Nail*)—Long galvanized nails with flat or curved heads for securing corrugated roofing to *I-beams*. They bend around, but do not penetrate, the I-beam.

"T" NAIL—A specially designed nail for use in *Nailing Machines*.

Nail Set A special tool, generally 4" long and with a tapered end, used to drive certain nails, usually *Finishing Nails*, below the surface.

National Building Code This model building code is published by the American Insurance Association and is adopted by some cities and counties in the eastern part of the United States but is not as widely used as other model codes. It is the oldest model code and is produced by insurance underwriters.

National Electrical Code This is the only model electric code in the United States. It is published by the National Fire Protection Association, with the National Electrical Manufacturers Association participating.

National Plumbing Code A model plumbing code issued by the American National Standards Institute. It is used to a limited degree only; its last publication was in 1955.

Natural Cement See *Cement*.

Natural Draft The flow of air through a *Chimney*, heater, or building caused by unequal air densities (weights) caused by differences of temperature, pressures, or wind, as opposed to *Forced Draft* (such as by a *Fan*).

Natural Gas See *Fuel*.

Neat Cement See *Cement*.

Neat Plaster A *Plaster* mix without any sand, consisting only of *Cement* and water.

Needle Beam A beam passing through or under a structure to provide support during alterations or moving.

Neon Lamps Glass tubes from which air is evacuated and neon gas added. The tube glows when a sufficiently high voltage is applied to cause current to flow through the tube.

Neoprene A synthetic *Rubber*, less resilient than natural rubber, having a high resistance to chemical corrosion.

Neoprene Washers *Washers* made from *Neoprene*, usually to provide insulating protection such as to prevent *Galvanic Corrosion* between dissimilar materials, as between a steel *Screw* fastening an aluminum sheet.

Net Area The area remaining after deducting holes, cut-outs, etc.; the actual usable area, as opposed to *Gross Area*.

Neutral Axis In structural design, an imaginary line through the cross section of a *Beam* which, during *Bending*, divides the areas of the beam which are in *Tension* and *Compression*, such line theoretically having zero *Stress*.

Neutral Conductor In electrical work, a *Conductor* which is usually *Grounded* and which has the lowest measured voltage between it and any other conductor in the system. It is also called the Common Conductor and is normally white in color.

Newell Post On a stairway, an end post supporting a run of *Handrail*.

Nickel One of the most commonly used alloying elements for *Steel* to increase strength without reducing ductility.

Nipple See *Pipe Fittings*.

Nitrocellulose Cotton or wood fibers treated with nitric acid; it is the major ingredient of most *Lacquers* (see *Paint*).

Noble Fir See *Softwood*.

No-Hubb Joint See *Pipe Joints*.

Noise Reduction Coefficient A measurement of the degree of sound control offered by a material or combination of materials. Most acoustical surfacing materials have established Noise Reduction Coefficients.

Nomogram Also called a Nomograph, it is a chart-like means of solving problems with several variables by laying a straight-edge across parallel rows of *Scales*, each of which represents a variable, and by the point of intersection on one of the scales, an unknown value can be determined.

Non-Bearing Wall See *Walls*.

Non-Combustible See *Incombustible*.

Non-Destructive Testing A means of testing the quality of materials without breaking or otherwise damaging them. Examples of non-destructive testing include *Dye Penetrating Testing, Magnetic Particle Testing, Ultrasonic Testing*, and X-Ray Inspection.

Non-Metallic Sheathed Cable An electrical cable having an insulated covering composed of a flame and moisture retardant rubber or plastic braid. It is refered to as ROMEX in the trade.

Non-Shrink Grout A *Grout* mix containing an additive which causes the mixture to expand during hardening and thereby offsetting the natural shrinkage.

Non-Sparking Floor See *Conductive Floor*.

Normalizing A general type of *Annealing* that usually involves air cooling of *Ferrous* materials after first heating to a specific temperature for a designated period of time. It is generally done to soften and increase *Ductiblity* but not to as great a degree as full annealing. The process is sometimes carried out on large or thick parts as a preliminary step in a multi-stage *Heat Treating* procedure.

Norman Brick See *Bricks*.

Nosing The front edge of a stair *Tread*.

Notice of Completion A notice filed by the owner of a building with the *County Recorder* indicating that, in his belief, all construction bills have been paid. A sub*Contractor* or material supplier has a set time thereafter in which to file a *Lien* in the event he has not been paid.

Nylon A tough, abrasion resistant *Thermoplastic* used for brush bristles, carpet fibers, appliance housings, coating metal, etc.

O

Oak See *Hardwood*.

Oakum Untwisted *Hemp Rope* impregnated with oil, used to caulk *Bell and Spigot* type *Pipe Joints*.

Obelisk A pyramid-topped four-sided pillar, usually of stone or concrete.

Obscure Glass See *Glass*.

Obsolescence The loss in value of an asset (buildings, tools, equipment, fixtures, etc.) due to such factors as change in technology, lack of demand, out-of-date, etc. See also: *Depreciation*.

Occupancy In *Building Codes*, the term refers to the use to which a building will be put, such as office building, warehouse, classroom, etc. Such occupancy classifications, which are usually given letter designations, have varying requirements as to degree of *Fire Resistive* construction, exits, and related safety requirements. In those occupancies with a large number of occupants or with other special hazards such as storage of explosive materials, the requirements are more severe.

Occupational Safety and Health Administration (O.S.H.A.) A government agency devoted to the development and regulation of occupational safety and health standards throughout the country. It was created in 1970 as the Williams-Steiger Occupational Safety and Health Act with the Department of Labor established as the federal agency responsible, and with certain research and investigatory functions under the jurisdiction of the Department of Health, Education and Welfare. The organization issues regulations, conducts investigations and inspections to insure compliance with those standards, and suggests penalties in cases where regulations are not adhered to. It maintains regional offices in the following areas: Boston, New York, Philadelphia, Atlanta, Chicago, Dallas, Kansas City, Denver, San Francisco, and Seattle. Copies of the O.S.H.A. handbook are available through the Government Printing Office, Washington, D.C. or regional offices.

Octagon An eight-sided geometric figure.

Ohm In electricity, the unit of electrical resistance. It is measured by an Ohmmeter whose two leads are connected to points in a *Circuit* between which the resistance is to be determined.

Ohm's Law A basic electrical formula relating *Volts, Amps,* and *Ohms*: E = I x R(Volts = Amps x Ohms), or in transposed form, Amps = $\dfrac{\text{Volts}}{\text{Ohms}}$

Oil Base Paint See *Paint*.

One-Way Glass A misnomer since there is no true "one-way" glass. It is properly called a *Transparent Mirror* and is described under *Glass*.

One-Way Slab See *Concrete Slabs*.

Open-End Block See *Concrete Blocks*.

Open Grain (Wood) See *Grain (of Wood)*.

Open Web Beam (or Joist) Steel beams resembling *Trusses* which are manufactured in many sizes, which have been standardized through the Steel Joist Institute to which most of the manufacturers belong. They usually consist of top and bottom *Chords* consisting of double (back-to-back) *Angles*, and *Web* members consisting of round rods.

Optimum Design The best or most favorable design.

Optimum Moisture Content In *Soil Mechanics*, the water content at which a given soil can be compacted to its maximum density. See also: *Compaction*.

Orangeburg Pipe See *Pipe*.

Orange Peel

1. A film surface, usually referring to *Paint*, which is pebbled, resembling the skin of an orange.

2. See *Grapple*.

Orbital Sander See *Sanders*.

Ordinary Hazard (Fire Sprinklers) See *Automatic Fire Sprinkler System*.

Organic Derived from living organisms, animal or vegetable, and generally containing carbon.

Organic Coatings Coatings produced from *Resins* derived from carbon compounds. Such raw materials are obtained from petroleum, coal, *Natural Gas*, and agricultural wastes. By comparison, *Inorganic* coatings are based on materials with a lack of carbon compounds.

Ornamental Iron Fabricated *Steel* or *Iron* items such as railings, gates, and ornamental grillworks, as opposed to structural steel items such as *Beams, Columns*, etc.

O.S.H.A. See *Occupational Safety and Health Act*.

Ounce A unit of weight equal to 1/16th of a *Pound*.

Outlet In electrical work, the termination of an electrical *Circuit* for connection to a *Receptical, Switch, Lighting Fixture*, appliance, or machine. See also: *Electric Box*.

Outlet Box See *Electric Box*.

Overbreak Excavation beyond an intended limit line or surface.

Overcurrent Protective Device In electrical work, a device, either a *Fuse* or *Circuit Breaker*, provided to open (interrupt) an electric *Circuit* to prevent damage to an electric *Conductor* in the event of an excessive current.

Overflow Scupper A hole through a *Parapet* wall which is slightly above the roof level and adjacent to a *Downspout* or *Roof Drain*. Its purpose is to relieve a water buildup in case the downspout or roof drain becomes clogged.

Overhead Overhead costs on a construction project include temporary field office facilities, insurance, job site telephone, superintendent's salary, and related items not directly a part of the construction cost.

Overhead Door See *Door Types*.

Overtime Extra pay—normally one-and-one-half the usual rate—for work done in excess of eight hours a day or 40 hours a week. Some work, such as on Sundays and holidays, requires (by union contact) double time payment.

Overturning Moment A structural design term meaning a horizontal force times a vertical distance, which tends to overturn a structural element, such as a retaining wall, tank, or component of a building. The horizontal force can be wind, *Seismic,* or *Earth Pressure.* The MOMENT OF STABILITY is the counter-acting force (vertical weight) times the horizontal distance from the center of gravity of the vertical forces to the edge about which the overturning rotation would occur.

Oxidation The chemical combining of an element with oxygen.

Oxyacetylene Welding and Cutting See *Gas Welding and Cutting.*

Oxychloride See *Cement (Magnesium Oxychloride).*

P

Pacific Silver Fir See *Softwood*.

Packaged Air Conditioning A factory assemblage of *Air Conditioning* equipment within one housing.

Package Deal A method of building whereby one company contracts, usually for a fixed sum, to provide a complete one-responsibility design, construction, and financing service for a client. Package deals differ from a TURN KEY JOB in that the latter generally involves design and construction only.

Paint A surface covering material to protect and/or decorate, applied in liquid form by brush, roller, or spraying, which hardens upon drying. It consists of a *Vehicle* which is the liquid (or binding) portion, such as oil, water, etc., and the *Pigment* which is generally in the form of fine powders. *Paint Finishes* range from *Flat* to *Gloss* indicating their degree of luster or shine. *Solvents* are a functionally related category of chemicals used to control the *Viscosity* or the thickness of the type of paint being applied. They are occasionally used as cleaners for removal of paint in undesired areas. Types of paints include the following:

ACOUSTIC PAINT—A coating specifically prepared for use on acoustical material but which does not significantly change the acoustical material's sound-absorbing quality.

ACRYLIC PAINT—A water-thinned *Vinyl Paint* deriving its name from the chemical substance making up the film.

ALUMINUM PAINT—A mixture of finely divided aluminum particles, in flake form, combined with a vehicle such as thin *Varnish*. It is used as a protective coating on metals.

ANTICORROSIVE PAINT—A first application of paint to *Iron* or *Steel* to inhibit or retard corrosion (*Rust*). See *Metal Primers (Paint)*.

BITUMINOUS PAINT—Originally referring to *Asphalt*, bituminous paint is a black, pitchy material which is the natural product of the oxidation of high boiling point petroleums. It is used on roofs to patch with, and as a water repellent between earth and *Concrete Blocks* to retard moisture.

CALCIMINE—An older type of water-thinned paint (rarely used today) consisting of *Glue, Calcium Carbonate, Clay, Pigments,* and water. It is sometimes spelled *Kalsomine*.

CASEIN PAINT—A type of water-thinned paint using proteinaceous material (*Casein*) as the binder.

CEMENT BASE PAINT—A special-

257

ized paint containing *Portland Cement*, a *Pigment* in an oil *Vehicle*, and other ingredients. It is in powder form and is mixed with water for application. Its primary use is on galvanized iron fences for animal enclosure since it contains no *Lead*.

ENAMEL—Paint having a hard, gloss, eggshell, or semi-gloss finish and consisting of a color added to *Varnish*.

EPOXY PAINT—An extremely hard, durable finish paint, having epoxy as its *Vehicle*. Epoxy paints consist of two or more components requiring mixing just prior to using. In clear form it can be used to cover enamel or vinyl for added durability. Its fumes are highly toxic and require special precaution.

FIRE RETARDANT PAINT—Paints applied to wood to increase resistance to combustion (it does not make the wood fire-proof). When the paint is heated there is a foaming action which serves as an *Insulation* to exclude the oxygen necessary for combustion.

FLAT PAINT—See *Paint Finishes*.

KALSOMINE—See *Calcimine*.

LACQUER—A flat, semi-gloss, or high gloss coating for wood, paper, or metal which is similar to *Enamel* but harder and faster drying. It consists of a *Resin* such as *Nitrocellulose* mixed with a fast drying *Solvent*. It is highly resistant to weather or sunlight.

LATEX PAINT—The term refers to water-thinned paints and finishes whose principle vehicle is *Latex*. Water may be used to clean the painting equipment. Different types of latex are known by the names of the chemical substances which make up the film, such as Styrene-butadiene, *Polyvinyl Acetate,* or *Acrylic*.

LUMINOUS PAINT—Coatings giving off light instead of just reflecting it; they must be rejuvenated by sunlight or artificial light.

METALLIC PAINT—Paints containing finely ground silver or gold particles to give a shiny finish, such as those used on cars.

METAL PRIMERS—Paints used for metals as a first coat designed to retard corrosion (*Rust*) and which may, or may not, be covered with additional decorative coats of paint. Metal primers require a surface free of foreign materials: loose and scaling rust must be removed. They include:

RED LEAD—Paint (red in color) containing an oxide of lead as the *Pigment*. It is also the source of lead in *Driers*. These primers are used for steel and iron surfaces.

RED OXIDE—Paint (brown-orange in color) containing zinc chromate and red iron oxide and used for steel and iron surfaces.

WHITE ANTI-CORROSION—An oil base paint used on iron and steel surfaces, and *Aluminum* when pretreated. Its advantages is that being white, an intermediate coat of paint is not required.

ZINC CHROMATE—Paint (yellow in color) containing zinc chromate as the *Pigment* and used on *Aluminum, Iron* and *Steel*.

OIL BASE PAINT—Paint in which the vehicle is oil (usually *Linseed Oil*). Oil base paints penetrate the surface to which they are applied.

RUBBER BASE PAINT—See *Latex Paint*.

SHELLAC—A clear coating material used to seal and give luster finish to wood. It consists of a *Resin* called "Lac" which is mixed with alcohol. After application the alcohol evaporates rapidly leaving the hard finish. Shellac is used in preference to varnish because of its rapid evaporation—*Varnish* requires 24 hours or more to dry. Shellacs are primarily used for floor coatings.

SPAR VARNISH—See *Varnish* below.

SPIRIT VARNISH—See *Varnish* below.

STAIN—Similar to paint but with a lesser amount of pigment; it can be transparent, semi-transparent, or opaque, but does not obscure the grain of the wood. It is lower in viscosity than paint and has high penetrating properties.

VARNISH—A clear protective coating for wood. It consists of *Resins* dissolved in alcohol or a quick drying oil. When alcohol is the drying agent it is called SPIRIT VARNISH; when an oil is the drying agent it is called SPAR VARNISH and has extremely high weather resistance.

VINYL PAINT—A general term indicating a *Plastic* in the paint. Vinyls comprise one of the most versatile groups of plastics with a wide range of properties dependent upon specific attachment to other chemical groups, i.e., when coupled with a chlorine group vinyl it becomes *Polyvinyl Chloride*. Vinyl paints adhere to the surface to which they are applied rather than penetrating as oil paints do.

WATER BASE PAINT—Paint having a vehicle that requires water for setting, solubility, or *Emulsion* purposes.

See also: *Color, Paint Finishes, Paint Thinner,* and *Solvents*.

Paint Finishes The final appearance of a painted surface with respect to its shine. The dull to shiny progression ranges from FLAT, SATIN, EGGSHELL, SEMI-GLOSS to GLOSS, and

is controlled by chemicals which break up the surface shine.

Paint Thinner *Volatile* liquids used to reduce or regulate the consistency of *Paint* and *Varnish*.

Panelboard In electrical work, a metal (usually steel) cabinet-like housing for electrical equipment usually mounted on, or recessed into, a wall and used to divide a single incoming *Feeder* into separate distribution *Circuits* (called branch circuits) such as to lights, receptacle outlets, motors, etc., such panelboards being called Light Panels or Power Panels. Each outgoing circuit is connected to an *Overcurrent Protective Device (Fuse* or *Circuit Breaker)*. See also: *Switchboard*.

Panel Door See *Door Types*.

Panelized Plywood Roof A method of constructing wood roofs comprising pre-assembled plywood panels, generally 4' x 8', having attached longitudinal framing members, usually 2 x 4's, called "sub-purlins," such that the panel is placed to span between *Purlins*, generally 4 x 12's or 4 x 14's spaced 8' on center. (See Figure P-1.)

Panic Hardware A type of quick-acting door opening *Hardware* consisting of a horizontal bar on the inside of a door. Pushing against the bar will, by a leverage mechanism, unlatch and open the door. Such hardware is legally required as a safety factor on certain exits such as from theaters and similar areas occupied by a large number of people.

Pan Joist Concrete Floor A type of concrete floor where a series of metal pans are used as forms to produce a floor system comprising closely spaced *Beams* and intervening *Slabs*. (See Figure P-2.)

Paper-Backed Metal Lath See *Metal Lath*.

FIGURE P-1 *Panelized Plywood Roof*

Parabola The curve formed by a vertical "slice" through a cone. Such a curve is said to be "parabolic."

Parallel Circuit See *Circuit*.

Parallelogram A four-sided geometric shape where opposite sides are of equal length and parallel.

Parallel Rule In drafting, a straightedge device for drawing horizontal lines and used in lieu of a "T" square. It extends across the drawing board and is guided vertically by wires attached to each corner of the drawing board. See also: *Drafting Machine*.

Parapet The top part of an exterior wall which is above the roof line.

Parcel In surveying, a term used to denote a small piece of *Property*.

Parent Metal See *Welding*.

Parkerizing Trade name (Parker Rustproof Corporation, a division of Oxy-Metal Finishes) for a zinc phosphate coating process for *Steel*. It is a corrosion protector for internal parts such as screws and other fasteners—particularly in the automotive field, calculating equipment, typewriters, etc. It transforms the steel surface into a non-metallic, non-conducting surface. Colors range from gray to black; the process is usually followed by an oil dip or wax finish, and can also be painted.

FIGURE P-2   Pan Joist Concrete Floor Forming with Reinforcing in Place

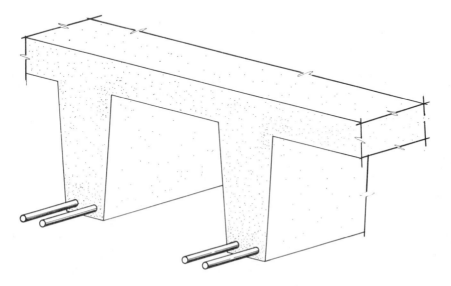

Pan Joist Construction

Parking For parking data, see Figure P-3.

Participation Agreement In financing, an agreement made as a condition to a real estate loan whereby the lender —such as an insurance company —receives a percentage of the profits and in some cases an ownership interest in addition to interest on the loan, and may have a voice in the management of the project being financed.

Particleboard A building panelboard consisting of small pieces or chips of wood bonded together with heat and pressure using synthetic *Resin* or other suitable binder. An example is Novoply, manufactured by U.S. Plywood.

Party Walls See *Walls*.

Passageway Set See *Lockset*.

Passive Earth Pressure See *Earth Pressure*.

Pattern Block See *Concrete Blocks*.

Pay Line In *Earthwork*, the outside line of an excavation for which the excavator is paid. No pay is given for excavating beyond the pay line even though sometimes it cannot be avoided.

Pedology The science that considers the nature, properties, formation, behavior, and use of *Soils*.

Peeling In painting, the process whereby a paint film detaches from its base surface.

Minimum Dimensions For Parking Layouts

a angle	*b* stall	*c* aisle	*d* overall	*e* layout
30°	16'	12'	44'	16'-0"
45°	19'	14'	52'	11'-3 3/4"
60°	20'	20'	60'	9'-4 1/4"
✱ 90°	18'	26'	62'	8'-0"

✱ *End stalls parallel to walls or fences should be a minimum of 10' wide.*

Parallel Parking
Minimum aisles are 10'
Minimum size parking space 8' x 24'

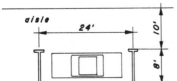

FIGURE P-3 Parking

Penalty Clause A clause sometimes put into a construction *Contract* indicating the amount to be paid as a penalty for nonfulfillment of a provision of the contract. It usually refers to a penalty for each day's delay beyond a specified completion date.

Pendant Light Fixture See *Light Fixtures*.

Penetrometer In *Soil Mechanics*, a device which measures the relative density and sometimes the bearing capacity of a soil by the resistance to penetration into soil of a cylindrical plunger or shaft. A *Procter Needle* is one type of penetrometer.

Penny Abbreviated "d", the term indicates *Nail* lengths. The abbreviation originally came from the Roman word "denarius," meaning coin, which the English adapted to penny and formerly meant the weight of 100 nails.

Pentagon A five-sided geometric figure.

Percolation Test A test to determine the permeability of *Soil* or rock by measuring the timed rate of change in the level of a column of water allowed to percolate into the soil or rock.

Perforated Gypsum Lath See *Gypsum Lath*.

Perimeter Warm Air Heating See *Heating Systems*.

Perlite A lightweight *Aggregate* used in acoustical and insulating plaster and *Lightweight Concrete*. It is a volcanic glass (obsidian) which, upon heating, expands greatly in size, weighing from 7 to 15 lbs. per cubic foot. Its use as an aggregate offers greater fire resistance, insulation value, and reduced weight.

Permafrost A permanently frozen zone of soil.

Permeability The rate at which water is transmitted through a material. It is most commonly used in reference to *Soils*, where its magnitude is expressed as the COEFFICIENT OF PERMEABILITY, and is measured by a *Permeameter*.

Permeameter An instrument to measure *Permeability*, generally for *Soils*.

Personal Property See *Property*.

Perspective A three dimensional drawing giving the illusion of depth by the convergence of lines in each plane toward varnishing points. (See Figure P-4.)

Phased Construction See *Construction Management*.

Phenolic A synthetic *Resin* forming the basis of a family of *Thermosetting* plastics. They are also used as impregnating agents and components of *Paints* and *Glue*.

Phenol-Resorcinol See *Glue*.

Philippine Mahogany See *Hardwood*.

Photogrammetry In *Surveying*, the use of aerial photography to measure land distances and elevations—the latter by using three-dimensional photographs and special viewing lenses. See also: *Aerial Surveying*.

Pi A mathematical (symbol: π) constant equal to the ratio of the circumference of a circle to its diameter, having the approximate value of 3.1416.

Pier An isolated column of *Masonry* or *Concrete*, and generally having a low ratio of height to width. See also: *Belled Pier*.

Pig Iron See *Iron*.

Pigment The coloring material in *Paint*, usually dry substances, mixed with the *Vehicle*.

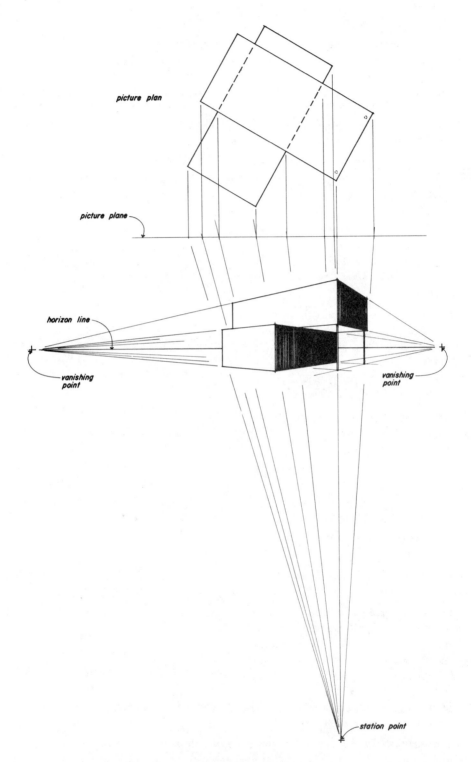

picture plan

picture plane

horizon line

vanishing
point

vanishing
point

station point

FIGURE P-4 Perspective

Pilaster A portion of a *Masonry* or *Concrete* wall which is thickened to act as a *Column* to support a *Beam* or *Truss*. (See Figure P-5.)

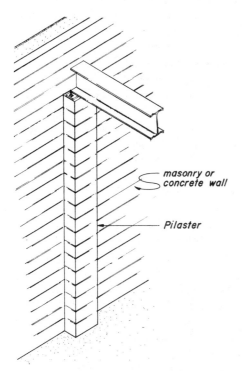

masonry or concrete wall

Pilaster

FIGURE P-5   *Pilaster*

Pilaster Block See *Concrete Blocks*.

Pile Cap A concrete mass at the top of a *Pile* or group of piles, to act as a means to transfer load from a structure to the piles.

Pile Driver A machine used to drive *Piles* into the ground. The guided weight which hits the top of the pile is the PILE HAMMER and can be steam, diesel, pneumatically, hydraulically, or vibratorily driven. The latter can extract as well as drive piles.

Pile Hammer See *Pile Driver*.

Pile Load Test A performance test

for the supporting capacity of *Piles*. The pile is loaded in increments up to 150 percent of the load it is designed to support. No appreciable settlement after 24 hours at maximum load is usually satisfactory evidence of its capacity.

Piles Shaft-like structural members placed vertically into the soil to support a building or structure, as opposed to *Footings, Belled Piers,* or a *Floating Foundation*. Piles support *Loads* by either the frictional resistance between the surface of the pile and the soil—called FRICTION PILES—or by *Column* action where the pile penetration is in poor soil and bears, as a column, upon a hard stratum at a deep level—called BEARING PILES. Piles are inserted into the ground by several methods, including driving by repeated blows with a heavy weight, or *Pile Hammer*, vibrating with an oscillating *Pile Driver* attached to the upper end of the pile, JETTING, whereby high pressure water discharged ahead of the tip displaces the soil permitting easy penetration with only nominal driving, and pouring *Concrete* in predrilled holes—called CAST-IN-PLACE PILES. Types of piles include:

BATTER PILES—A pile placed into the ground at an angle, such as to provide increased resistance to *Lateral Loads* or *Uplift*.

CONCRETE PILES—These are either precast (made in a plant and shipped as a unit for driving at the job site) or cast-in-place as described above.

"H" PILES—Steel "H" *Beams*, which are available in a number of standard rolled shapes, which are driven into the ground by a pile driver.

STEP-TAPER PILE—This is a cast-in-place pile where successively reduced diameter *Steel* casings are driven into the ground and subsequently filled with concrete such that a ''step-tapered'' pile results.

WOOD PILES—Round wood piles driven into the ground. They are usually tapered and are of *Pressure Treated Wood* to impede deterioration.

Pile Shell A metal casing driven into the ground and filled with *Concrete* to make a *Pile*.

Piling See *Sheet Piling*.

Pilot (or Man-) Door See *Door Types*.

Pilot Light A small flame, used in gas-heating or gas-cooling devices, which burns constantly to ignite the main gas supply when it is turned on.

Pine See *Softwood*.

Pipe Pipe is used to carry liquids, gases, or solids suspended in liquid, and is manufactured in a wide range of materials, some of which are better suited than others for specific uses. Pipe is roughly classified as either PRESSURE PIPE, which can be subjected to internal pressure such as for water supply systems, or DRAINAGE PIPE, which is also called non-pressure or gravity pipe and is used primarily for drainage, waste lines, and venting. Pipe is specified in terms of its internal diameter as opposed to *Tubing* which is specified by the outside diameter. The term CLASS refers to pressure pipe and indicates pressure resisting characteristics; the term SCHEDULE refers to wall thicknesses. Types of pipe and descriptive information about each are listed below. See also: *Pipe Fittings, Pipe Joints*.

ASBESTOS CEMENT PIPE—Pipe made from *Portland Cement* and *Asbestos* fibers; it is primarily used for *Drainage and Vent Piping*. It is generally available in 3″ to 18″ diameters and in lengths of 10 to 13 feet. It is most often connected by *Hubless (Pipe) Joints*.

BITUMINIZED FIBER PIPE—Pipe manufactured from *Cellulose* and asbestos fibers combined with coal tar pitch. It is primarily used for below grade drainage and sewer lines. It is available in diameters of 3″ to 6″ and generally in 8′ lengths. The tapered ends are friction fitted into a corresponding tapered sleeve. Bituminized fiber pipe is supplied with its own couplings.

BRASS PIPE—Pressure pipe used primarily for water service and air service. It is generally available in 1/2″ to 2″ diameters and in 20′ lengths, and is connected by *Brazed (Pipe) Joints* or *Screwed Joints*. Thinwall brass pipe is used for drainage and venting where pressure presents no problems. Brass tubing is also available.

CAST IRON PIPE—Used primarily for water supply, drainage, and *Sewer* lines. Most commonly used diameters range from 1-1/2″ to 15″ and standard lengths are 5 and 10 feet. Cast iron pipe is available uncoated or coated, the latter with asphalt or bitumastic material to prevent corrosion in underground installations. Ends of pipe are either of the *Bell and Spigot* type for joining by caulking with lead and *Oakum* (in declining use), or by *Compression (Pipe) Joints*. It is also manufactured with flanged ends for *Mechanical (Pipe) Joints*.

CEMENT ASBESTOS PIPE—See *Asbestos Cement Pipe*.

CERTAIN-TEED PIPE—An asbestos cement pipe manufactured by Certain-Teed Corporation under the trade name Fluid-Tite. It is used both as a pressure pipe for water supply systems and as drainage and sewer pipe. The standard diameters are from 3″ to 24″ and 13′ lengths. It uses a *Hubless (Pipe) Joint* including a lubricating paste in the connection.

CLAY PIPE—Used primarily for sewer and storm drain lines. It is not used under buildings due to its limited structural capacity. Most commonly used diameters are 4″ through 12″, although larger diameters are available. Standard available length is 5 feet. Clay pipe is joined by *Bell and Spigot (Pipe) Joints*, or, more commonly, with plain ends for hubless joints. Clay pipe is also vitrified to provide an impervious surface, such pipe being termed VITRIFIED CLAY PIPE (VCP).

CONCRETE PIPE—Concrete pipe is available either reinforced, in diameters of 12″ to 144″ and 8, 12, and 16 foot lengths, or non-reinforced, in diameters of 4″ to 72″ and in 30″ to 8′ lengths. It is used for water supply systems, drainage pipe, and sewers. Joints may be either tongue-and-groove or *Bell and Spigot (Pipe) Joints*, using cement mortar and a rubber ring to effect the seal.

COPPER PIPE—Copper pipe is used for water supply systems, waste lines, (primarily for medical and industrial use), and vent lines. It has high corrosion resistance. Standard available diameters range from 1/4″ to 6″ and 20′ lengths. Copper pipe is connected by *Sweat (Pipe) Joints* or can be threaded for screwed joints. Copper tubing is available in 60′ coils or 20′ straight lengths.

CORRUGATED METAL PIPE—Corrugated metal pipe is formed from galvanized sheet metal, plain or asphalt coated, with the corrugations running around the perimeter of the pipe for added stiffness; aluminum is also available but not as widely used. It is generally used for storm drains and *Culverts* and is available in diameters of 6″ to 54″ and 20′ lengths. Corrugated metal pipe is joined by bolt-

ing. An arch configuration is also available for uses requiring a lower height. Dimensions for this type are from 28″ span x 20″ rise to 83″ span x 57″ rise.

GALVANIZED STEEL PIPE—See *Steel Pipe*.

LEAD PIPE—Lead pipe is now primarily used in the telecommunication field in the splicing of underground cables (also called sleeving) in diameters of 1/2″ to 7″ and 6′ lengths. As tubing it is available in 3/8″ and 1/4″ and lengths up to 200 feet. Lead pipe is connected by a *Flanged (Pipe) Joint* or *Wiped (Pipe) Joints*.

ORANGEBURG PIPE—Trade name (Flintkote Corp.) for a *Plastic Pipe* (PVC or polyethlene) used principally for cold water service systems. It is available in 3/4″ to 2″ diameters and 20′ lengths. As tubing its diameter and length dimensions correspond to that of *Copper Tubing*. Orangeburg pipe is connected by *Solvent Weld (Pipe) Joints*.

PLASTIC PIPE—Also called *PVC (Polyvinyl Chloride)* pipe. It is used for water lines, waste lines, vent piping, and in the chemical industry for acid wastes. It is available in diameters from 1/2″ to 12″ and in 20′ lengths. Connections are most often by *Solvent Weld (Pipe) Joints*.

STAINLESS STEEL PIPE—Pipe used primarily for corrosion resistance such as in chemical, petroleum, and food industries (the latter use for sanitation requirements). It is available in 1/8″ to 6″ diameters connected by threaded or *Welded (Pipe) Joints* and in 8″ to 24″ diameters connected by flanged joints or welded joints. It is available in 18′ to 20′ lengths. Stainless steel tubing is also available in sizes corresponding to copper tubing.

STEEL PIPE—Steel pipe is avilable as black (non-galvanized) or *Galvanized*. It is used for water supply lines, fire protection water systems, vents, and storm drains. It is available in standard diameters of 1/4" to 12" and lengths of 21 feet. Connections are by flanged joints, screwed joints, or welded joints. Steel pipe is also used for structural purposes such as for fence posts and columns. For such use, it is available in three wall thicknesses for each nominal diameter: standard, extra strong, and double extra strong.

TRANSITE PIPE—Trade name (Johns-Manville) for a type of *Asbestos Cement Pipe*.

VITRIFIED CLAY PIPE (VCP)—See *Clay Pipe*.

Pipe Color Coding Industrial piping can be color coded by the following accepted identifying colors:

RED—Fire protection piping, such as a sprinkler system.

YELLOW—(or orange)—Dangerous, such as gas or steam.

GREEN—(or white)—Safe materials.

BLUE—Protective materials, such as chilled water

DEEP PURPLE—Extra valuable materials.

Pipe Dope Any material used to lubricate and seal screwed type *Pipe Joints*.

Pipe Fittings Connecting devices placed between sections of *Pipe* to serve various function such as to change direction, etc. Principal types of pipe fittings include:

BULL HEADED TEE—See *Tee*.

CAP—A cover screwed onto the end of a pipe to close it.

CALDER COUPLING—See *Pipe Joints*.

COUPLING—A short length of internally threaded pipe used to connect two

pipe ends to form a straight run. See also: *Nipple*.

CROSS—Similar to a *Tee*, but joining four pipe sections, each 90° apart.

ELL (or "L")—A fitting, also called an ELBOW or "bend," used to change the direction of a run of pipe. They are available for angle changes of 11-1/2, 22, 45, and 90 degrees.

INCREASER—See *Reducer*.

NIPPLE—A short piece of externally threaded pipe which is screwed into and between two sections of longer pipe or fittings to effect a connection.

PLUG—An externally threaded solid piece screwed into the end of a pipe to close it.

REDUCER (or INCREASER)—A fitting joining pipes of different sizes.

SADDLE—A pipe straddling device used to tap into the side of an existing run of pipe.

TEE—A coupling between straight runs but having a side outlet to connect a pipe joining perpendicularly. If the side outlet is larger than the run ends it is called a BULL HEADED TEE.

UNION—A pipe fitting used in place of a *Coupling* which is internally threaded in such a way that turning the union draws in both pipe ends. It is used where future dismantling and reconnections are anticipated and the pipe ends cannot be turned.

WYE (or "Y")—A fitting used to divide a single run of pipe into two diverging runs, resembling the letter "Y". The two types are "Lateral Y's" (Y) and "True Y's" (Y).

Pipe Flange A round, collar-like plate around the end of a *Pipe* used to join sections of pipe together by bolting a matching flange on one pipe to another flange on the other pipe. A BLIND

FLANGE is one without a hole for passage of liquids, used to close the end of the pipe.

Pipe Joints A means of connecting ends of *Pipe*, including the following types:

BALL JOINT—A connection in which a ball is held in a cuplike shell that allows movement in any direction in the joint except along the *Axis* of the pipes that are joined. Also called a SWIVEL JOINT.

BELL AND SPIGOT JOINT—A connection in which one end of the pipe is manufactured with a flared out "bell" or hub into which an adjoining plain end fits. The overlapped space was originally intended to be filled with *Oakum* fiber and sealed with molten *Lead*. This is declining in use, being largely replaced by the COMPRESSION type joint which uses a *Neoprene* or rubber gasket inserted into the hub with the seal effected when the plain end is pushed into the gasketed hub. Bell and spigot type joints are used for *Cast Iron, Clay* and *Concrete Pipe*. The latter is generally caulked with *Cement Mortar*. (See Figure P-6.)

BRAZED JOINT—Similar to a *Sweat Joint* but using brass instead of *Solder*, requiring a high melting point (over 1100°F).

CALDER COUPLING—Trade name (Joints, Inc.) for a type of *Hubless Joint*.

CAULKED JOINT—In general, any joint made by caulking, such as in a *Bell and Spigot Joint*.

COMPRESSION JOINT—See *Bell and Spigot Joint*.

DUAL JOINT—Same as *Compression Joint*.

FLANGED JOINT—A bolted connection whereby matching holes in face-to-face flanges are bolted together with a metal or rubber gasket between. It is primarily used for larger diameter *Cast Iron Pipe, Lead Pipe*, and *Steel Pipe*.

FLARED JOINT—A joint for *Tubing* where a threaded coupling, beveled to receive the flare, draws the ends together as the coupling is tightened.

HOT-POURED JOINT—A joint, such as of the *Bell and Spigot* type, where either molten *Lead* or heated *Bitumastic* material is used to seal the joint.

HUBLESS JOINT—A means of joining sections of plain end pipe. A neoprene sleeve under a steel band is placed around the joint and tightened by two *Stainless Steel* screw clamps. It is used on *Drainage and Vent Piping*.

LEADED JOINT—A *Hot-Poured Joint*, using molten lead and *Oakum* fibers.

MECHANICAL JOINT—A general term meaning a pipe joint made by mechanical means, requiring nuts and

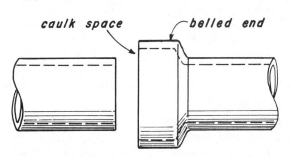

FIGURE P-6   Bell and Spigot Joint

bolts. It is most often used for *Cast Iron Pipe*.

NO-HUB JOINT—A trade name (California/Alabama Pipe Co.) for a type of *Hubless Joint*.

SCREWED JOINT—A joint using threaded couplings. They are used primarily for *Steel Pipe*.

SLIP JOINT—Any connection in which one pipe slides into another and is held in place by a gasket or some kind of *Sealant* or *Adhesive*.

SOLVENT WELD JOINT—A joint used for *Plastic Pipe* whereby a solvent "glue" is applied at the pipe connection which dissolves a portion of the plastic material and fuses the parts together.

SWEAT JOINT—A soldered joint, used for *Copper Pipe*. The heat to melt and fuse the *Solder* is applied by a torch or a *Soldering Iron*.

SWIVEL JOINT—See *Ball Joint*.

TYTON JOINT—Trade name (U.S. Cast Iron Co.) for a type of *Slip Joint* using a special type of rubber ring.

WELDED JOINT—Joining of steel pipe by *Welding*.

WIPED JOINT—A means of joining *Lead Pipe* whereby molten lead or *Bitumen* is wiped around the end perimeters of the pipe before joining.

Pipe Wrench A wrench with slightly curved, serrated jaws, designed to tighten the grip on the pipe as the handle is turned. It is also called a STILLSON WRENCH. See also: *Chain Wrench*.

Pitch

1. Black, sticky, gummy substance made from the distillation (turning to a vapor by heating and then condensing back to a liquid) of the oil constituents found in coal—also called *Tar* pitch.

2. A *Resin* found in certain evergreen trees (conifers).

3. The angle of the threads of a *Screw, Bolt*, or similar configuration.

4. The slope of the sides of a roof, expressed by a ratio of height to span.

Pitch Pocket In *Lumber* terminology, a well defined opening between *Annual Rings* which usually contains, or has contained, *Pitch*—either liquid or solid.

Pith In *Lumber* terminology, the small soft core in the center of a *Log*.

Pivot Hinge See *Hinge Types*.

Plain Gypsum Lath See *Gypsum Lath*.

Plain Sliced Veneer See *Plywood*. *(Hardwood.)*

Plane Table Surveying A means of plotting and mapping in the field, by the use of a drawing board mounted on a tripod. Sightings are made through an instrument called an ALIDADE which is basically a sighting telescope mounted on a straight edge and which sets on the board and can be moved about such that sighting can be made and a line drawn parallel to its axis which will represent the direction of the sighting on the drawing.

Plank Generally, a large board used as a walking or bearing surface. See also: *Lumber*.

Planning In a general sense, the arranging of a number of activities or elements in such a way that their interaction most efficiently achieves a desired result; an orderly program to achieve a desired result.

Planning Commission A citizens' committee working within a governmental entity whose members are appointed or elected and are legally authorized to set

Zoning policies for a city or county and pass on appeals for *Variances*.

Plans

1. Generally the term "plans" means a complete set of *Working Drawings* for a project.

2. The term "plan" means a drawing showing, in horizontal projection (i.e., "looking down"), a part of a building, a total building, or larger area.

Types of such plans include:

FLOOR PLAN—A drawing showing the arrangement of rooms, walls, partitions, doors, and windows, giving dimensions, names or room numbers, and other information.

FRAMING PLAN—Primarily a structural drawing similar to a *Floor Plan* but showing, in plan view, structural framing members such as *Joists*, *Rafters*, *Beams*, and *Girders*. Such a drawing also shows *Bearing Walls*, *Columns*, and other points of support and is dimensional correlation to the *Floor Plan*.

FOUNDATION PLAN—A plan view drawing showing the layout of the *Foundations*.

GRADING PLAN—Similar to a *Site Plan* but primarily showing desired *Contours* and related *Earthwork* information to show how the site is to be graded.

PLOT PLAN—Same as a *Site Plan*.

SITE PLAN—A drawing showing a complete *Property* (site), indicating the location of buildings, walkways, paved areas, street frontages, etc.

Plaster A mixture of *Cement*, sand, *Lime Putty* or *Hydrated Lime*, and water which, when freshly mixed, can be applied in thin layers (1/8″ to 1″) over a

backing, called *Lath*, and which hardens to form a rigid surface. The two basic types of plaster are *Gypsum Plaster* (made with *Gypsum* cement) and *Cement Plaster* (made with *Portland Cement*). See also: *Metal Lath*, *Plaster Coats*, *Plaster Finishes*.

Plasterboard Same as *Gypsum Board*.

Plaster Coats Plaster is applied in two or three coats depending upon the surface to which it is applied. For three coat application, generally used for *Cement Plaster* over *Metal Lath*, the coats are called, in order of their application; SCRATCH COAT which is applied to the lath and "scored" with grooves for better bonding of the next coat, the BROWN COAT. The final coat is called the FINISH COAT, and can be given a number of surface treatments as described under *Plaster Finishes*. Two coat application is generally used for *Gypsum Plaster* over *Gypsum Lath* and consists of the brown coat and a finish coat.

Plaster Finishes Commonly used plaster finishes include:

ACOUSTICAL PLASTER FINISH—A sprayed-on finish plaster coat resulting in a textured and relatively cushioned surface to reduce sound wave reverberation. (See Figure P-7.) It consists of a lightweight, pourous *Aggregate* such as *Perlite*. A simulated acoustical effect can also be obtained by a *Spray Finish*.

COLOR FINISH—Obtained by integrally mixed color pigments or colored aggregates.

DASH COAT—A fine or coarse texture applied by hand or machine.

FLOAT FINISH—Obtained by floating with a shingle, carpet, wood, rubber, etc., to bring the aggregate to the surface.

Courtesy National Gypsum Co.

FIGURE P-7   Application of Acoustical Plaster

GLITTER FINISH—Mica or metallic flakes blown onto the wet surface.

MARBLECRETE—Surface embedded materials such as marble chips, crushed ceramic tile, etc., thrown forcibly onto a bedding coat either by hand or machine, and then tamped lightly to give uniform embedment.

PUTTY COAT—A *Trowelled Finish* coat consisting of *Lime Putty* or *Hydrated Lime* gaged with gypsum *Gaging Plaster* or *Keene's Cement*.

SAND FINISH—A finish having a sand-like texture; it may or may not have color added.

SKIM COAT—A very thin coat (3/32″-1/16″) applied on the surface, usu-ally to cover imperfections or to add extra thickness to plaster.

SPRAY FINISH—A sprayed-on finish plaster coat resulting in a textured sur-face. It is applied by machine and is popu-larly used on exteriors and interior ceil-ings.

TEXTURE FINISH—Any finish coat of plaster applied in an artistic manner, using uneven designs, etc.

TROWELLED FINISH—A smooth, dense finish using a steel *Trowel*.

Plaster Fireproofing *Plaster* used as a fireproofing material, over structural steel and wood beams and columns by either spraying the plaster directly onto such members and building up a thickness in accord with established fire resistive standards, or by a method where *Lath* and plaster are applied around, but separated from, the structural steel or wood members. The thickness of plaster required for a given fire resistance may be decreased when higher insulating value *Aggregates* are used, such as *Perlite*.

Plaster Ground   See *Base Screed*.

Plastering Machine   A mechanical device by which wet *Plaster* is conveyed under pressure through a flexible hose to deposit the plaster by blowing it onto a surface. It is also known as a "plaster pump" or "plaster gun." It is distinct from *Gunite* in that for the latter the plaster or *Concrete* is conveyed dry through the hose and water is added at the nozzle.

Plaster of Paris   A variety of *Gypsum* used (decreasingly) for plaster-like castings, such as for ornamentation.

Plastic Foam   See *Foamed Plastic*.

Plasticity

1. Descriptive of improved workability characteristics given to *Plaster* and *Mortar* mixes by such agents as *Lime Putty*, *Hydrated Lime*, an *Air Entraining Agent*, or other *Admixture*.

2. In structural design, and as distinguished from *Elasticity*, the stressing of a material beyond the *Elastic Limit*, where permanent deformation will result when the stress is removed.

Plasticizer

1. An *Admixture* for *Mortar* or *Plaster* mixes to give a more plastic (workable) quality.

2. In painting, materials used to blend with brittle *Resins* to produce better flexibility of the *Vehicle*.

Plastic Limit   In *Soil Mechanics*, the water content at which soil changes from semi-solid to a plastic state.

Plastics   A group of synthetic materials (meaning man-made rather than occurring in nature) which can, by the application of heat and pressure or both, be caused to be in a flowable (plastic) state during manufacture into their end product. Plastics are the result of petroleum technology *(Organic* in substance) consisting of carbon compounds, hydrogen, oxygen, and nitrogen. The various types of plastics displaying widely different physical characteristics are the result of chemical formulation of the above materials together with various modifying additives. The two basic categories of plastic are THERMOPLASTICS which can be repeatedly heated and reshaped to another hardened state when cooled, and THERMOSETTING plastics which, once formed, cannot be reheated without destroying their properties. Examples of the former include *Acrylics, Polyethylenes, Polypropylenes, Polystyrenes,* polyacetals, cellulosics, *Vinyls, Polyesters,* and *Nylon*. The latter includes *Epoxy, Melamine,* phenolics, *Polyester,* and *Silicone*. Plastics used in the building industry include acrylics, *Alkyds, Elastomers, Epoxies, Foamed Plastics, Polycarbonate, Polyethlene, Polypropylene, Polystyrene, Polysulphide, Polyurethane, Polyvinyl Acetate (PVA),* and *Polyvinyl Chloride (PVC)*. Plastic materials are usually shipped to the fabricator in granular or powder form. Methods used by the manufacturer to fabricate plastics into finished products include: molding (for example, injection molding and blow molding), extruding, foaming, production of films (blown film) and sheets, rotational casting, laminating into sheet form with other materials such as paper, cloth, or wood, and coating whereby plastic materials are applied as a thin membrane over other materials such as paper, wood, metal, or fabrics.

Plat A map or drawing showing a parcel of land.

Platform Framing In *Carpentry*, a two story *Stud* wall framing system whereby the stud walls are constructed one story at a time with the floor *Joists* set onto the top of the wall thereby serving as a "platform" for the next higher wall section. See also: *Balloon Framing*.

Plate Glass See *Glass*.

Plates In structural *Steel*, a rectangular shape defined according to rolling procedure as either "Sheared" or "Universal." The former are rolled between horizontal rolls and trimmed on all edges by shearing or gas cutting. The latter are rolled between horizontal and vertical rolls and trimmed on the ends only. Plates are defined as being over 8″ in width and .230″ in thickness.

Plate Washer See *Washers*.

Plenum In *Air Conditioning*, an air holding compartment connected to one or more distributing *Ducts*.

Plexiglas Trade name (Rohm & Haas Co.) for transparent plastic sheets.

Plinth Block A wood finish piece placed at the base of a door casing and against which a *Baseboard* abuts.

Plot Plan See *Plans*

Plug Receptacle See *Receptacle*.

Plug Strip An assembly of wire *Receptacle* outlets. They are uniformly spaced and mounted in a metal strip enclosure for fastening to a wall or work bench. Some trade names are Wiremold and Plugmold.

Plumb Bob In *Surveying*, a pointed weight in the shape of an inverted cone, which is hung by a string below a surveying instrument so that the instrument can be precisely centered over a reference point. It is also used in *Chaining* and for giving sights on a point.

Plumber Union plumbers belong to and are represented by the various local District Councils of the United Association of Pipe Trades, affiliated with the AFL-CIO.

Plumber's Friend A cup-shaped device of rubber on the end of a wood or metal handle used for unplugging stoppages in pipes by the action of siphonage and/or compression. It is also called a *Pneumatic* plunger.

Plumbing A system of *Pipes, Plumbing Fixtures*, and related accessories for the on-site supply and distribution of water. It also includes *Drainage and Vent Piping*.

Plumbing Fixture A device for using water and discharging the used water, such as sinks, *Lavatories, Water Closets*, drinking fountains, etc., as opposed to *Pipe Fittings, Valves*, etc.

Ply One complete layer, as of a *Veneer* in *Plywood*, or a layer of roofing *Felt* on a roof.

Plyclip A trade name (American Plywood Association) for a specially extruded aluminum "H" clip used as a substitute for *Blocking* where edges of sheets come together.

Plyform See *Plywood*.

Plyscord See *Plywood*.

Plywood A *Wood* panel product made from a number of thin layers of wood *Glued* together under pressure, with the *Grain* of each layer generally at right angles to adjacent layers, to provide strength and dimensional stability in both directions. The standard size sheet is 4′ x 8′, although other sizes are available such as 4′ x 9′ and 4′ x 10′. The two basic classifications of plywood are:

CONSTRUCTION AND INDUSTRIAL PLYWOOD—Construction and industrial plywood is manufactured from over 70 different species of wood, 15 of which are *Hardwoods*. This type of plywood has many uses in all phases of frame construction and in concrete forming. These 70 species are divided into five basic stiffness groups, from Group 1 (the stiffest) to Group V. Available thickness range from ¼" to 1-1/8". Construction and industrial plywood is manufactured in two types: "interior," which has moisture resistant glue bonds between layers, and "Exterior" plywood having waterproof glue bonds. Plywood is also divided into two catagories: "Engineered" grades and "Appearance" grades.

ENGINEERED GRADES include C-D Interior, Structural I C-D, and Structural II C-D, along with PLYFORM. C-D Interior is the grade used for sheathing and similar applications, replacing the old PLYSCORD and "Standard" grades. Structural panels have restrictions on species which make them suitable for use in engineered applications, such as highly stressed *Diaphragms*. THE IDENTIFICATION INDEX number shown on grade stamps for Engineered grades has two numbers separated by a slash. The number on the left indicates the spacing in inches for supports when the panel is used for roof sheathing with the grain of the face layer across the supports. The number on the right shows the support spacing when used for a floor, again with the face again across the supports. PLYFORM is manufactured with properties expecially suitable for concrete *Formwork*.

APPEARANCE grades include sanded panels, which are specified with a two-letter designation describing the appearance of each exposed face; for example, grade A-D meaning an "A" appearance on one side and a "D" appearance on the other side. These designations are summarized as follows:

N—"Special Order" *Veneer* with a smooth sanded surface free from essentially all defects, such that it is suitable for a natural finish.

A—Smooth sanded surface veneer where a limited number of defects have been patched.

B—Similar to A but permitting more patched defects. Tight *Knots* permitted.

C—Allows some open (unpatched) defects such as 1" knot holes and *Pitch Pockets*. Minimum permitted in Exterior plywood.

C-PLUGGED—Improved C veneer with plugged knotholes and other open defects limited.

D—Lowest appearance grade. Permits a higher degree of unpatched defects than any of the above.

Exterior type panels are required to contain all "C" grade or better veneers. Any panel that permits a "D" grade veneer is interior type even though many interior type panels are bonded with exterior glue. Another classification of Appearance grade panels are those with textured surfaces, usually sold as "303 Siding" panels. Such textures include re-sawn surfaces, relief grain, and numerous combinations involving various depths, widths, and spacing of grooves. Also available are "High Density Overlaid" and "Medium Density Overland" plywood, each with surfaces of plastic-impregnated fiber. The high density surface is hard and glossy; medium density is opaque, masks all grain characteristics of the plywood, and provides a premium quality base for paint.

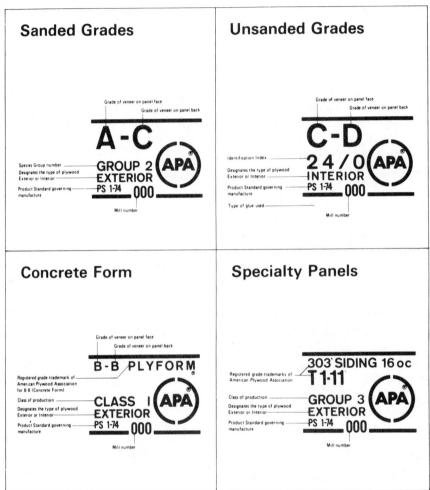

FIGURE P-8   Plywood Grade Stamps

The principal grading and standardization association for the construction and industrial plywood industry is the American Plywood Association, which has established grade stamps, samples of which are illustrated in Figure P-8. The stamps shown are appropriate to *Product Standard* PS 1-74. The stamps show the important qualities of the particular sheet. The standard source of information for plywood grades and thicknesses available is "Guide to Plywood Grades" available from the APA.

HARDWOOD PLYWOOD—Hardwood plywood is also called PREFINISHED PLYWOOD and is used in construction essentially only for paneling. This is generally ¼" plywood with a surface veneer of a wide range of *Hardwoods* such as mahogany, birch, walnut, etc. In addition to the wood species used, it is also classified according to the way it is cut from the log. ROTARY CUT VENEER is cut around the circumference of the log; PLAIN SLICED VENEER is made by cutting parallel to a diameter; QUARTER SLICED VENEER is made by cutting perpendicular to a diameter. RIFT CUT VENEER is used only for oak and is obtained by cutting at an angle of approximately 45° to a diameter. The principal grading and standardization agency for American decorative hardwood plywood is the Hardwood Plywood Manufacturers Association.

Plywood Patches Insertions of sound wood in plywood veneers for replacing defects. BOAT PATCHES are oval shaped with the size tapering in each direction to a point, or to a small rounded end; ROUTER PATCHES have parallel sides and rounded ends; SLED PATCHES are rectangular with feathered ends.

Pneumatic Operated by air pressure.

Pointing In *Masonry* construction, the filling or packing of *Mortar* into a joint after the masonry unit is laid.

Points In *Real Estate* finance, an additional charge made by a lender to a buyer for making a loan. One "point" equals one percent; for example, 3 points means 3 percent of the amount of the loan.

Poisson's Ratio In structural design, the ratio of sidewise elongation to lengthwise elongation when a material is subjected to *Stress*. For *Steel*, the value is approximately 0.3.

Pole Structure This term usually means a roof structure supported upon round wooden *Columns* (poles). The term is generally applicable to open-sided agricultural type sheds.

Polycarbonate A *Thermoplastic* used for molded parts and in sheet form where high impact strength and heat resistance are required. Its uses include safety helmets, machinery guards, and appliance components.

Polyester A class of *Thermoplastic Resins* used primarily with *Fiberglass* for laminating or molding products such as corrugated roofing panels, flat sheets for room dividers, or *Glazing*. Polyesters have outstanding weather resistance, transparency, and toughness.

Polyethylene A *Thermoplastic* produced from ethylene. It is furnished in flexible, semi-rigid and rigid grades. It is characterized by light weight, chemical resistance, and low cost. Its uses include film for *Vapor Barriers* and packaging, trash cans, and coatings for paper.

Polyphase In electrical terminology, other than *Single Phase*—that is, *Two Phase*, *Three Phase*, etc.

Polypropylene A *Thermoplastic* obtained from petroleum. It is stiffer than rigid *Polyethylene* and can withstand boiling water. It is used for packaging, indoor-outdoor *Carpets*, and upholstery fabric.

Polystyrene A *Thermoplastic* with exceptional clarity and chemical resistance. When blended with synthetic rubber, it is tough and used in refrigerator parts, and appliance housings. In crystal form it is widely used for containers and disposable glasses.

Polysulfide A family of *Elastomers*. They are available in liquids or solids and are impermeable to gases and vapors. The cured products have excellent resistance to oil, *Solvents,* and weather.

Polyurethane A large family of *Plastics* furnished as *Resins* for coatings,

Elastomers, and as flexible and rigid *Foamed Plastics.*

Polyurethane Foam See *Urethane Foam.*

Polyvinyl Acetate A transparent *Thermoplastic* whose major uses are in *Water-Based Paints* and *Glues* such as white glue. In comparison to *Vinyl Paint,* the polyvinyls (such as polyvinyl acetate and *Polyvinyl Chloride*) have a more complicated chemical formula and are considered "exotic" coatings.

Polyvinyl Chloride (PVC) The most important *Vinyl* plastic. It can be modified with *Plasticizers* to produce a wide range of plastics ranging from elastic to rigid. It is used for coatings, floor coverings, pipes and fittings, shower curtains, and wire insulation, as well as hospital items such as I.V. tubing, blood collecting systems, etc. It is also used in *Latex Paints,* particularly for use on swimming pools, and other facilities requiring water resistant paints.

Pop Rivet See *Blind Rivet.*

Porcelain Enamel A hard, glossy finish used on metals whose end result is similar to *Baked Enamel.* However, porcelain enamel is baked at a higher temperature (1000°F.) and the raw ingredients consist primarily of sand and glass.

Porosity The degree to which water can pass through the inter-connected pores within a material.

Portland Cement See *Cement.*

Portland Cement Plaster See *Plaster.*

Port Orford Cedar See *Softwood.*

Post Indicator Valve See *Automatic Fire Sprinkler System.*

Post-Tensioned See *Prestressed Concrete.*

Potable A drinkable supply of water.

Potential Energy See *Energy.*

Pot Life The period of time during which a material, requiring two or more compounds to be mixed prior to application, remains suitable for use.

Poured Gypsum Roof Deck See *Gypsum Roof Deck.*

Pouring Wool Insulation See *Loose Fill Insulation.*

Pour Strip In *Precast Concrete Wall* construction (tilt-up), a concrete perimeter strip to close the space between the wall panels and floor slab, such space being necessary for erection of the panels. (See Figure P-9.)

Powder Actuated Tools A fastening tool firing a specifically designed cartridge to drive a nail-like, stud-like fastener (also called a DRIVE PIN) into or through a variety of materials but primarily into *Concrete* and *Steel,* such as to fasten a wood *Sill* to a concrete *Slab.* Interchangeable barrels can be used for the different sizes of fasteners and the penetration depth can be accurately controlled.

Power The rate of performing work. Common units are *Horsepower, British Therman Units,* and *Watts.* See also: *Electric Power.*

Power Activated Tools A general term meaning a wide variety of portable or stationary tools using electric or *Pneumatic* power. Some examples are drills, *Saws, Sanders, Routers, Rivet* guns, etc.

Power Factor The ratio between actual electrical *Power* consumption in *Watts* and the apparent power obtained by multiplying *Volts* time *Amperes.*

Power Panel See *Panelboard.*

POUR STRIP

FIGURE P-9   Pour Strip (to effect closure between pre-cast wall panels and floor slab. Protruding dowels will be bent so as to be embedded within the pour strip.)

Power Shovel   See *Shovel*.

Pozzolan   A siliceous (containing *Silica*) or siliceous and aluminous material which in itself possesses very little cementitious value but will, in finely divided form and with moisture, chemically react with calcium hydroxide *(Lime)* at ordinary room temperatures to form compounds having cementitious properties. As an additive to a *Concrete* mix, pozzolans will reduce heat generation and thermal volume change.

Pratt Truss   See *Trusses*.

Pre-Action System   See *Automatic Fire Sprinkler System*.

Precast   A term generally referring to concrete panels, slabs, or structural shapes manufactured or constructed in a place other than its final intended location. An example might be precast concrete panels made in one place and trucked to the job site.

Precast Concrete Walls (also called Tilt-Up Walls)   A method of constructing *Concrete* walls where a concrete slab is poured in a flat position over another

FIGURE P-10 *Temporary Bracing and Pilaster Forms Pre-Cast Concrete Wall Panels in Place Showing*

FIGURE P-11 *Pre-Cast Concrete Wall Panel Just Lifted and Being Carried to Position*

slab—usually the floor of the building-to-be—and after hardening is lifted by a *Crane* into a vertical position and set on a foundation to form a finished wall. A special parting agent is used to prevent the slab from sticking to the one below, and inserts are embedded into the panel for attachment of the lifting cables. After being set into place, the panels are braced until the roof is in place. Panel thicknesses are usually from 6″ to 8″ and are joined by cast-in-place *Pilasters*, welded connectors, or similar means. (See Figures P-10 and P-11.)

Pre-Engineered Steel Building A steel building, generally of the industrial type, usually consisting of *Rigid Frames, Girts, Purlins,* and having *Metal Decking* for roofs and walls. Manufacturers of such buildings generally have standardized designs (hence the term "pre-engineered") and prepared drawings for a variety of sizes, spans, etc. The components of the building are usually fabricated in the manufacturer's shop and delivered to the job site for erection. The term "Pre-engineered Steel Building" is also used interchangeably with STANDARD STEEL BUILDING, and is a more accurate replacement for "Pre-fabricated Steel Building" since the standardization occurs in engineering rather than fabrication. (See Figure P-12.)

Pre-Fabricated   Housing See *Factory Built Housing, Industrialized Building*.

Prefinished Plywood See *Plywood*.

Preheating In *Air Conditioning*, to heat air in advance of other processes.

Pressure Grouting Pumping a cement *Slurry* or other chemical into the ground under pressure, under a *Concrete Slab*-on-grade to stabilize the underlying soil or to raise the slab in order to correct settlement. It creates a barrier to the

Courtesy Soulé Steel Co.

FIGURE P-12   Pre-Engineered Steel Building

movement of water and/or fills voids in soil or rock. It is also termed MUD JACKING.

Pressure Reducing Valve See *Valves*.

Pressure-Relief Valve See *Valves*.

Pressure Welding See *Resistance Welding*.

Prestressed Concrete Embedding within a concrete *Beam* or *Slab*, high strength *Steel* cables, called *Tendons*, which are tensioned to cause compressive forces within the beam or slab. By so placing and tensioning the tendons, the resulting compressive forces can be so applied as to induce a bending opposite to that caused by the applied load thereby increasing the efficient use of the concrete. If the tendons are first stretched (tensioned) between unyielding supports and the concrete then placed around them, it is said to be PRE-TENSIONED. If the tendons are placed within cables and tensioned after the concrete has hardened, it is said to be POST-TENSIONED. (See Figures P-13 and P-14.)

Pre-Tensioned See *Prestressed Concrete*.

Prime Coat In painting, a first coat of paint applied on bare material. It is used to seal pores (as when applied on wood) and/or otherwise serve as a base for additional coats.

Prismoidal Descriptive of a solid shape having parallel ends in the shape of irregular polygons and sides in the shape of *Trapezoids*. The volume within such a shape can be computed approximately by the *Prismoidal Formula*.

Prismoidal Formula A formula to determine with close (but not exact) accuracy, the volume of earth (or water or other material) contained within a

Prismoidal shape. It is:

$$V = (A_1 + 4A_2 + A_3)(h/6)$$

where V = Volume, A_1 and A_3 = area of each end,

A_2 = area midway between A_1 and A_3, and

h = distance or height between A_1 and A_3.

Privacy Set See *Lockset*.

Private Sewage Disposal System An on-site means of disposing of *Sewage* as opposed to connection to a public *Sewer*. It usually consists of a *Septic Tank* to remove solid matter from the liquid *Effluent* (liquid discharge) which goes either to a *Seepage Pit* or *Disposal Field* for seepage into the soil. The use of such a system is governed by the local *Building Code* and soil conditions.

Proctor Needle In *Soil Mechanics*, a type of *Penetrometer* having varying diameter tips which can be put on the plunger. It is used for measuring the relative density of soil by the resistance to penetration of the tip.

Production Welding See *Automatic Welding*.

Product Standard Replacing the term "Commercial Standards," these are produced by the U.S. Department of Commerce, National Bureau of Standards to establish standard quality requirements, methods of testing, rating, certification, and labeling of commodities, and to provide a basis for fair competition developed by a balanced cooperation among manufacturers, distributors, consumers, and other interests. Copies may be obtained from the Superintendent of Documents, U.S. Government Printing Office, Washington, D.C. 20402.

Professional Liability Insurance See *Insurance*.

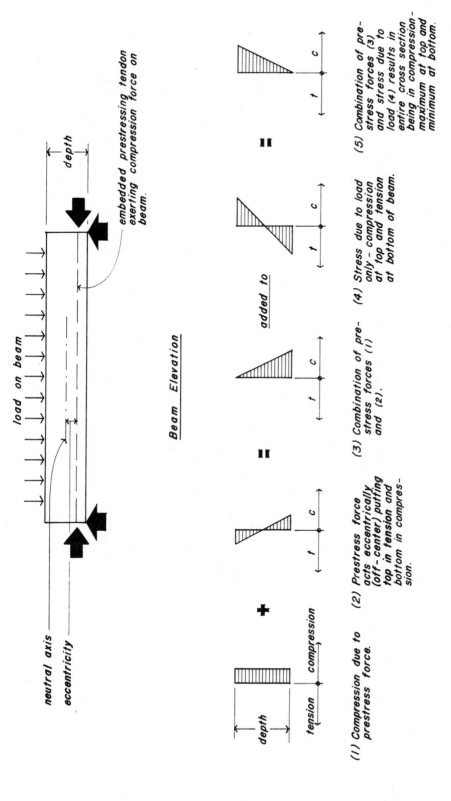

load on beam

neutral axis
eccentricity

depth

embedded prestressing tendon exerting compression force on beam.

Beam Elevation

depth

tension | compression

(1) Compression due to prestress force.

+

(2) Prestress force acts eccentrically (off-center) putting top in tension and bottom in compression.

=

t   c

(3) Combination of pre-stress forces (1) and (2).

added to

t   c

(4) Stress due to load only- compression at top and tension at bottom of beam.

=

t   c

(5) Combination of pre-stress forces (3) and stress due to load (4) results in entire cross section being in compression- maximum at top and minimum at bottom.

Diagram of Stress Distributions from Top to Bottom of Cross Section

FIGURE P-13   Prestressed Concrete Principle

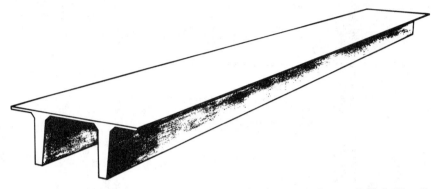

Courtesy Rockwin West Corp.

FIGURE P-14 *Prestressed Concrete Double Tee*

Profile A plotted line showing the *Elevation* along the length of something, such as a curb, pipeline, or the centerline of a roadway.

Program See *Computer Program.*

Programming

1. In *Computer* terminology, the preparation of a PROGRAM.

2. In *Planning* terminology, the series of steps or events by which a desired result can be achieved.

Progress Payment A payment made to a contractor as the work of his contract proceeds. They are usually at predetermined, progressive stages, such as when the foundations are completed, walls completed, etc. Also, they are often monthly payments proportional to the work completed.

Projected Window See *Window Types.*

Propane See *Liquified Petroleum Gas.*

Propane Torch See *Gas Welding and Cutting.*

Property All property is classified into two types: REAL PROPERTY, which is land, buildings, and other things that the "attached" to the ground and PERSONAL PROPERTY, which are all other things such as machines, furniture, etc.

Proposal See *Bid.*

Protractor A drafting device for measuring and plotting angles.

Psychrometer An instrument for measuring the amount of moisture in the air by comparing *Dry-Bulb Temperature* and *Wet-Bulb Temperature*. The evaporation is less in moist air, hence the difference between the thermomenter readings is greater when the air is dry.

Puddling To contain a "puddle" of water within an area, such as to consolidate *Soil.*

Pull Box See *Electric Box.*

Pullman A built-in cabinet containing a countertop with a *Lavatory, Sink,* etc.

Pulpboard Similar to *Particleboard*

but made from the smaller findings (particles) of wood waste.

Pulpwood *Wood* used for ''pulp'' (ground up *Cellulose* material) for the manufacture of paper and related products.

Pumice A lightweight, porous volcanic glass used as an *Aggregate* in *Lightweight Concrete, Plaster,* and as an *Abrasive.*

Punch List A list prepared by an *Architect* and given to the *Contractor* when a building is substantially complete, enumerating those items necessary for total completion, or defective items requiring correction.

Punching Shear In *Structural Design,* the *Shear* forces resulting from the tendency of a *Column* to pierce through an end support, such as through a concrete slab upon which it rests. The *Stress* thus caused is the total force divided by the area of the perimeter of the hole that would be punched.

Purlin In an industrial type building, a horizontal roof member spanning between *Beams* and *Trusses* and to which the roofing is attached.

Push Plate A metal plate attached to the face of a door to be pushed against to open the door. See also: *Kick Plate.*

Putty A doughy mixture (usually *Whiting* and *Linseed Oil*—sometimes mixed with white lead) whose uses include filling small cracks and nail holes and setting glass in window frames. See also: *Spackling Compound.*

Putty Coat See *Plaster Finishes.*

P.V.C. Pipe (Polyvinyl Chloride) See *Pipe.*

Pyrometer An instrument for measuring high temperatures.

Q

Quantity Surveying A listing ("take-off") of quantities of materials from a set of *plans*.

Quarry An open excavation in the earth for mining stone, such as road *Aggregate, Limestone, Marble,* and *Granite*.

Quarry Tile A shale, clay type of hard burned tile which is unglazed and most commonly supplied 6″ x 6″ x ½″ thick, although other available sizes are also frequently used. It has a very low water *Absorption Rate*.

Quarter Sliced Veneer See *Hardwood Plywood*.

Quartz The chief constituent of *Sand*; a crystalline silicon dioxide mineral.

Quay Wall A retaining wall separating a water area from the shore.

Queen Post Truss See *Trusses*.

Quenching A process of controlled, usually rapid, cooling of heated parts in liquid such as water, brine, oil, or molten salt, or in a gas such as air, nitrogen, hydrogen, argon, or various mixtures of hydrocarbons.

Quickline See *Lime*.

Quicksand A phenomena of sandy soil whereby pressure caused by upward flowing water balances the downward weight of soil particles causing a "suspension-like" condition of the soil mass.

R

Rabbet Joint See *Woodworking Joints*.

Raceway In electrical work, a general term for any continuous channel or pipe-like enclosure to house and protect electric *Conductors*. The term can include *Conduit*, underfloor metal raceways, *Wireways*, and *Bus Ways*.

Racking In *Masonry* construction, a method of building the end of a wall by stepping back each successive course.

Radial Saw See *Saws*.

Radian A unit of angular measurement (approximately 57.3°) equal to the angle to the center of a circle formed by two radii cutting off an arc whose length is equal to a radius.

Radiant Heating See *Heating Systems*.

Radiation Transfer of heat by electromagnetic waves, whereby an object in the line-of-sight path of such waves (as from an *Infra-Red* heater) is heated by the dissipation of the waves at the contact surface.

Radiator See *Heating Systems*.

Radius Half the *Diameter* of a circle; the distance from the center of a circle to its perimeter. For illustration see *Circle Nomenclature*.

Radius of Gyration In structural design, a mathematical property of a cross-section shape, such as of a *Column*, which is related to its *Moment of Inertia* and which is used as a measure of resistance to buckling of a member in compression. Its symbol is "r" and for any cross-section its value is $r = \sqrt{I/A}$, where I is the moment of inertia and A is the area of the cross-section.

Rafter Framing member supporting a roof (as opposed to a *Joist*) which is relatively closely spaced and which usually frames into a *Beam or Bearing Wall*. For illustration see *Wood Framing*.

Raft Foundation See *Floating Foundation*.

Rag Felt A *Building Paper* manufactured primarily from compressed rag fibers as opposed to a cellulose base paper.

Rail (of a Door) In a *Panel Door* the horizontal pieces joining the vertical side *Stiles*.

Railroad Terminology Rail lines serving industry include the following types:

MAIN LINES—The principal lines used by the railroad and built on railroad property.

BRANCH LINES—Those deviating from the *Main Lines*, running for shorter distances, and also built on railroad property.

DRILL TRACK—Those lines connecting *Branch Lines* and *Spur Tracks*. The land on which they are built may either be owned by the railroad or private industry. If the land strip of the railroad's *Right-of-Way* is 50' wide, it can also accomodate the spur track. Generally, however, the right-of-way is 30'.

SPUR TRACK—Tracks on the land of the customer being served , and which are usually privately owned.

Rails Rolled structural steel rails used for railroad equipment and craneways are available in various weights and dimensions. Railroad rails vary from 75 lbs. to 90 lbs. per yard of length. Rails are bolt joined and of an average length of 33'. Craneway rails are available in a variety of lengths, generally from 30 to 60 ft., and range in weight from 40 lbs. per yard of length to 175 lbs., and vary from 3½ to 6 inches in depth.

Rainfall For designing drainage systems, *Culverts, Catch Basins, Gutters, Roof Drains,* etc., statistics are available from the National Weather Bureau regarding anticipated rainfall intensities in inches per hour and are indicated as to their probability of occurrence every 1, 5, 10, 25, 50, and 100 years. Intensities are also indicated for periods of 5 minutes, one hour, and twenty-four hours. In the design of drainage it is necessary to select the probability (e.g., a "10 year rain") together with inches per 5 minutes (or longer). Such rainfall information is available from the National Weather Bureau.

Rake The edge of a roof on a sloping side.

Raked Joint See *Masonry Joints*.

Rammed Earth Construction A method of constructing *Adobe* walls by placing plastic soil into forms, much like placing *Concrete*, and ramming it into place, after which the forms are removed.

Range Lines In the Government Survey method of *Land Description*, range lines are imaginary reference lines running in a north-south direction and spaced six miles apart to the east and west of a principal *Meridian*. See also: *Land Description, Township Lines*.

Ratchet Nail See *Nails*.

Reaction In structural design, the force required to hold up a *Beam* or *Column* at a support point; also, the force transferred to one structural member by another.

Ready-Mix Concrete Concrete mixed at a *Concrete Batch Plant* and delivered to the job in specially built trucks with rotating drums. Concrete is mixed en-route and discharged through a chute. Capacities of ready-mix trucks vary from 5 cubic yards to 9 cubic yards.

Real Estate Land (including buildings and improvements) and its natural assets such as minerals, water, etc. See also: *Property*.

Real Estate Syndicate A general term meaning a group ownership of *Real Property* which can take several forms, often that of a *Limited Partnership*.

Real Property See *Property*.

Ream To enlarge a hole.

Rear Dump Truck See *Truck Types*.

Rebar Contraction of *Reinforcing Bar*.

Receptacles An electrical device to receive the prongs of a plug and which is connected to an electric *Circuit*. They are available in several types to receive two or more prongs and with or without a *Ground wire*.

Recessed Light Fixture See *Light Fixtures*.

Reciprocating Compressor See *Compressor*.

Reciprocating Saw See *Saws*.

Recorder See *County Recorder*.

Record of Survey A map recorded by a surveyor to incorporate *Surveying* information into public records often to clarify an ambiguous or uncertain descriptive situation.

Rectangle A four-sided geometric figure having opposite sides of equal length and parallel, and with 90° angles at all corners.

Rectifier In electricity, a device to convert alternating current to direct current.

Red Oak See *Hardwood*.

Red Spruce See *Softwood*.

Reducer See *Pipe Fittings*.

Reducing Valve See *Valves*.

Redwood See *Softwood*.

Reflective Insulation See *Insulation*.

Refractory Any material that will withstand high heat, such as a lining of a kiln or furnace, but usually referring to *Fire Bricks*.

Refrigerant A substance which produces a refrigerating effect by its absorption of heat while expanding or vaporizing. *Freon* is the most commonly used refrigerant in the Mechanical Refrigeration System described under *Refrigeration Equipment*.

Refrigeration Cooling a space by removing heat from it.

Refrigeration Cycle A method whereby heat is removed, or absorbed, from one space and discharged outside the space by a continuously repeating heat transfer process. The two basic means by which this is accomplished are called Mechanical Refrigeration, utilizing electrical energy, and Absorption Refrigeration, which utilizes energy in the form of heat (example: *Gas Air Conditioning*). See: *Refrigeration Equipment*.

Refrigeration Equipment There are two basic types of refrigeration equipment, the processes of which are as follows:

ABSORPTION REFRIGERATION (also called GAS REFRIGERATION)— A cycle using *Ammonia* as the refrigerant, and heat, usually in the form of burning *Natural Gas* (or *Liquified Petroleum Gas*), as the energy source. The operating cycle is as follows: A combination of water and ammonia is boiled by heat thus removing the ammonia from the water. The ammonia vapor is then condensed and passed to the *Evaporator* where its pressure is reduced by the separated water absorbing ammonia vapor. As it absorbs heat from the evaporator, the water absorbs ammonia and is passed to the *Generator* where the ammonia is separated and the process repeated. Larger capacity units (over 15-25 *Tons of Refrigeration*) use, instead of ammonia, *Lithium Bromide* as the *Absorbent* and water as the refrigerant, in a similar cycle, but one which is under a vacuum rather than a positive pressure as in the ammonia cycle.

MECHANICAL REFRIGERATION —A *Refrigeration Cycle* where a *Refrigerant*, usually *Freon*, goes through a cycle of alternately absorbing and giving off heat. Freon as a gas is compressed by a *Compressor* where the very high pressure raises its boiling point to where it condenses into a liquid in the *Condenser*. The freon, then in liquid form, is allowed to enter an *Expansion Coil* (also called an EVAPORATOR which is within the space to be cooled. After going through an expansion *Valve* the liquid is permitted to expand since it is no longer under pressure and thereby "boils" and in so doing absorbs heat from the surrounding space. The cycle begins again as the resulting gas enters the compressor.

Register An opening into a room for the passage of conditioned air, as from a *Furnace* or *Air Conditioner*. It has a *Grille* covering and *Dampers* to control the air flow. See also: *Diffuser*.

Reglet A continuous slot built into a wall such as to provide a means for attaching the upper edge of *Flashing* to the wall in a weather-tight manner.

Rehabilitation The remodeling or otherwise improving of a run-down or out-of-date building. A building that has been rehabilitated is sometimes called a "rehab." See also: *Renovation, Restoration*.

Reinforced Grouted Masonry See *Grouted Masonry*.

Reinforcing Bar (also called Rebar) A round steel bar embedded in *Reinforced Concrete*. Their purpose in *Beams* is to increase the tension value of *Concrete* and in *Columns* to supplement the bearing capacity. They are available in diameters from ¼" to 1½". Almost all bars are *Deformed Bars*, meaning they have a surface deformation pattern to increase their bond to concrete to prevent slippage. Plain bars are those without such deformations and are usually only available in small diameters (usually ¼") and are used for *Ties* and *Stirrups*. Bar

sizes are designated by a number system where the number of the bar is equal to the number of eighths of an inch in the diameter—for example, a #6 bar is 6/8 of an inch in diameter, or ¾".

Relative Humidity The ratio, expressed as a percentage, of the actual weight of moisture in the air to the maximum possible weight of moisture that the air could hold at the same temperature and pressure.

Rendering An artist's *Perspective* drawing of a building.

Renovation To repair or remodel a building to a like-new condition and generally meaning a greater degree of improvement than *Rehabilitation* but less than *Restoration*.

Repose, Angle of See *Angle of Repose*.

Residual Stress See *Stress*.

Resilient Clips In a *Plaster* or *Drywall* partition, a means of reducing sound transmission through the wall by using formed metal devices to connect *Gypsum Board* or *Gypsum Lath* to *Studs* in a way that results in a cushioned or "floating" attachment.

Resilient Flooring A floor covering material in the form of tiles or sheets made from a variety of materials which are press-formed, rolled, or calendered (special rolling-type process) under heat and set in *Mastic* or other *Adhesive* over a wood or concrete floor to form a finished wearing surface. Resilient flooring is non-rigid as opposed, for example, to *Ceramic Tile* or *Concrete*. Types of flooring included under resilient flooring are the following:

ASPHALT TILE—Consists of asphaltic *Resins*, *Asbestos* fibers, color *Pigments*, and inert fillers. These tiles are available in a standard thickness of ⅛″(1/16″ and 3/32″ are also available) and 9″ x 9″ square. This is generally the least expensive type of resilient flooring.

BATTLESHIP LINOLEUM—A type of *Linoleum* characterized by its color and no longer manufactured.

COMPOSITION FLOORING—A general term meaning any resilient flooring.

LINOLEUM—Comprised of cork and linseed oil with a felt backing and generally available in 6′ wide rolls, with a standard thickness of 3/32″ but also available in ⅛″ thickness.

SHEET VINYL—Same as *Vinyl Tile* except available in sheet form, in a range of thicknesses from 1/16″ to 1/4″. Sheet vinyl has a backing of either asbestos or felt.

VINYL ASBESTOS TILE—Consists of vinyl resins, asbestos fibers, pigment, and inert filler and is available in thickness of 1/16″, 3/32″, and ⅛″, and either 9″ x 9″ or 12″ x 12″ square. It differs from *Vinyl Tile* in its asbestos fiber content.

VINYL TILE—Consists of *Polyvinyl Chloride* resins, pigment, and inert fillers (which are clay based) and is available in thicknesses of 1/16″, 3/32″ and ⅛″, and 9″ x 9″ or 12″ x 12″ square tiles.

Resin A semi-solid or solid material varying from yellow to amber to dark brown, being transparent or translucent. Resins can be both natural occurring or synthetic. The former are used in the manufacture of *Varnish* and certain *Driers* and the latter are principally used in the manufacture of *Plastics, Glues,* and as protective coatings. Natural resins are obtained from the sap of cone-bearing trees; synthetic resins are produced from simple chemicals found in petroleum,

natural gas or coal. Both types are soluble in alcohol, ether, and other alkali solutions, and soluble in water.

Resistance The property of a material to resist or impede conduction of *Electric Current,* measured in *Ohms*. High resistance means poor conductivity, and vice versa.

Resistance Welding A method of *Welding* based upon the principle that heat is generated by *Electric Current* flowing through a resistance. The most common type of resistance welding is SPOT WELDING where two thin sheets of metal are clamped together and electrodes are placed on opposite sides of the sheets. A low voltage-high ampere current flowing for a short period between the electrodes produces enough heat to melt the metal, and by applying pressure the sheets will fuse together at the localized area. SEAM WELDING is a series of spot welds.

Resisting Moment See *Moment of Resistance.*

Restoration Repairing or remodeling a building to its original appearance, usually referring to the preservation of historical buildings. See also: *Rehabilitation, Renovation.*

Resultant Force A single imaginary force having a magnitude and direction which could replace, and be equivalent to, two or more separate forces.

Retainer A payment made in advance for services that are to be performed, such as an architect receiving partial payment of his fee before starting to work on a project.

Retaining Wall Any wall intended to retain earth at a higher elevation on one side. For stability, it usually employs a wide *Footing* such that the *Resisting Mo-*

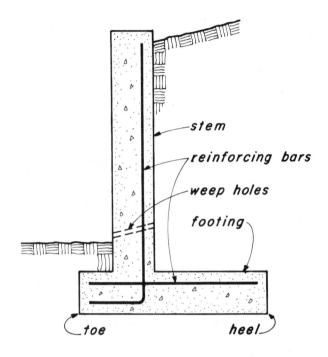

stem

reinforcing bars

weep holes

footing

toe   heel

FIGURE R-1   Retaining Wall

ment is greater than the *Overturning Moment*. It should contain *Weep Holes* near the base to prevent *Hydrostatic Pressure* build-up. The corner of the footing under the retained earth side is called the HEEL of the footing and the front corner, the TOE. (See Figure R-1.)

Retarder   An *Admixture* put into a *Concrete*, *Mortar*, or *Plaster* mix to slow the rate of setting.

Retempering   Adding water to a partially hardened *Concrete* or *Mortar* mix to restore *Plasticity*.

Return   Any surface turned back from the face of another surface.

Return Air   See *Air, Return*.

Reveal   A continuous recessed, slot-like space, usually for decorative effect, as around a window or door opening.

Reverberation   A reflecting or "echoing" of sound waves off a surface. It can be an acoustically disturbing effect since it occurs after the original sound is produced, but can be minimized by the proper selection of materials which absorb rather than reflect sound waves.

Reversible Window   See *Window Types*.

Revolving Door   A door pivoted about its center *Axis* to allow entrance or exit on either side by rotation of the door.

Rheostat   An electrical device to vary the amount of *Resistance* in a *Circuit* and thereby change the quantity of current flowing.

Ribbon   A horizontal board nailed onto, or recessed into, the side of a stud wall to support the ends of *Rafters* or *Joists*.

Rib Metal Lath See *Metal Lath*.

Richter Scale A quantitative measurement of the magnitude of an *Earthquake*. The magnitude is the logarithm of the relative amplitude of the earth movement as would be recorded on a standard *Seismograph* 100 kilometers from the center of the quake. Since this is a logarithmic scale an increase of one in magnitude indicates an increase of ten of the earth's movement and an even greater increase in the amount of energy released by the earthquake. On this scale a magnitude of 5 is potentially destructive whereas the San Francisco earthquake of 1908 had a magnitude of 8.3.

Ridge The line formed where two upward sloping roof surfaces meet, as opposed to a *Valley*.

Rift Cut Veneer See *Plywood*.

Rigging The use of cables and related accessory equipment, usually meaning their application to moving materials and equipment by a *Crane*.

Right Angle An angle of 90 degrees; being perpendicular to another line or surface.

Right Hand Door See *Hand (of a Door)*.

Right-of-Way Generally, any right of use to a strip of land but usually meaning a strip of land owned by, and/or used by, a railroad.

Rigid Frame A structural framework in one plane made up of pieces fastened rigidly together at their joints. (See Figure R-2.)

Rigid Metal Conduit See *Conduit*.

Rigid Pavement A *Concrete* pavement, as opposed to *Asphaltic Concrete* pavements, which are called FLEXIBLE PAVEMENTS.

Rim Joist A *Joist* along the side of an opening and parallel to adjacent joists.

Rim Lock A lock which is applied to the surface of a door, not *Mortised* into it.

Ring Pull A cupboard door opening

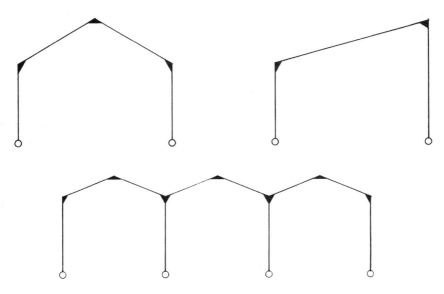

FIGURE R-2 Rigid Frames

device consisting of a hinged ring which can be pulled open by inserting a finger through the ring.

Ringer Crane A crane whose base is set in a ring device which allows the crane to rotate a full 360 degrees.

Ring Shank Nail See *Nails*.

Riparian Right The right to utilize the fluid in a body of water, such as the right to withdraw water from a stream.

Ripper In earthmoving, a large pick-like, or plow-like, implement mounted behind a *Tractor* which is pulled through dense or rocky material in order to loosen it.

Rip Rap Rocks or boulders placed along a shoreline embankment to prevent erosion by wave action.

Ripsaws See *Saws*.

Rise and Run Refers to the steepness of a stairway by stating the amount of "rise" (vertical distance for each step) and the amount of "run" (net horizontal dimension of each step.)

Riser

1. In a stairway, the vertical part of a step.

2. In plumbing, a vertical water supply pipe.

Rivet A *Bolt*-like fastening device without threads, with a head similar to a thickened nail head. Rivets are placed in plain holes and the rivet's reverse side is "hammered" to flatten it, creating another head. Rivets are used primarily for assembling metal parts in lieu of bolts. Their use is declining however, in favor of *High Strength Bolts* which have equivalent holding power.

Rock A consolidated, hard, naturally formed mixture of minerals, usually distinguished from *Soil* by the difficulty of excavation. See also: *Igneous Rock, Metamorphic Rock, Sedimentary Rock*.

Rock Gun In plastering, a device for throwing *Aggregate* (usually marble chips) into a still wet plaster surface to provide a decorative finish.

Rock Wool See *Mineral Wool*.

Rod

1. In *Surveying* a stick-like measuring device usually marked in feet and tenths of a foot which is used with a *Level* to measure vertical distances. It is held upright upon a point from which a measurement is to be taken and, by sighting through a level, a reading can be made to determine the vertical distance from the base of the rod to the height of the level. The rod can then be moved around to determine relative heights of other points.

2. The term is also used in *Welding*, particularly *Arc Welding*, to mean the source of the filler metal deposited.

Rod Buster Slang for a workman who places *Reinforcing Bars* for concrete work.

Rolled Beams Any metal *Beam* formed by passing flat bar stock through successive rollers until the desired shape is achieved.

Roller A self-propelled earth or pavement leveling machine having a front roller-wheel and one or two rear roller-wheels. It is different from other *Compactors* in that it leaves a smooth surface.

Roll Felt See *Building Paper*.

Roll-Up Door See *Door Types*.

Roman Brick See *Bricks*.

Romex Wiring A trade name (Rome Cable Co.) for a non-metallic sheathed electric *Cable*.

Roof Diaphragm See *Diaphragm*.

Roof Drain A round, sink-like receptacle at a low point of a roof to collect rain water. An outlet pipe at the bottom leads to an outside *Leader* (or *Gutter*), or to inside drainage piping. Such drains have a grate-like domed covering to prevent leaves or other material from clogging them.

Roofing The waterproof covering over a roof, or the act of placing such a covering.

Roofing Bond See *Bond (Roofing)*.

Roofing Felt An *Asphalt* impregnated rag fiber paper manufactured in rolls and used on built-up roofs.

Roofing Nail See *Nails*.

Roof Pitch See *Pitch (of a Roof)*.

Roof Shapes Nomenclature for various roof shapes are illustrated in Figure R-3.

Room Air Conditioning A self-contained unit intended for a through-the-wall installation whereby the heat ab-

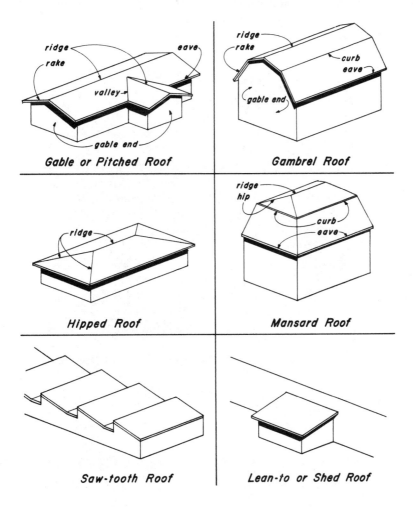

FIGURE R-3   Roof Shapes

sorbed from cooling the room is dissipated out the exterior side of the unit. Such units can also provide heating by forcing the air over hot wires or by the *Heat Pump* principle. See also: *Air Conditioning*.

Rope Rope is made by twisting strands of a material together. The most commonly used types of rope, in order of increasing price, are MANILA ROPE (made of natural fibers), POLYPROPYLENE ROPE, and NYLON ROPE. The latter two are made from synthetic fibers. For the same diameter, Manila rope is less strong; for the other two the strength is about equal. WIRE ROPE is made by braiding strands of wire; a CABLE is a term meaning a larger size and usually indicates several strands of wire rope braided together.

Rosewood See *Hardwood*.

Rosin A residue in the distillation of crude turpentine or from chemically treated pine stumps. It is hard and brittle and is used in making *Varnish* and insulation. The color varies from light yellow to almost black.

Rotary Compressor See *Compressor*.

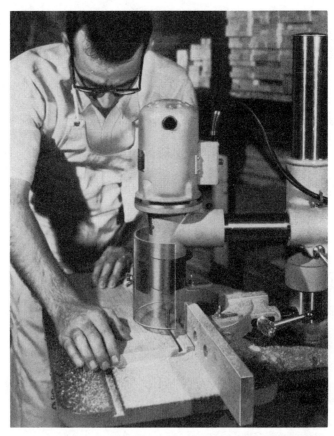

Courtesy Rockwell Manufacturing Co.

FIGURE R-4 **Router**

Rotary Cut Veneer See *Plywood*.

Rough Grading See *Grading*.

Rough Lumber See *Lumber*.

Rough Hardware See *Hardware*.

Rough-In Plumging The installation of concealed and permanent parts of the piping up to, but not including, the fixtures.

Rout To cut away material as for a groove or slot. See also: *Router*.

Router A power driven woodworking tool used for cutting grooves, slots, and for rounding edges, fluting, etc. With special accessories, a router may be used for cutting slots for door frame hinges and for *Butt* and lock *Mortising*. (See Figure R-4.)

Rowlock In *Masonry* construction, a brick placed crosswise in a wall with only its end showing. For installation see *Brick Joinery*.

Rubber Natural rubber is an elastic material derived from a gum-like *Resin* obtained from certain trees and plants. All natural rubbers, except adhesive rubbers, are vulcanized. "Synthetic" rubbers are alike only in their degree of elasticity. They are also called ELASTOMERS in the chemical industry and are more resistant to chemicals and weathering, and are superior to natural rubber in most uses such as gaskets, tires, oil seals, and hoses.

Rubber Base Paint See *Paint*.

Rubble Masonry *Masonry* work composed of irregular or random sized stones.

Running Bond See *Masonry Bond Patterns*.

Rupture, Modulus of See *Modulus of Rupture*.

Rust A reddish-brown surface which forms on *Iron* and *Steel* through oxidation when exposed to moist air. Rust can be impeded by coating the surface, as with a *Metal Primer*.

S

S 4 S See *Surfaced Lumber*.

Sabre Saw See *Saws*.

Sacking In *Concrete* work, a means of patching irregularities on a concrete surface by using a soupy mixture of cement, sand, and water which is slapped on and smoothed out with a burlap sack or similar material.

Safety Glass See *Glass*.

Sag Rod In a *Steel* framed industrial type building, a round rod perpendicular to and between *Purlins* or *Girts* to prevent their sagging.

Sampling Spoon In soil testing, a metal sleeve, usually lined with thin brass rings, which is forced into the ground to retrieve a core of soil.

Sand Siliceous (containing *Silica*) mineral grains ranging in size from *Silty Soil* (.002 inches) to *Gravel* (0.2 inches) in diameter. See also: *Sandy Soil*.

Sandblasting The use of compressed air mixed with sand to discharge a high velocity stream of sand. The sand particles act as an *Abrasive* to remove paint or to provide a decorative texture to a smooth concrete wall.

Sand Displacement Method See *Compaction Test*.

Sanders Power sanding machines include the following rypes:

BELT SANDER—Utilizes band shaped paper and a continuously moving belt motion. They sand in one direction only. (See Figure S-1.)

CIRCULAR SANDER—Same as a *Disc Sander*.

DISC SANDER—Utilizes disc shaped paper and a rotary motion. Circular sanders are also available as special attachments to power drills.

FINISHING SANDER—Utilizes rectangular shaped paper and either orbital (around) or vibrating (back-and-forth) motion. It has a rectangular pad plate to which the sandpaper is attached.

ORBITAL SANDER—See *Finishing Sander*.

VIBRATING SANDER—See *Finishing Sander*.

Sand Finish See *Plaster Finishes*.

Sandpaper Paper which is coated with a glue or other bonding material and into which a mineral *Abrasive* such as flint, emery, garnet, aluminum oxide, or silicon carbide is embedded. Sandpaper is primarily used for resurfacing or cutting wood, metal, plastic, glass, etc. It is available in various sizes and grades ranging

Courtesy Black & Decker Manufacturing Co.

FIGURE S-1　Belt Sander

from # 12 (coarsest) to #600 (finest); the latter is used for polishing purposes. Some manufacturers use a number classification such as 2½ - 3 for extra coarse to 3/0 - 4/0 for extra fine sandpaper.

Sand Trap　See *Interceptors*.

Sandwich Panel　A building panel having two smooth finish surfaces which are separated by a filler material such as plastic foam, corrugated paper, or similar material being lightweight and having fairly high compressive strength. The surfaces are bonded to the core with *Adhesive* usually requiring high temperature and pressure to effect the bond.

Sandy Soil　A type of soil characterized by a predominance of *Sand* sized particles. Sandy soils usually have low compressibility and high *Permeability*.

Sanitary Sewer   See *Sewer*.

Sap   The juice that circulates through a living tree, from the roots to the leaves, carrying nutrients and other chemicals to the cells.

Sapwood   The outer layers of growth in a tree located between the *Bark* and *Heartwood*, which contain living elements; it is usually lighter in color than heartwood.

Sash   The framework into which panes of glass are set.

Satin Finish   See *Paint Finishes*.

Saturated Air   See *Air, Saturated*.

Sawed Joint   In a concrete slab-on-grade, a saw cut (made with a *Concrete Saw*) part way through a slab to act as a weakened *Contraction Joint*. To be effective, they are usually made within 48 hours after the slab is poured.

Sawing of Lumber (from a Log)   As illustrated in Figure S-2, the three principal means of sawing lumber from a log

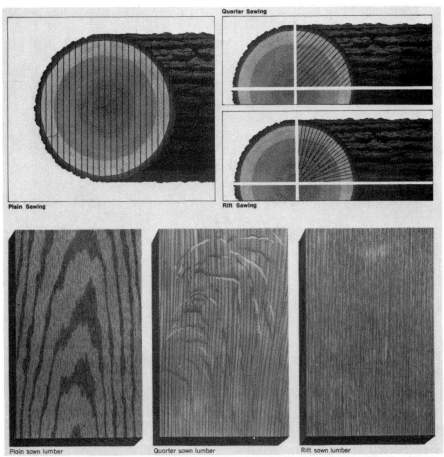

Courtesy Architectural Woodwork Institute

FIGURE S-2   Sawing of Lumber from a Log

are "plain sawn lumber," "quarter sawn lumber," and "rift sawn lumber."

Saw Kerf A groove or notch made by cutting with a saw.

Saws Basic types of saws, both manual and power operated are:

BAND SAW—A stationary power saw used primarily for cutting curves. The blade is in the form of a continuous belt (band) such that the cutting edge moves in one direction. (See Figure S-3.)

BACKSAW—A manually operated saw with a metal stiffening along the back. This type of saw is usually used in a *Miter* box.

BAYONET SAW—See *Sabre Saw*.

CHAIN SAW—A power driven (usually by a small gasoline engine), portable saw in which the blade is the outer side of a continuous loop of chain. They are used primarily for cutting trees and logs.

CIRCULAR SAW—A power saw in which the blade is a circular disc with the teeth (cutting edge) around the perimeter. It is usually table mounted and the mate-

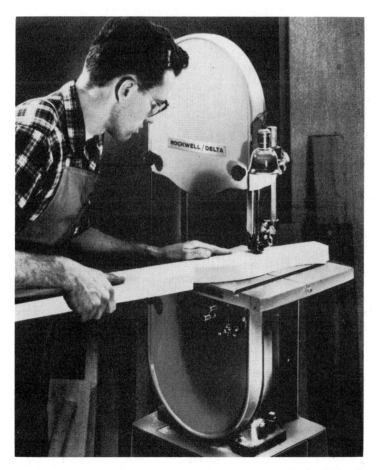

Courtesy Rockwell Manufacturing Co.

FIGURE S-3   Band Saw

rial is fed to it. **A SKILSAW** (trade name of Skil Corp.) is a portable circular saw.

COMPASS SAW—A manually operated saw principally used for cutting holes and openings. It has detachable blades and is similar to a *Keyhole Saw* which is smaller.

CONCRETE SAW—Same as a *Masonry Saw*.

COPING SAW—A manually operated saw used for cutting sharp angles and curved surfaces, having a narrow blade stretched across the open end of a "C" frame.

CROSSCUT SAW—A standard carpenter's hand saw, used to cut across the grain of wood.

HACK SAW—A manually operated saw similar to a *Coping Saw* but having a wider blade and used to cut metal.

HANDSAW—Manually operated and of two types—*Ripsaw* and *Crosscut Saw*.

KEYHOLE SAW—See *Compass Saw*.

MASONRY SAW—Any of several types of power saws having specially hardened steel blades for cutting *Bricks*, *Concrete*, or *Concrete Blocks*.

RADIAL SAW—A power saw supported above the work—it is pulled back and forth over the support table with the

Courtesy Rockwell Manufacturing Co.

FIGURE S-4   Radial Saw

material remaining stationary. (See Figure S-4.)

RECIPROCATING SAW—A portable power saw which cuts with an in-and-out action.

RIPSAW—A manually operated saw used to cut wood with the grain (as opposed to a *Crosscut Saw)*.

SABRE SAW—Also called a *Bayonet Saw*. It is a portable power saw with special blades available for cutting a variety of materials. It has an orbital blade-cutting action with the blade cutting on the upward stroke.

SKILSAW—See *Circular Saw*.

TABLE SAW—See *Circular Saw, Radial Saw*.

Sawtooth Roof A roof which in cross section has a series of peaks resembling sawteeth, whereby the steeper slope generally has windows or ventilators and the flatter slope is covered with roofing material. For illustration see *Roof Shapes*.

Scab Slang for a worker who takes the place of a striking union worker.

Scaffold A temporary platform-like structure upon which workmen may work above ground level.

Scale The proportional relationship between an actual object and its visual representation on paper.

Scarf Joint In wood construction, an end-to-end splice between boards, usually in *Glued Laminated Lumber*, whereby the ends of the joining pieces are taper cut so they can overlap but remain in the same plane, thus increasing the contact surface between them. For illustration see *Woodworking Joints*.

Scarifier A rake or plow-like implement attached to and drawn by a *Tractor* to *Scarify* soil. It is usually referred to as a "scarifier" when it is attached to a Motor Grader; it is referred to as a RIPPER on all other units.

Scarify To plow up or otherwise loosen a top layer of soil, usually by means of a *Scarifier*.

Scheduling The sequential planning of items of work on a construction project, relating to lengths of time and planned completion dates.

Schematic Drawing A diagramatic or preliminary type drawing concept (scheme) as opposed to the definitive *Working Drawings*.

Scissor Truss See *Trusses*.

Score A mark, notch, or groove, or the act of making one.

Scored Block See *Concrete Blocks*.

Scouring The lowering of the ground surface around a foundation by the erosive action of moving water or wind.

SCR Brick See *Bricks*.

Scraper In earthmoving, a large scoop-like piece of grading equipment used to excavate, haul, and deposit earth. It is supported by rubber tires and is pulled by a *Tractor* and sometimes also pushed by a *Bulldozer*. Some models also have assisting, power driven wheels. As it moves, the earth is scraped up into the bowl-like cavity. Some types have chain operated paddles within the bowl to assist in loading and unloading of the material. (See Figure S-5.)

Scratch Coat See *Plaster Coats*.

Screed In plastering or placement of *Concrete Slabs*, a straight piece of wood or metal set into the surface to act as a thickness and leveling guide.

Screen Block See *Concrete Blocks*.

Screwed Joint See *Pipe Joints*.

Screw Nail See *Nails*.

Courtesy Caterpillar Tractor Co.

FIGURE S-5   *Scraper (Self Loader)*

Screws Wood screws are conically tapered metal fastening devices. They have greater holding power than nails due to gripping of the threads and can be withdrawn with less damage to the material than with nails. Most wood screws are steel, having either a natural finish, bright finish (unplated), zinc plated (galvanized), blued (special heat treatment), lacquered, chrome plated, or plastic coated. Screws are also available in brass, bronze, copper, and stainless steel (for corrosion resistant applications). The ridge between the grooves around the *Shank* is called the "thread" and the angle of the threads is the "pitch." Screws are sized by gage number and length: the higher the gage number, the larger the diameter. Screws are generally available from #4 gage x ¼" up to #24 gage x 8". Types of screw heads include flat head which is countersunk so as not to project above the surface; round head which is usually used for attaching metal over a washer; oval head which is partly each of the preceeding two; hexagonal which has a bolt-like head without a slot for a screwdriver; and pan head which projects above the surface and has a flat top with rounded edges. A Phillips screw is one having crossed slots within the head of the screw rather than one slot extending the full width, and requiring a special screwdriver. Other types of screws include the following:

DRIVE SCREW—Actually a type of nail. See *Nails*.

DRYWALL   SCREW—A   bugle headed screw (resembling a golf tee) with high threads and a special head to grip the material to pull through so it will not tear the drywall paper. It makes its own hole.

LAG SCREW—Also called a *Lag Bolt*. It has a hexagonal head such that it is inserted with a wrench rather than a screwdriver—it does not require a nut.

MACHINE SCREW—Available with

both fine and coarse threads. They are used for the assembly of metal parts such as to fasten *Butt Hinges* to metal *Jambs*. They have a straight rather than a tapered shank.

SELF-TAPPING SCREW—A hardened steel screw with a special, partially slotted shank which, as it is screwed into a plain hole, will cut or form its own threads.

SHEET METAL SCREW—A heat treated, short screw (up to 2″ long) with steeply inclined, enlarged threads. It is used to fasten lapping sheet metal sheets.

Scribe To mark and cut the edge of a sheet of material such that it will fit tightly against an adjoining irregular surface.

Scrubber See *Air Washer*.

Scupper An opening in a *Parapet* wall, just above the roof line, for drainage of water from the roof. See also: *Overflow Scupper*.

Scuttle An access opening from a ceiling into an attic space or through a roof.

S-Dry See *Seasoning*.

Sealant A compound used to fill and seal a *Joint*, as opposed to a *Sealer* which is a liquid used to seal porous surfaces. It is intended to remain elastic and bonded to the sides of a joint such that watertightness is retained with minor movement of the joint.

Sealer Generally, a liquid applied to a relatively porous surface to seal it and improve the adherence of the finish material, or for waterproofing. See also: *Floor Sealer*.

Seam Welding See *Resistance Welding*.

Seasoning (of Lumber) Seasoning of

lumber refers to the *Moisture Content* at the time the lumber is graded and involves the weight of the tested sample. Seasoning is divided into four catagories: green lumber (abbreviated S-GRN), having in excess of 19% moisture content; air dried lumber (abbreviated S-DRY) not exceeding 19% moisture content; and KILN-DRIED lumber having a moisture content not exceeding 15% (abbreviated MC 15). PAD (meaning Partially Air Dried) is a form of seasoning done to retard mold and stain and reduce shipping weight—it is not the same as S-Dry.

Second A unit of angular measurement equal to 1/3600 of a *Degree* or 1/60 of a *Minute*.

Section A drawing showing a sidewise view at an imaginary "cut" through a building or portion thereof, or of any assembly or component, to clarify its manner of construction. It is also called a *Cross Section*.

Section (of Land) In the "Government Survey" method of *Land Description*, a section is a one mile square area (640 *Acres*) comprising 1/36th of a *Township*. Each section of land within a township is thus numbered from 1 to 36. For illustration see: *Land Description*.

Sectional Door See *Door Types*.

Sectional House A house which is put together at its site from prefabricated parts (sections) which have been made in a factory and shipped to the job site. It differs from a MODULAR HOUSE in that the latter is a factory built complete room or group of rooms.

Section Modulus In structural design, a mathematically obtained property of a given cross-sectional area which is used to compute *Stresses* in *Beams*; it is abbreviated "S". Steel and wood handbooks usually give values or formulas for various standardized shapes. The *Bending Stress* in a beam equals the *Bending Moment* divided by the *Section Modulus* (not applicable for reinforced concrete).

Sector (of a circle) See *Circle Nomenclature*.

Sedimentary Rock One of the three catagories of rock (the others being *Igneous Rock* and *Metamorphic Rock*). It is formed from consolidation of fragments of rocks, minerals, and organisms, or as deposits from sea water. Examples of sedimentary rock include sandstone and *Limestone*.

Seepage Pit Similar to a *Leach Field* where solid wastes are held in a separate enclosed area *(Septic Tank)* and only the *Effluent* passes to the pit from where it is dispersed to the soil. They are compact in area but rather deep and must not perforate the *Water Table*. Size and location are governed by local *Building Codes*.

Segment (of a circle) See *Circle Nomenclature*.

Seismic Refers to *Earthquake* action.

Seismology The study of earth vibrations or movements caused by *Earthquakes*.

Seismograph An instrument used to measure activity within the earth—the intensity and duration of *Earthquakes* and other earth tremors.

Self-Climbing Crane A type of crane, usually hydraulically operated, which is used on the erection of tall buildings and which is moved upward as the building is constructed. In effect, the building is built around the crane which is supported by each successive floor.

Self-Closing Door See *Door Types*.

Self Extinguishing A material which will stop burning once a flame is re-

moved. It usually applies to *Plastic* materials.

Self-Furring Metal Lath See *Metal Lath*.

Semi-Gloss Finish See *Paint Finishes*.

Semi-Tractor-Trailer See *Truck Types*.

Separator See *Interceptors*.

Septic Tank A part of a private *Sewage Disposal System* consisting of an underground *Concrete* (usually *Precast*), *Masonry,* or *Cast Iron* device to separate the solid matter, which is biologically digested, and discharge only the liquid *Effluent* for final disposal into the soil by a *Seepage Pit* or *Leach Field*. Septic tanks are compartmentalized, have inlet and outlet pipes, and must be cleaned periodically. Decomposition of the solid matter takes place during the retention period. Their use, construction, and size is governed by local *Building Codes*.

Series Circuit See *Circuit*.

Serrated Having a saw-tooth like edge.

Service (Electrical) See *Electric Service*.

Service Head In electrical terminology, a device for connecting overhead open service *Conductors* to a *Conduit* system; that point or place on a building where the electric company makes an overhead connection to the building from a power pole. See also: *Electric Service*.

Service Sink In plumbing, a sink usually used for mops and other janitorial purposes.

Set Back The legal distance a structure or building must be from a property line.

Setting Time For any *Concrete, Mor-*

tar, Grout, Plaster, or similar cementitious mixture, the time measured from when water is first added until the mixture has hardened so that it cannot be worked further. See also: *Initial Set.*

Settlement The downward movement of a foundation due to consolidation of the underlaying soil mass. Although some settlement always occurs, its magnitude is limited by design so undersirable distortion or cracking of the structure will not occur.

Sewage Liquid waste from a building, which can also include rain water, and *Industrial Wastes*, and other solid materials.

Sewage Disposal *Sewage* is disposed of either by discharge into a public sewer for collection, treatment, and ultimate dispersal in a non-hazardous manner, or by a private sewage disposal system which usually consists of a *Septic Tank* followed by either a *Seepage Pit, Disposal Field* or *Leach Field.*

Sewer A pipe to carry away *Sewage.* If it carries human waste it is called a SANITARY SEWER (or DOMESTIC SEWER), and if it carries only rain water it is called a STORM SEWER. If it carries both it is called a COMBINATION SEWER. See also: *Drainage and Vent Piping.*

S-GRN See *Seasoning.*

Shade See *Color.*

Shadow Block See *Concrete Blocks.*

Shake In *Lumber* terminology a separation along the grain, the greater part of which occurs between the *Annual Rings.*

Shakes See definition under *Shingles.*

Shank The cylindrical part of a *Screw* or *Bolt* (other than its head).

Shaper A woodworking machine,

usually of the floor model variety, used for grooving and shaping edges of boards and for combinations of decorative cuts.

Shatter Resistant Glass See *Glass.*

Shear In structural design, a force acting perpendicularly to the length of a structural member tending to cause a sidewise displacement. Scissors cut paper by "shearing" action. The design of nearly all structural members must, in addition to designing to resist *Bending* forces, be checked for resistance to shear. In terminology related to *Bolts*, the term SINGLE SHEAR means the bolt connects only two members such that the tendency is to shear the bolt in only one plane. DOUBLE SHEAR refers to a bolt through three or more members where there is a tendency to shear the bolt in two or more planes. See also: *Walls (Shear), Stress (Shear Stress), Horizontal Shear.*

Shear Connector A timber connector joining two members on a lap-type joint that has such a design or configuration that the shearing area is increased over that of, for example, a bolt.

Shear Face Block See *Concrete Blocks.*

Shear Panel Similar to *Shear Walls* but usually referring to one portion (panel) of a longer wall which, for bracing purposes, is made more rigid than the rest of the wall.

Shear Wall See *Walls.*

Sheathing Material used for covering a wall, floor, or roof surface, or the act of installing such material. It usually refers to such covering consisting of thin boards or *Plywood.* The term is often used synonymously with, but is preferred to, *Sheeting.* DIAGONAL SHEATHING has boards laid diagonally across the supports. When laid perpendicular, or straight across their support, it is called STRAIGHT SHEATHING or HORIZONTAL SHEATHING.

Shed A shelter type building usually open on one or more sides.

Sheepsfoot Roller In earthwork, a heavy roller having many blunt-ended protrusions which is pulled behind a *Tractor* to compact filled ground (See Figure S-6). It is one type of *Compactor*.

Sheet Glass See *Glass*.

Sheeting Vertical planks or steel *Sheet Piling* driven into the ground along a line where an excavation is to be made. Its purpose is to act as a retaining wall during excavation. See also: *Sheathing* (which has a different meaning).

Sheet Metal A comprehensive term for light gage *Steel* used for forming

Gutters, Downspouts, Flashing, etc. It is available in gages ranging from #30 (.102″) to #12 (.1046″) (the smaller the gage number, the heavier the material).

Sheet Piling A continuous, fence-like curtain of wood boards or interlocking steel sections which are driven into the ground to retain earth during excavation.

Sheetrock Trade name (United States Gypsum) for *Gypsum Board* wall and ceiling systems.

Sheet Vinyl See *Resilient Flooring*.

Shellac See *Paint*.

Shell (of a Building) The basic minimum enclosure of a building consisting of the structural framework, roof covering, exterior walls, floor and exterior doors, but exclusive of interior partitions, plumbing, electrical work, etc.

FIGURE S-6   *Sheepsfoot Roller*

Sheet Light gage material either in flat pieces or coils consisting of material over 12″ in width and under .203″ in thickness.

Shielded Arc Welding See *Arc Welding*.

Shielded Electrode See *Arc Welding*.

Shielding In *Welding* terminology, the protection from oxidation of molten metal.

Shim A thin wedge of wood or metal used to raise the end of a *Beam* or *Column* by driving it into the space underneath.

Shiner A nail driven through a board such that it shows from the other side.

Shingle Nail See *Nails*.

Shingles Shingles are used as a roof covering but are suitable only for roof slopes steeper than 4 in 12. They are also used for siding. Shingles are manufactured in a variety of materials, including plastic, clay tile, and aluminum (to simulate "shingles"), but most shingles are of the following types:

ASBESTOS CEMENT SHINGLES —Made from *Asbestos Cement* in sizes similar to *Wood Shingles*.

ASPHALT SHINGLES—Manufactured from asphalt impregnated felt. They are available in short strips with varying sizes, pattern, and mineral color coatings.

CEDAR SHINGLES—See *Wood Shingles*.

COMPOSITION SHINGLES—Same as *Ashalt Shingles*.

SHAKES—These differ from *Wood Shingles* in having textured grooves and a rough, or "split" appearance.

WOOD SHINGLES—Also called CEDAR SHINGLES since they are chiefly manufactured from Western Red Cedar. They are tapered and manufac-

tured in standard lengths of 16, 18, and 24 inches, and packaged in "bundles." For use as a decorative siding, they are also manufactured as plywood strips.

Shiplap Siding Horizontally laid wood siding applied so the lower edge of one board overlaps the upper edge of the board below, for water tightness.

Shop Detailer See *Detailer*.

Shop Drawings As opposed to the *Working Drawings* for a construction project, shop drawings are prepared by a sub-*Contractor* (for example, a steel fabricator) for use within his shop. They are usually submitted to the *Architect* or *Engineer* for approval to assure proper interpretation of the working drawings, insofar as show fabricating methods, sizes, etc., are concerned.

Shoring The temporary support of earth banks during construction.

Shortleaf Pine See *Softwood*.

Shotcrete See *Wet Mix Shotcrete*.

Shoved Joint See *Masonry Joints*.

Shovel A self propelled excavating machine, either *Crawler* mounted or wheel mounted, which is revolvable (capable of excavating in directions different from its line of travel). It is also called a DIPPER SHOVEL.

Shrinkage

1. In *Concrete*, the decrease in volume caused by losing its moisture content during the hardening process. The control of the resulting cracks is of concern in the design of concrete structures and is minimized by the use of *Admixtures*, specially proportioned mixes, and proper placement of *Contraction Joints*.

2. In *Earthwork*, the loss of volume of a soil from its excavated state to its in-

place compacted state, due to densification.

3. In *Wood*, the closing of the distance between wood fibers due to the slow drying out of wood. The distortion effect of such shrinkage on an individual board can be predicted by knowing (by the *Grain* pattern) from what part of the log the board was cut and knowing that the loss of moisture causes a log to shrink in diameter and circumference. See also: *Warping*.

Shrinkage Crack　A crack occurring in *Concrete*, *Plaster*, *Masonry*, or similar material caused by the failure of the material in tension due to tensile *Stresses* within the material caused by drying shrinkage.

Shrinkage Limit　In *Soil Mechanics*, the water content at which a soil has minimum volume; a further decrease in moisture does not result in a further decrease in volume.

Shut-Off Valve　See *Valves*.

Shutter　A louvered covering over an opening which is either removable or hinged.

Shutter, Fire　See *Fire Shutter*.

Siamese Connection　In plumbing, a "Y" type hose connection outside a building usually for fire department connection. See also: *Pipe Fittings*.

Sieve Analysis　Same as *Mechanical Analysis*.

Sieve Size　Refers to the number of openings per inch in each direction in a sieve, used to classify *Soils* or *Fine Aggregate* by grain size. A #4 sieve means four spaces (and four wires) per inch in each direction such that all grains larger than .25 inches will be retained on the sieve.

Silica　The most common solid mineral known; quartz and sand are largely silica. Chemically it is silicon dioxide, SiO_2.

Silica Gel　A form of aluminum oxide which *Absorbs* moisture readily and is used as a drying agent.

Silicon Carbide　A very hard crystalline mineral used as an *Abrasive*, as on grinding wheels.

Silicone　A group of synthetic *Resins* with a wide variety of uses, designated as inorganic *Plastics* since silicone takes the place of carbon. Properties include high heat resistance (500°F.), weatherability, and excellent water resistance. Silicones are used for *Paints* (when temperatures are above 300°F.); *Masonry* treatment to prevent water absorption and staining; for *Caulking*, for *Sealants* and for *Cements*.

Sill

1. The bottom horizontal part of a window opening.

2. The bottom horizontal member of a *Stud* wall upon which the studs rest and which is bolted to the foundation slab —also called a *Mudsill*.

Sill Block　See *Concrete Blocks*.

Sill Course (Masonry)　Same as *Belt Course*.

Silo　A round storage bin resting on the ground used for storing sand, cement and similar materials.

Silt　A fine-grained, unconsolidated *Soil* with particles between *Sand* and *Clay* in size. It is usually carried or put down as sediment by moving water.

Silty Soil　A type of *Soil* having *Silt* as the predominant grain size and properties between *Clay* and *Sand*. It has very little

Cohesion between particles, and is relatively *Permeable*.

Simple Span Beam See *Beam*.

Single Phase Power In electrical work, a 2 or 3 wire distribution system commonly used for residences and small commercial buildings. See also: *Electric Service*.

Single Shear See *Shear*.

Single Strength Glass See *Sheet Glass*.

Sink In plumbing, a shallow fixture, ordinarily with a flat bottom, that is commonly used in a bathroom *(Lavatory)*, kitchen, or in connection with the preparation of food, and for certain industrial processes.

Sintering Bonding of metal particles which are shaped and partially fused by pressure and heating below the melting point.

Site Plan See *Plans*.

Site Planning The design processes involved in determining the most functional and visually pleasing arrangement of buildings, parking areas, walkways, landscaped areas, and other features for a particular piece of land.

Sitka Spruce See *Softwood*.

Skew At an angle.

Skids Timbers placed on the ground, upon which a structure is built such that it can be moved by attaching cables to the skids and pulling the structure to a new location.

Skilsaw See *Saws (Circular)*.

Skim Coat See *Plaster Finishes*.

Skin Friction In *Soil Mechanics*, the resistance to movement between the surface of a friction *Pile* and the soil in contact with it.

Skiploader See *Loader*.

Skylight A window built into a roof to provide natural light for the area below.

Slab Any thin layer of material but usually referring to *Concrete Slabs*.

Slabs-On-Grade See *Concrete Slabs*.

Slag A metallic residue from the *Blast Furnace* production of *Iron*. It is a lightweight porous material consisting of *Silica*, *Alumina*, and *Lime*. Its chief uses are as an *Aggregate* for *Concrete* such as in the manufacture of *Concrete Blocks*, as a coating for some roofing materials, and as an ingredient in some *Portland Cements*.

Slaked Lime See *Hydrated Lime*.

Slaking

1. In *Soil Mechanics*, the complete disintergration or loss of structure of dry soil or rocks when immersed in water.

2. In *Masonry* work, the act of adding water to *Quicklime* to produce *Hydrated Lime* or *Lime Putty*.

Slate A *Metamorphic Rock* derived from shale and clay which has been transformed by natural processes of heat and pressure and which can be split into thin layers for decorative paving.

Sleepers Wood strips nailed or otherwise fastened over a concrete surface to provide a nailable means for attachment of other materials.

Sleeve A short piece of tube or pipe acting as protection around a pipe or *Conduit* which passes through a wall. See also: *Expansion Shield*.

Sleeve Nut An elongated *Nut* having internal threads such that two threaded rods can be joined.

Slenderness Ratio A *Structural Design* term related to the resistance of a structural member to *Buckling* under compressive loads. It is usually expressed as a ratio of the free length divided by the *Radius of Gyration* of the compression member.

Slide Rule A hand-held, sliding graphic *Scale* used chiefly for multiplying and dividing but can be also used to obtain square roots, cube roots, and trigonometric functions. It is based on a logarithmic scale, and is accurate to about three digits.

Sliding Door See *Door Types*.

Slip Form Construction A method for constructing high concrete walls whereby concrete is deposited between the forms which are continuously moved upward and supported by the partially hardened concrete emerging below the moving form.

Slip Joint See *Pipe Joints*.

Slope (of a Roof) A measure of the steepness of a roof measured in the number of inches of rise in a horizontal length of 12 inches. Example: a 4 in 12 slope means a rise of 4″ in a horizontal length of 12″.

Slope Ratio In *Earthwork*, the steepness of an embankment expressed as the ratio of its horizontal distance to one foot of vertical distance. For example, a 2:1 slope means a 2 foot horizontal distance for each 1 foot of vertical distance.

Slop Sink See *Service Sink*.

Slotted Hole An elongated hole to permit slippage of a *Bolt* within the hole.

Slot Weld A means of *Welding* two plates together where a hole in one plate is filled with metal and is bonded at the bottom to the backing plate. For illustration see *Weld Types*.

Sludge In *Sewage* terminology, the solid matter in various stages of decomposition, as opposed to the *Effluent*.

Slumped Block A variety of *Concrete Blocks* having a rough surface resembling *Adobe Bricks*.

Slump Test A measure of the "sloshiness" (or more exactly, *Plasticity*) of a *Concrete* mix. In a standard slump test, the concrete is placed into a 12″ high cone-like container with an 8″ diameter top. When full and stuck flush, the container is lifted and the amount, in inches, by which the top surface "slumps" from the original shape is the "slump" of the sample of concrete. "Stiff" concrete with a low *Water-Cement Ratio* may have a slump of only 3″ while very "wet" concrete may have a slump as high as 5″.

Slurry A soupy mixture of *Cement*, sand and water.

Smoke Detectors See *Fire Detectors*.

Smoke Vent A covered opening in a roof which, in the event of a fire, will open automatically when a *Fusible Link* melts and will let smoke and heat escape, thereby helping slow the spread of the fire. Some types also act as a skylight, and others have a specially designed plastic dome which will melt away to let the smoke escape. (See Figure S-7.)

Soap A *Brick* split through the middle of its height; also called a *Split*.

Sodium Vapor Lamp See *High Intensity Discharge Lamps*.

Soffit The underside of a horizontal surface which projects beyond a wall line, such as an overhanging roof.

Software In *Computer* terminology, the *Computer Programs* and similar data, as opposed to the physical equipment, which is termed HARDWARE.

FIGURE S-7   Smoke Vent

Softwood   One of the two botanical catagories of *Wood*, the other being *Hardwood*. Softwood is derived from conifers, having needle-like leaves, and one of the primary uses is for structural *Lumber*. Most lumber for contruction, for *Sheathing*, framing, *Beams*, and posts, are softwoods; however, a few hardwoods are used, for example, alder and aspen. Species with similar and sometimes indistinguishable wood characteristics are often grouped for convenience under *Commercial Group Names*, as listed in Figure S-8. The basic standard for softwoods is the *American Softwood Lumber Standard* which operates under the auspices of the U.S. Department of Commerce. Based upon this standard, various grading agencies, such as the Western Wood Products Assn., establish allowable design *Stresses*, grade classifications, and other marketing and consumer standards.

Soil   An aggregation of loosely bound mineral grains, organic material, water and gas resulting from the mechanical and chemical decomposition or dissolution of *Rock*. Soil is differentiated from rock, by its ability to be readily excavated by mechanical processes. (See Figure S-9.)

Soil Cement   A low strength *Portland Cement* concrete in which soil is the *Aggregate*. It is used when the strength of the soil alone is not satisfactory, such as under pavement.

Soil Mechanics   The application of the laws of mechanics and hydraulics to *Soils*.

FIGURE S-8   Commercial Group Names

COMMERCIAL NAME	SPECIES INCLUDED	GRADING AGENCY *
Aspen (Bigtooth-Quaking)	Bigtooth aspen, Quaking aspen	1, 2, 6
Balsam Fir	Balsam fir	1, 2
Black Cottonwood	Black cottonwood	7
California Redwood	Redwood	3
Coast Sitka Spruce	Sitka spruce	7
Coast Species	Douglas fir-larch, Western hemlock, Pacific silver fir (Amabilis), Grand fir, Sitka spruce	7
Douglas Fir South	Douglas fir from Utah, Colorado, Arizona, and New Mexico	6
Douglas Fir-Larch	Douglas fir from anywhere in the U.S. except the four states named above, and Western larch	5, 6
Douglas Fir-Larch (North)	Douglas fir, Western larch	7
Eastern Hemlock-Tamarack	Eastern hemlock, Eastern larch	1, 2
Eastern Hemlock-Tamarack (North)	Eastern hemlock, Eastern larch	7
Eastern Spruce	White spruce, Red spruce, Black spruce	1, 2
Eastern Spruce-Balsam Fir	White spruce, Red spruce, Black spruce, Balsam fir	1
Eastern White Pine	Eastern white pine	1, 2
Eastern White Pine (North)	Eastern white pine	7
Engelmann Spruce	Engelmann spruce	6
Engelmann Spruce-Lodgepole Pine	Engelmann spruce, Lodgepole pine	6
Hem-Fir	Western hemlock, California red fir, Grand fir, Noble fir, Pacific silver fir, White fir	5, 6
Hem-Fir (North)	Western hemlock, Pacific silver fir (Amabilis), Grand fir	7
Idaho White Pine	Western white pine	6
Lodgepole Pine	Lodgepole pine	6
Mixed Species	Douglas fir from all states except Utah, Colorado, Arizona, and New Mexico, all hemlocks from West, all true firs from West, Western larch, all cedars from West, all spruces from West, all pines from West	5
Mountain Hemlock	Mountain hemlock	5, 6
Mountain Hemlock-Hem Fir	Mountain hemlock, Western hemlock, California red fir, Grand fir, Noble fir, Pacific silver fir, White fir	6
Northern Aspen	Quaking aspen (Trembling), Balsam poplar, Bigtooth aspen (Largetooth)	7
Northern Pine	Jack pine, Red pine, Pitch pine	1, 2

FIGURE S-8 Commercial Group Names (cont.)

Northern Species	Douglas fir, Western hemlock, Pacific silver fir (Amabilis), Grand fir, Eastern hemlock, Tamarack, White spruce, Black, spruce, Red spruce, Lodgepole pine, Jack pine, Subalpine fir (Alpine), Balsam fir, Sitka spruce, Ponderosa pine, Alaska cedar (Pacific Coast Yellow), Western redcedar, Western white pine, Red pine, Eastern white pine.	7
Northern White Cedar	Northern white cedar	1
Ponderosa Pine	Ponderosa pine	7
Ponderosa Pine–Sugar pine	Ponderosa pine, Sugar pine	6
Ponderosa Pine–Lodgepole Pine	Ponderosa pine, Lodgepole pine	6
Red Alder	Red alder	6
Red Pine	Red pine	7
Sitka Spruce	Sitka spruce	5
Southern Pine	Shortleaf pine, Loblolly pine, Longleaf pine, Slash pine	4
Southern Pine (minor)	Pitch pine, Pond pine, Virginia pine	4
Spruce-Pine-Fir	White spruce, Red spruce, Black spruce, Engelmann spruce, Lodgepole pine, Jack pine, Subalpine fir (Alpine), Balsam fir	7
Subalpine Fir	Subalpine fir	6
Western Cedars	Western redcedar, Incense cedar, Port Orford cedar, Alaska cedar	5, 6
Western Cedars (North)	Western redcedar, Alaska cedar (Pacific Coast Yellow)	7
Western Hemlock	Western hemlock	5, 6
Western White Pine	Western white pine	7
Western Woods	All Douglas fir, Western hemlock, Mountain hemlock, White fir, Noble fir, Pacific silver fir, California red fir, Grand fir, Subalpine fir, Western larch, Engelmann spruce, Lodgepole pine, Ponderosa pine, Sugar pine, Idaho white pine, Incense cedar, Western redcedar	6
White Woods	Engelmann spruce, All true firs from West, all hemlocks from West, all pines from West	6

* 1   Northeastern Lumber Mfrs. Assn., 13 South St., Glens Falls, N. Y. 12801.
  2   Northern Hardwood & Pine Mfrs. Assn., Suite 501 Northern Bldg., Green Bay, Wis. 54301.
  3   Redwood Inspection Service, 617 Montgomery St., San Francisco, Calif. 94111.
  4   Southern Pine Inspection Bureau, Box 846, Pensacola, Fla. 32502.
  5   West Coast Lumber Inspection Bureau, Box 23145, 6980 S.W. Varns Rd., Portland, Ore. 97223.
  6   Western Wood Products Assn., 1500 Yeon Bldg., Portland, Ore. 97204.
  7   National Lumber Grades Authority, 1055 W. Hastings St., Vancouver 1, B.C. Canada.

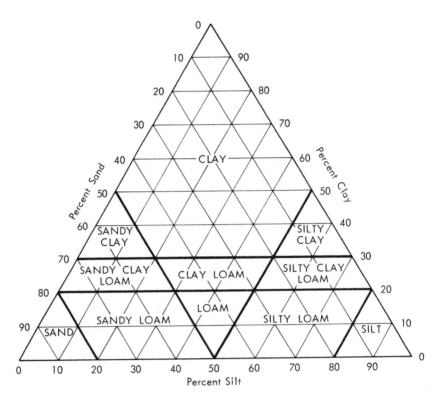

FIGURE S-9   Soil Textural Classification Chart, U.S. Bureau of Soils

Soil Pipe See *Drainage and Vent Piping*.

Soil Pressure See *Bearing Capacity (of Soils.)*

Soil Stack See *Drainage and Vent Piping*.

Soil Testing Laboratory A laboratory especially equipped to test soil samples, usually to determine *Shear* strength, degree of *Compaction,* expansiveness and other characteristics necessary for design of foundations.

Solar A term referring to the sun; for example, a "solar heat load" is the heat from the sun.

Solar Angles The angle the sun makes with the horizon, measured verti-

cally. Also called the "altitude," it varies with the time of day, time of year, and with the *Latitude*. This information is used in the design of roof overhangs and for other studies where the location of the sun is a design consideration. Publications are available which give information pertaining to solar angles for varying latitudes and times of the day and year.

Solar Glass See *Glass*.

Solder See *Soldering*.

Solder Gun An electric tool which imparts heat at a metal tip. It is used to solder electric or electronic components where instant heat is required.

Soldering Soldering is used most often for joining sheet metal to provide water-tightness or air tightness rather than strength, and for joining electric *Conductors*. Heat is applied by a soldering iron (declining in use), a *Solder Gun* or a *Propane* or *Butane* torch. The wire-like solder, which is a mixture of lead and tin, has a relatively low melting point and, as heat is applied, melts over the joint and bonds to the base metal.

Solderless Connection A mechanical means (as opposed to *Soldering*) of joining two electric wires whereby they are pushed into a threaded and tapered device which can be turned to tightly engage the wires, or a special slotted fitting into which the wires are pushed and the fitting is then crimped with a special tool to bind the wires together.

Soldier Course In masonry, a course of *Brick* in which the bricks are set on end with their long dimension up and down. For illustration see: *Brick Joinery, Masonry Bond Patterns*.

Soldier Piles *Piles* driven at regular intervals between which timbers are placed for temporary retainment of earth during excavation.

Solenoid An electrically activated mechanical switching device whereby an electric current flows through an electromagnet having a movable core. Movement of the core, caused by current flowing in the electromagnet, activates a *Switch*, *Valve*, or other device.

Solid Core Door See *Door Types*.

Solid State In electrical terminology, electronic devices which are "solid" in structure but have electrical properties which enable them to function in place of heated filaments in tubes and/or moving parts.

Solvents Liquids, generally *Volatile*, used in *Paints* and similar coating materials to dissolve various parts of the *Vehicle*, thus giving greater control of the consistency. Common types of solvents include:

ACETONE—A highly flammable solvent used in the manufacture of *Lacquers* and *Paint Removers*. It is also used as a thinning agent.

BENZENE—A powerful solvent for many organic materials, sometimes called "Benzol" and often confused with *Benzine*. It is formed from the distillation of coal tar and is highly flammable and toxic, requiring proper precautions for its use. Its use is generally restricted to industrial finishes for spray application.

BENZINE—A highly flammable solvent derived from the distillation of petroleum. It is often used as a *Lacquer* thinner, but involves high fire hazard in shipping and storage. It has low solvent power and evaporates rapidly.

MINERAL SPIRITS—A solvent used as a thinner for *Enamel* and *Varnish*, and as a brush cleaner; they are generally derived from petroleum.

TURPENTINE—A colorless liquid which is distilled from *Resin* and is used as a *Paint Thinner*.

Sound Level Meter An instrument for measuring the intensity of sound in *Decibels*.

Southern Pine See *Softwood*.

Southern Standards Building Code Published by the Southern Building Code Congress in Birmingham, Alabama. The code is used principally in the southern states and is not as widely used as some of the other building codes.

Space Frame A three dimensional *Truss*-like framework used to span a rectangular area whereby the individual

FIGURE S-10   Space Frame

members are so interconnected that a truss effect is achieved to carry imposed loads to all four support sides. (See Figure S-10.)

Space Heaters   See *Heating Systems*.

Spackling Compound   A powder mixed with water which dries very hard and is used primarily on *Wallboard* to cover seams, nail holes, and fill cracks. See also: *Putty*.

Spall   To break off in chunks, usually referring to *Concrete*.

Span   The distance between supports, such as of a *Beam*, *Girder*, or *Truss*.

Spandrel   A *Beam* spanning between *Columns* on the exterior of a building.

Spar Varnish   See *Paint*.

Specifications   Written instructions accompanying *Drawings*. Generally, they describe material choices and quality, workmanship, performance and installation procedures, and other information more easily presented in written form. By convention, specifications are separately bound in book form and issued with the project *Working Drawings,* which together form complimentary portions of the *Contract Documents.* Specifications contain *General Conditions* followed by technical divisions, such as site work, concrete, masonry, finishes, etc.

Specific Gravity The ratio of the weight of a material to the weight of the same volume of water.

Spigot

1. In plumbing, the end of a pipe that fits into a *Bell.*

2. A word used synonymously with *Faucet.*

Spikes Nails over 6″ long. See also: *Nails.*

Spirally Reinforced Column See *Concrete Columns.*

Spiral Shank Nail See *Nails.*

Spirit Varnish See *Paint (Varnish).*

Splash Block A tile-like concrete or masonry piece placed on the ground under a downspout to prevent erosion from the discharging water.

Splay To bevel a corner such that there is not a 90° angle.

Splice The end-to-end joining of two members.

Spline A strip of *Wood* or metal fitting into a slot, usually to form a longitudinal side-to-side connection between two members

Split In *Lumber* terminology, a lengthwise separation of the wood due to the tearing apart of the wood cells.

Split Beam The longitudinal cutting of an "I" type beam to form two separate *Tee Beams.* (See Figure S-11.)

Split (Brick) See *Soap.*

Split-Face Block See *Concrete Blocks.*

Split Level Two or more adjoining floor levels separated vertically by less than a full story.

Split-Ring Connector A type of *Timber Connector* used to join two wood

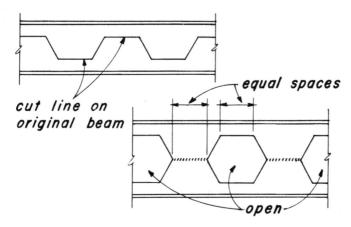

cut line on
original beam

equal spaces

open

FIGURE S-11   Split Beam

members, consisting of a steel ring fitted into circular grooves in lapping members such that the depth of the ring is half-way into each member. A *Bolt* through the two members holds the ring in place. Its purpose is to provide increased *Shearing* resistance at the connection over what could be obtained by a bolt alone.

Split System (Air Conditioning) See *Central Air Conditioning*.

Spot Welding See *Resistance Welding*.

Spray Finish See *Plaster Finishes*.

Spread Footings See *Footings*.

Spring Line An imaginary line connecting the end points of an *Arch* or similarly curved member or structure.

Springwood See *Annual Rings*.

Sprinkler System See *Automatic Fire Sprinkler System*.

Spruce See *Softwood*.

Spud Wrench An open end, non-adjustable wrench primarily used by *Ironworkers* for erection of structural steelwork.

Spur Track See *Railroad Terminology*.

Square (of a number) The product of a number multiplied by itself. For example, the "square" of 4 is 16 (4 x 4 = 16).

Square (of roofing) A unit of measure used in roofing terminology. It is a coverage of 100 square feet. For example, a roof area of 1800 square feet comprises 18 "squares."

Square Root (of a number) A number which, when multiplied by itself, will equal the number for which the square root is desired. The square root of 9 = 3 (3 x 3 = 9).

Stabilization See *Soil Stabilization*.

Stack In plumbing, a general term used for any vertical line of *Drainage and Vent Piping*.

Stacked Bond See *Bond (Masonry)*.

Stadia In surveying, a means of measuring distances using a *Transit* and a *Rod*. By sighting the rod through the transit, the length of the rod contained between two horizontal cross lines in the telescope can be noted and using an appropriate proportional scale, the distance from the transit can be determined. It is a fast means of measuring but it is not exact.

Staggered Stud Wall A framing method for wood *Stud* walls whereby studs on one side do not extend through to the other side. The method is usually used for sound insulation since *Wallboard* or *Plaster* on one side is not directly attached (through a single stud) to the other face of the wall. (See Figure S-12.)

Staging Same as *Scaffolding*.

Stain See *Paint*.

Stainless Steel Steel manufactured with *Chromium* and added *Nickel* to provide high resistance to corrosion. Oxygen reacts with the chromium to form a microscopic protective film on the surface.

Stake Truck See *Truck Types*.

Stanchion A *Column*-like member, usually between window openings.

Standard Penetration Test In soil testing, a test in which a *Sampling Spoon* is driven 12 inches into the ground using a 140 pound standard weight dropped from a height of 30 inches on the sample. The resistance to penetration as measured by the number of blows required is used as a measure of the *Soil Bearing Capacity*.

Standard Steel Building See *Pre-Engineered Steel Building*.

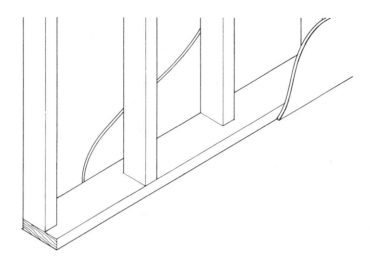

FIGURE S-12   Staggered Stud Wall

Standby Generator See *Emergency Generator*.

Standing Seam Roof A metal deck roof consisting of flat sections joined by vertical, overlapping seams.

Standpipe Piping within a building used solely for fire protection. They are divided into two classifications: DRY STANDPIPES which have outlets at each floor and whose source of water supply depends upon a fire department hook-up; and WET STANDPIPES which are under pressure and connected to the building water supply so that it is available for fire protection at all times.

Staples See *Nails*.

Star Drill A drill bit with a specially hardened star-shaped tip used for drilling holes in *Concrete* or *Masonry*.

Starter (Fluorescent) A now obsolete device in a pre-heat type *Fluorescent Lamp* to provide a high starting voltage when the lamp is turned on.

Stat Short for *Thermostat*.

Static Not subject to change, as opposed to dynamic.

Statically Indeterminate Structure A *Structural Design* term meaning a framework having a complexity such that the reactions and forces in individual members cannot be computed by the basic laws of *Statics* and more complex mathematical methods are required for their solution. Also termed an INDETERMINATE STRUCTURE.

Statics A fundamental concept of structural design having three basic laws:

a. The summation of all horizontal forces acting on a framework or structural member must equal zero.

b. The summation of all vertical forces acting on a structure or structural member must equal zero.

c. The rotation causing *Moment* about any point must equal zero.

Steam the vapor resulting from heating water to its boiling point.

Steam Heating See *Heating Systems*.

Steel Steel is made from *Iron*, coal which has been reduced to almost pure carbon, *Limestone*, and various trace alloying elements such as *Manganese*, phosphorous, and silicon. The chief means of producing steel is the basic oxygen furnace, which is largely replacing the open hearth process, followed by the electric furnace process. CARBON STEEL is the term used to distinguish lower strength steels from ALLOY STEEL, or steel containing special alloying elements to suit special requirements, such as to increase strength, increase abrasion characteristics or to make *Stainless Steel*. The carbon content generally has the largest effect on the strength of steel. Lowering the carbon content increases *Ductility* and raising it increases strength and hardness. Based upon carbon content, steels are classed as LOW CARBON STEEL (.06 to .30% carbon), also called MILD STEEL, and the most commonly used steel in construction; MEDIUM CARBON STEEL (.30 to .50% carbon), which includes most high strength structural steel; and HIGH CARBON STEEL (.50 to .80% carbon) which is used chiefly for tooling *High Strength Bolts* and similar uses. Properties of steel can be changed by *Cold Working* and *Heat Treating*. Steel is specified according to the properties desired (strength, corrosion resistance, etc.) and the most common means is to specify its ASTM (American Society for Testing and Materials) number. The most commonly used steels in construction are as follows:

ASTM A36—Probably the most commonly used carbon steel in the construction industry, having a minimum yield stress of 36,000 p.s.i. It is suitable for *Welding, Bolting,* and *Riveting.*

ASTM A440—A high strength, low alloy steel having a minimum yield stress of from 42,000 to 50,000 p.s.i. depending upon the thickness. It is undesirable for welding but is used economically for riveted and bolted structures.

ASTM A441—Similar to A440 but with alloy modification (principally adding vanadium rather than increasing carbon) to improve weldability.

ASTM A570—The most commonly used light gage steel used primarily for sliding, decking, and light gage structural members. This steel is available in yield strengths of 25,000, 30,000, 33,000, 40,000, and 42,000 p.s.i.

ASTM A572—Probably the most economical steel for general structural applications. It is a high strength, low alloy steel suitable for welding and having minimum yield stresses of 42,000, 45,000, 50,000, 55,000, 60,000, and 65,000 p.s.i., designated as Grades 42, 45, 50, 55, 60, and 65 respectively.

ASTM A588—A general construction steel with increased atmospheric corrosion resistance (four times carbon steel) used primarily in welded bridges and buildings. It is available in yield strengths from 42,000 to 50,000 p.s.i..

Steel Pipe See *Pipe.*

Steel Studs See *Metal Studs.*

Steel Wire Steel wire is most commonly available in gage (U.S. Steel and Wire Gage) of from 18 to 000 gage, the former being approximately .0475″ in diameter and the latter, .362″ in diameter. The most commonly used finishes are black (ungalvanized), *Galvanized*, or bright—the latter for plating purposes. Fine wire is purchased in spools and larger wire by coils, both being sold by the pound. See also: *Cable.*

Step-Taper Pile See *Piles.*

Stiff-Leg Derrick See *Derrick.*

Stiffness The capacity of a structural assembly or individual member to resist loads without excessive *Deflection* or distortion.

Stile On a wood panel or frame, a vertical piece into which the interior or secondary members fit.

Stillson Wrench Same as *Pipe Wrench*.

Stinger Slang for a truck mounted *Crane*, usually having a telescopic boom, and usually hydraulically operated. It can handle up to 10 to 15-ton loads.

Stippling A textured finish on a newly painted surface before it is dry. It is achieved through the use of a special stippling brush or roller stippler.

Stirrup In *Concrete* and *Masonry* construction, a hoop-shaped *Reinforcing Bar* wrapped around the longitudinal bars of a beam to hold them in place and to increase resistance to shearing forces. (See Figure S-13.)

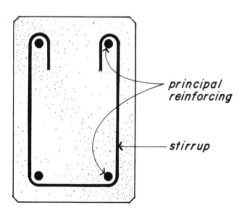

Concrete Beam

Cross - Section

FIGURE S-13 **Stirrup**

Stitch Bolt (or Rivet) A *Bolt* used to hold two longitudinal side-by-side members together.

Stochastic A term descriptive of a trial-and-error method for solving a problem as contrasted to a fixed, established step-by-step method.

Stockpile Earth or other material which is piled and stored for future use.

Stone Masonry *Masonry* laid up with natural stones.

Stop See *Door Stop*.

Storm Sewer See *Sewer*.

Story Pole In *Masonry* work, a marked pole used to establish and control vertical heights during the construction of a wall.

Stove Bolt See *Bolts*.

Straight Sheathing See *Sheathing*.

Strain Deformation. See also: *Stress*.

Strain Gage An instrument to measure movement (strain) over a given gage length. The device is affixed to the surface of a structural member and records the movement by the electrical principle that the resistance of a wire varies with its tension—stretching it decreases its cross section area and increases its electrical resistance. The magnitude of resistance is converted to micro-inches per inch.

Strap Hinge See *Hinge Types*.

Straw Nail See *Nails*.

Stress Force per unit area. For example, a force of 1,200 lbs. acting on an area 4 in. x 5 in. would result in a stress of $\frac{1200}{4 \times 5} = 60$ lbs. per square inch. Designation of types of stress include:

ALLOWABLE STRESS—The maximum permitted stress established by an

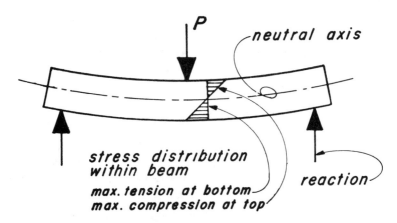

FIGURE S-14   Bending Stress

applicable *Building Code*. It is less than. the *Ultimate Stress* by a *Factor of Safety*.

BENDING STRESS—Stress resulting from *Bending*: same as FLEXURAL STRESS. For a *Beam* of a homogeneous material such as *Steel* or *Wood* as opposed to REINFORCED CONCRETE which consists of both concrete and steel) such stress is highest at an outer edge and reduces to zero at an imaginary line, called the NEUTRAL AXIS, which is at the *Center of Gravity* of the beam's cross section. When the beam bends, one edge is in TENSION—the fibers tending to pull apart; and the other edge is in COMPRESSION—the fibers tending to crush. The magnitude of the stresses is assumed to vary uniformly from zero at the neutral axis to a maximum *Tensile Stress* and maximum *Compressive Stress*, which points are also called EXTREME FIBER STRESS IN BENDING. (See Figure S-14.)

COMBINED STRESS—The cumulative effect of stresses caused by two or more loading conditions, such as when a column is subjected to both *Bending* and *Compression*, or when a *Bolt* is subjected to *Tension* as well as *Shear* forces.

COMPRESSIVE STRESS—Stress caused by compressive forces.

DESIGN STRESS—Usually means the same as *Allowable Stress* but sometimes means the actual stress existing in a member subjected to the design load.

EXTREME FIBER STRESS IN BENDING—See *Bending Stress*.

FLEXURAL STRESS—Same as *Bending Stress*.

RESIDUAL STRESS—A stress remaining in a structural member after application and removal of a force.

TENSILE STRESS—Stress caused by tension forces.

TORSIONAL STRESS—Stress caused by *Torsion*.

ULTIMATE STRESS—The maximum stress a material can withstand before failing. It is higher than the *Yield Stress*.

WORKING STRESS—The stress to which a member is subjected but which is always (or should be) below the *Allowable Stress*.

YIELD STRESS—A stress at which, for a given material, a permanent deformation results when the stress is removed.

Stressed-Skin   Panel In   *Wood*

construction, two layers of *Plywood* or similar panels separated by longitudinal wood framing members. The panels are glued or otherwise fastened to the framing members such that they act structurally as an integral unit.

Stress Relieving A process of *Heat Treating* to lessen or eliminate internal *Stresses* in metal parts that generally occur from fabrication operations such as *Welding*, machining, forming, or other processes. It is usually done at a lower temperature than *Annealing* or *Normalizing*.

Stress-Strain Curve A graph, usually applicable to metals, which shows the relationship between applied force *(Stress)* and resulting deformation *(Strain)*. Such a plot shows a constant proportionality between stress and strain until the *Yield Point* (also ELASTIC LIMIT is reached. For values below the yield point, the ratio of stress divided by strain is called *Modulus of Elasticity*. (See Figure S-15.)

Stretcher Course A course (or layer)

of *Masonry* in which the units are laid lengthwise in the wall. For illustration see *Brick Joinery*.

Strike (of a Lock) A metal plate which is a part of a door locking mechanism. It is fastened to the *Jamb* of a door and has a recess into which the door bolt engages.

String Course (Masonry) See *Belt Course*.

Stringer A longitudinal inclined beam supporting a stairway.

Strip Lath In plastering, a flat strip of *Metal Lath* used as reinforcement over joints between sheets of *Gypsum Lath* or where dissimilar materials join.

Strong Back A stiffening brace, usually meaning such a brace on *Precast Concrete Walls* to stiffen them during lifting, such as where door and window openings occur.

Struck Joint In *Masonry* construction, any *Mortar* joint which has been finished with a trowel. See also: *Masonry Joints*.

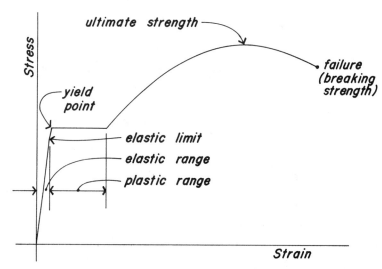

FIGURE S-15   *Stress-Strain Curve for Mild Steel*

Structural I, II See *Plywood*.

Structural Design The application of structural engineering principles and knowledge of construction methods to determine the materials, assemblage, and means by which a predetermined set of *Loading* conditions can be provided for to design an economic effective, and safe structure.

Structural Engineer See *Engineers*.

Structural Shapes For commonly used structural shapes, see Figure S-16.

Structural Steel Shape Designations Standard rolled structural *Steel* shapes are designated by three units of information, such as:

W 8 x 17

Indicates type of section as described below.

Depth of section in inches.
Weight of section in pounds per foot.

In 1970 the designation system was changed; for example, a 12″ deep wide-flange beam weighing 27 lbs. per foot was indicated as a 12 WF 27. The new designation is W 12 x 27 using the information sequence as in the example above. "W" shapes are *Wide-Flange Beams*; "S" shapes are "I" *Beams*; "M" shapes are light steel beams; and "HP" shapes are "H" *Beams* used generally for *Piles*: "C" shapes are for *Channels*; "WT" shapes are *Tees* cut from wide-flange beams; "MT" shapes are tees cut from light steel beams; and "ST" shapes are tees cut from "I" beams.

Strut

1. A compression member similar to a *Column*, but in a horizontal or inclined position and generally used as a lateral tie.

2. In *Earthwork*, a horizontal timber brace spanning between the two sides of a trench.

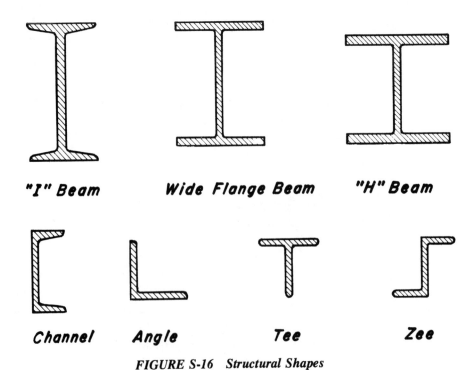

"I" Beam   Wide Flange Beam   "H" Beam

Channel   Angle   Tee   Zee

FIGURE S-16 Structural Shapes

Stucco *Portland Cement Plaster* applied to the outside of a building.

Stucco Mesh In plastering, a woven wire *Mesh* applied over *Building Paper* against a wood stud wall to serve as *Lath*. It is generally of 17 gage wire with hexagon shaped openings about 1-½ inch. It is also called STUCCO NETTING, and in smaller gage wire with smaller openings, *Chicken Wire*.

Stucco Netting See *Stucco Mesh*.

Stud A regularly spaced upright framing member (usually wood) in a wall to which the finish material, or other covering, is applied. See also: *Metal Studs*.

Stud Bolt See *Bolts*.

Studio Apartment See *Apartment Nomenclature*.

Studs, Powder Actuated See *Powder Actuated Tools*.

Stud Welding A means of attaching a short piece of round metal rod, usually threaded, onto another metal surface by using a gun-like device which holds the stud perpendicular to, and against, the surface. By a timing device, an intense electrical current flows from the stud to the *Base Metal* and, within a specially processed tip, fusion takes place to join the two.

Sub-Base In *Earthwork* or paving, the ground immediately below the *Base*.

Sub-Bids See *Bid*.

Subcontractor See *Contractor*.

Subdivision A leval division of a *Tract* of land into smaller *Parcels*. Common usage implies a land development including street improvements, lot grading, drainage, sanitary facilities, and may include new housing developments. A subdivision can also be used to combine several lots or parcels into a single lot or acreage.

Subdivision Map See *Land Description*.

Sub-Flooring A structural *Sheathing* material such as *Plywood* applied over *Joists* to provide a base for the installation of the finished floor.

Sub-Lease See *Lease*.

Submerged Arc Welding See *Arc Welding*.

Subsidence Downward movement of the ground surface due to consolidation of a subsurface zone. Subsidence may be caused by oil withdrawal, collapse of limestone caverns, decay of organic material in the subsoil, and similar phenomena.

Subsoil The soil zone below the *Top Soil*.

Substation A special electrical facility where the higher transmission voltages in an electrical system are stepped down (by *Transformers*) to distribution voltages.

Sub-Structure The foundation or that part of a structure below the ground level.

Sugar Pine See *Softwood*.

Summerwood See *Annual Rings*.

Sump A basin or pit-like depression at the low point of a floor or pit to collect waste water for easier removal.

Sump Pump A pump to remove water accumulated in a *Sump*. It usually operates automatically when the water level activates a switch to turn on the pump.

Sun Angle See *Solar Angles*.

Superstructure That portion of a structure which is above ground.

Superintendent This term usually means a person in charge of a total construction project and under whose supervision several *Foremen* may work.

Surcharge In foundation design, an added load placed on top of the soil, such as behind a retaining wall, as from an adjacent footing, storage loads, weight of additional earth, etc.

Surety A guarantor, usually a corporate insurance company, which, for a fee paid by a *Contractor* will guarantee performance of a *Contract*. Its purpose is to provide an owner with legal assurance that the obligations of a contractor will be fulfilled regardless of any misfortune occurring to him. The term is used synonymously with BONDING COMPANY.

Surfaced Lumber Lumber that has been surfaced by a planing machine for the purpose of attaining smoothness and uniformity of size. The designation S 4 S means "surfaced four sides." Similarly, S 2 S means surfaced two sides; S 2 S 1 E means "surfaced two sides and one end." Unless specified to be "rough sawn" or "rough lumber," most lumber purchased from a lumber yard is surfaced on each side. See also: *Lumber*.

FIGURE S-17   **Switchboard**

Surface Mounted Light Fixture See *Light Fixtures*.

Surface Water That portion of rainfall or other water which runs off over the surface of the ground.

Surge Drum (or Surge Tank) See *Accumulator*.

Surveying The science of land measurement, conveyancing, description, etc.

Suspended Ceiling See *Ceiling*.

Sway Bracing Inclined bracing in a vertical plane, similar to "X" bracing. It usually refers to bracing used to hold *Columns* or *Trusses* plumb during erection, being removed when the structure is completed.

Sweat Joint See *Pipe Joints*.

Sweets Catalog A standard reference of building products, published by F.W. Dodge Co. (a division of McGraw Hill). It is published annually in three classifications: Light Construction, Industrial Construction, and the larger Architectural series.

Swell (of a Soil) The increase in volume of a soil caused by the addition of moisture. It is most evident in *Clay Soils*.

Switchboard Similar in function to an electric *Panelboard* but considerably larger, usually standing from the floor and sometimes having access to the back. The service switchboard is the first point of entry of *Electric Service* to the building, containing the electric meter, main *Overcurrent Protective Device,* and disconnecting means, instruments (optional), and provision for dividing the incoming *Feeder* into outgoing sub-feeders for other switchboards and panelboards from which subsequent *Branch Circuit* distribution continues.

Large switchboards are also referred to as SWITCHGEAR. (See Figure S-17.)

Switch Box See *Electric Box*.

Switchgear See *Switchboard*.

Swivel Joint See *Pipe Joints*.

Symmetrical Identical on both sides.

Systems Engineering A systematic method for performing a complex engineering project by coordinating independent parts of the design such that their total interaction leads to an optimum overall end result. For example, a particular air conditioning design may not be the most economical, on an overall basis, if it requires more floor or *Duct* space (which would add cost) than an alternate system requiring less space.

T

Table Saw See *Saws*.

Tack The quality of stickiness of a *Paint* or varnish film during the drying period. Oil paints and spar varnishes may maintain a certain degree of tack for some weeks after they are basically considered dry.

Tack Weld A small, temporary weld to hold a member in position until permanent connections can be made.

Take-out Loan Final or permanent financing for a construction project.

Tail Joist A *Joist* framing perpendicularly into a *Header* joist.

Take-off See *Quantity Surveying*.

Tamarack See *Softwood*.

Tamper A one-man operated compacting tool either manually or power operated, which compacts soil by a hammer-like action of a blunted end against the ground.

Tap

1. A tool used for cutting inside threads.
2. To bore a hole in a pipe, tank, or other device.

Tape Tapes are available in a variety of types of material, widths, colors, and for many special purposes. Most tapes used in building construction conform to *Federal Specifications* and generally belong to one or more of the following categories:

PAPER BACKED TAPE—Having a crepe paper backing and generally used for interior protecting, holding, or sealing purposes, it includes the well known "Masking Tape" as well as drafting and freezer tapes.

DOUBLE COATED TAPE—Indicates adhesive application to both sides of the tape base (usually paper or plastic). These tapes are generally used for joining, mounting, splicing, etc., and are available with extended liners, that is, edges lacking adhesive.

FOAM TAPE—Available both as *Double Coated Tape* or with an adhesive on one side only; they have a cellular plastic foam base available in 2, 3, or 4 layer thicknesses. Foam tapes are generally used for sealing, holding, and padding applications—some provide thermal insulation and reduce vibration transmission.

REINFORCED TAPE—Containing glass yarn, polyester, nylon, or rayon fillament reinforcement and used for sealing corrugated boxes, bundling, etc., where additional tape strength is desired.

FILM BACKED TAPE—Usually *Plastic* backed, these tapes come in a variety of colors depending on adhesive or color coating. They are used for *Elec-*

troplating and labeling protection, *Duct* sealing, and include the "transparent" mending variety.

MISCELLANEOUS TAPES—Tapes having various properties for a specific use such as: grass cloth for solvent resistance; cotton cloth for curved or irregular surfaces; lead foil for radiation barriers; aluminum foil for reflection and metal patching; rubber to protect certain areas from sandblasting; Teflon (trade name of E.I. DuPont de Nemours Co.) for pipe wrapping and corrosion resistance.

Tapered Steel Girder A fabricated steel girder tapering in depth from a maximum at the center to a minimum at each end. Usually the purpose for the tapering is to provide a more economical design by proportioning the material used to the strength required throughout the length. (See Figure T-1.)

Tar Black, sticky, gummy *Bituminous* material from the distillation (turning to a vapor by heating and then condensing back to a liquid) of oil. It has a lower melting point than *Asphalt*, and is principally used as a roofing cement to bond plys of *Roofing Felt*.

Tare Weight The empty weight of a truck which can be subtracted from the loaded weight when driven onto a scale, to determine the weight of material transported.

Teak See *Hardwood*.

Tee A structural shape which is rolled, extruded, or otherwise shaped resembling a "T" in cross section. For illustration see *Structural Shapes*. See also: *Pipe Fittings*.

Tee Beam In *Concrete* construction, a *Beam* resembling the letter "T", making more efficient use of a given amount of concrete by providing a large area on the compression side (at the top) and a relatively small stem into which the

FIGURE T-1 **Tapered Steel Girders**   *Courtesy Delta Steel Corp.*

Tension resisting *Reinforcing Bars* are added.

Temper In *Masonry* construction, the moistening of *Mortar* and remixing it to the proper consistency to compensate for moisture lost between the time it is mixed to the time it is used. It is also called *Retempering*.

Temperature Scales The two basic temperature scales are: CENTIGRADE SCALE, where 0° is the freezing point of water and 100° the boiling point, and FAHRENHEIT SCALE, where 33° is freezing and 212° is the boiling point. See also: *Dry Bulb Temperature, Wet Bulb Temperature*.

Tempering A *Heat Treating* process that generally follows *Quenching* or *Normalizing*. Usually it is done to increase toughness and *Ductility* and is accompanied by decreases in strength and hardness.

Template A piece of wood or steel with accurately spaced holes through which bolts may be placed and temporarily held in correct position when they are built into a wall or foundation. It may also refer to plastic drafting templates with cutouts of letters, symbols (ie, furniture pieces, etc.), or mathematical designations (circles, rectangles, triangles, etc.).

Tendon A wire strand or *Cable*, usually referring to those used in *Prestressed Concrete*.

Tensile Stress See *Stress*.

Tension An area of a structural member is said to be in tension if the acting forces tend to cause lengthening of the particles.

Ten-Wheeler (Truck) See *Truck Types*.

Termites Whiteish, ant-like insects which, under certain conditions of mois-

ture and proximity to the ground, cause structural damage by eating wood members.

Terneplate A corrosion resistive coating applied over *Steel*, consisting of a mixture of *Lead* and *Tin*.

Terra Cotta

1. A term literally meaning "baked earth," it refers to hollow, molded clay building units. Use of this product is declining but it is still available in sizes and shapes roughly corresponding to those of *Concrete Blocks*.

2. In decorating, "terra cotta" is used to refer to an earthy, brownish-red color.

Terrazzo A floor finish made by sprinkling *Marble* chips onto a still plastic *Concrete* base. After hardening has taken place, the surface is ground smooth.

Tetrahedron A pyramid-like solid shape having four triangular shaped faces (including the bottom).

Texture Finish See *Plaster Finishes*.

Theodolite See *Transit*.

Therm A quantity of heat equal to 100,000 B.T.U.'s.

Thermal Refers to heat.

Thermal Expansion Expansion of a material caused by raising its temperature. It is expressed in inches per inch per degree of temperature change. See also: *Coefficient of Expansion*.

Thermocouple A temperature sensitive device using the principle that two dissimilar metals in contact at different temperatures will generate an electric *Voltage*. They are used for temperature indications, such as for gas fired water heaters, wall heaters, etc.

Thermodynamics The science of heat energy and its transformations to and from other forms of energy.

Thermopane Trade name (Libby Owens Ford Co.) for factory built *Insulating Glass* window units composed of two or more glass panes separated by hermetically sealed air spaces.

Thermoplastic See *Plastics*.

Thermosetting See *Plastics*.

Thermostat A switch-like device which senses the temperature and thereby can remotely control the flow or temperature of air to a space.

Thincoat High-Strength Plaster A high density, low consistency plaster applied in one or two coats directly to the *Lath* and to a thickness of 1/16″ to 1/8″. Also called VENEER PLASTER.

"T" Hinge See *Hinge Types*.

Thinner See *Paint Thinner*.

Thin Set A term descriptive of one method of setting *Ceramic Tile*.

Thin Shell Concrete A term for any roof structure consisting of a thin, shell-like surface, usually curved and usually of concrete, and generally depending upon arch-like "shell" action for its structural capability.

Thinwall Conduit See *Conduit*.

Thiokol Trade name (Thiokol Chemical Division of Thiokol Corporation) for *Concrete* coatings, *Sealants, Caulking,* and *Admixture* products. Thiokol is often used as the generic term for *Polysulfide*.

Threads See *Bolts, Screws*.

Three Phase Power See *Electric Service*.

Three-Way Switch System A convenience outlet switching system whereby any two switches can turn an outlet on or off. They are commonly used where it is necessary to switch lights from two different locations.

Three Wire Service See *Electric Service*.

Threshold A strip of metal or wood used under a door to cover the joint between differing flooring materials on each side of the door and to help seal the gap under the door.

Throat of a Weld The plane of least cross sectional area of a weld. In a *Fillet Weld*, this is the plane formed by an angle bisecting the corner angle.

Tie

1. A light *Tension* member connecting parts of a structure that would, without such a member, tend to spread.

2. In *Concrete* work, small size *Reinforcing Bars* bent into a rectangular shape and placed ("tied") around, and at right angles to, the principal reinforcing to hold it in place and keep it from spreading (see *Concrete Columns*), or to act as *Stirrups* to resist *Shear* forces.

Tied Column See *Concrete Columns*

Tier

1. A level, as several tiers (or levels) high.

2. In *Masonry* construction, a vertical section of a wall a single unit in depth. Also called a WYTHE.

Tie Rod A *Tension* rod resisting the thrust from opposing sides of an *Arch, Rigid Frame,* or similar framework.

TIG Welding See *Arc Welding (Gas Sheilded)*.

Tile A term usually referring to *Ceramic Tile* but which can apply to any thin, square, or rectangular piece of *Clay*, stone, or *Concrete* used for flooring, roofing, or other surface covering.

Timber See *Lumber*.

Timber Connectors Any of a number of types of metal devices to join wood members, such as *Framing Anchors*, *Shear Connectors*, and *Split-Ring Connectors*.

Tilt-Up Walls See *Precast Concrete Walls*.

Time Clock A clock which has a built-in time printing device usually used for stamping time cards to check workers in and out of a plant.

Time Switch A clock which has built-in switches to control the time at which lights or other electrical devices go on or off.

Tin A relatively soft metal used in making *Solder* and used as an alloy with *Copper* to form *Bronze*. Its chief source is from the ore cassiterite (SnO_2).

Tint See *Color*.

Tinted Glass See *Glass*.

Title Insurance See *Insurance*.

Title Insurance Company A company engaged in researching real estate records to discover encumberances, restrictions, etc., prior to a *Title Insurance* policy being issued.

Title Report A report issued by a *Title Insurance Company* prior to issuance of a policy of *Title Insurance*, containing information relative to the title to the property. It is usually subject to review by a potential buyer.

"T" Nail See *Nails*.

Toe (of a Retaining Wall) See *Retaining Wall*.

Toe Nailing Diagonal nailing through a corner.

Toggle Bolt See *Bolts*.

Tolerance A measure, usually of quantity, by which something can vary from a specified amount and still be ac-

ceptable. For example, a *Concrete Slab* may, by specification, be limited to being level within a tolerance of ¼" in any ten foot length.

Ton 2,000 pounds.

Tongue and Groove See *Woodworking Joints*.

Ton of Refrigeration In *Air Conditioning*, a cooling effect equal to 12,000 BTUH. It is equivalent to the amount of heat required to melt one ton of ice in one hour.

Tooled Joint In *Masonry*, the pulling or rubbing of a tool along the joint to form a tight, smooth junction of a desired shape. See also: *Masonry Joints*.

Toothing In *Masonry* construction, the temporary ending of a wall where the units in alternate courses project horizontally, giving a tooth-like effect. The purpose is to improve *Bonding* with a subsequent section of wall.

Topographic Map A map showing the surface features of a region such as hills, valleys, rivers, etc., and including man-made features such as bridges, canals, roads, etc.

Topography The accurate, detailed description of a region as would be shown on a *Topographic Map*.

Topping A thin layer generally of *Concrete*, placed over a slab or other *Sub-Base* to form a smooth, level finish.

Top Soil The surface zone of *Soil* which supports vegetation.

Torches See *Gas Welding and Cutting*.

Torque A twisting force which tends to cause rotation about some point.

Torque Wrench A wrench having a calibrated dial that will read in *Foot-Pounds* of torque applied to the tightening of a *Nut*, as for *High Strength Bolts*.

Torsion A twisiting action whereby load applied to a member causes such member to twist (or rotate) about its longitudinal axis.

Torsional Stress See *Stress*.

Touch Latch A type of *Catch* mounted inside a cabinet or cupboard door in such a manner that pressing on the door releases the catch and pushes the door ajar. This type of latch is used when a concealed door, or a door without pulls, is desired.

Tower Crane A type of crane, usually truck mounted, consisting of a vertical tower with a boom at the top which can rotate in a full circle. It depends for

stability on counterweights. Its significant feature is its ability to move materials within a wide radius and to a relatively great height. (See Figure T-2.)

Townhouse See *Apartment Nomenclature*.

Township See *Land Description (Government Survey)*.

Township Lines In the Government Survey method of locating land, township lines are imaginary reference lines running in an east-west direction, spaced six miles apart to the north and south of the *Base Line*. See also: *Land Description, Range Lines*.

Toxic Medically harmful; poisonous.

Tracing Paper A semi-transparent

Courtesy Tower Cranes of America, Inc.

FIGURE T-2   Tower Crane

paper upon which drawings can be made and reproduced by the *Blueprinting* process. See also *Vellum*.

Track See *Railroad Terminology, Rails*.

Tract An area of land set apart as a unique unit which is often subdivided into blocks and lots.

Tractor Tractors are of two general types: rubber tired such as are used for some types of *Loaders*, and *Crawler* tractors having, instead of tires, continuous track-like belts of interlocking steel treads for greater traction. The term "tractor" also applies to the highway truck-like motive unit used to pull a semi-trailer. See also: *Bulldozer, Loader, Scraper*.

Tractor-Trailer Truck See *Truck Types*.

Trade Unions For construction industry trade unions see Figure T-3.

Figure T-3

Trades (and included crafts) Unions

Boilermakers   1*
    Blacksmiths, Forgers, Iron ship
    builders, and Helpers.
Bricklayers   2
    Brick masons, Stone masons,
    Marble masons, and related Plas-
    ters and Gunite workers, Tile
    and Terrazzo setters.
Carpenters   3
    Joiners, Floor layers, Wood
    treaters, Drywall workers,
    Millwrights, Shinglers, Pile
    drivers, Acoustical installers, In-
    sulation and weather stripping
    installers.
Electrical Workers   4
Elevator Constructors   5

Hoisting Trades   6
    Portable building trades, Truck
    and Equipment operators.
Insulators   7
    Abestos workers.
Iron Workers   8
    Sheeters, Reinforced iron work-
    ers, Riggers, Heavy machinery
    movers.
Laborers   9
    Brick and Plaster tenders, Hod
    carriers, Pipe layers, Gardeners,
    Sandblasters, Diggers.
Lathers   10
Marble Polishers, Rubbers,
  and Sawyers   11
    Slate and Stone polishers, rub-
    bers, and sawyers, Tile and
    Marble setters helpers, Marble
    Mosaic and Terrazzo helpers.
Painters   12
    Glaziers, Scenic artists and sign
    painters, Carpet and Resilient
    floor layers, Decorative covering
    workers.
Plasterers   13
    Cement masons, Shop hands.
Plumbers   14
    Pipe fitters, Steam fitters.
Roofers   15
    Slate, Tile, and Composition
    roofers, Damp and waterproof
    workers.
Sheet Metal Workers   16
Teamsters   17
    Chauffeurs, Warehousemen,
    Helpers.

*Numbers indicate unions as follows:

1. International Brotherhood of Boilermakers, Iron Ship Builders, Blacksmiths, Forgers, and Helpers.

2. Brick layers, Masons, and Plasterers International Union of America.

3. United Brotherhood of Carpenters and Joiners of America.

4. International Brotherhood of Electrical Workers.

5. International Union of Elevator Constructors.

6. International Union of Operating Engineers.

7. International Association of Heat and Frost Insulators and Asbestos Workers.

8. International Association of Bridge, Structural, and Ornamental Iron Workers.

9. Laborers International Union of North America.

10. Wood, Wire, and Metal Lathers International Union.

11. International Association of Marble, Slate, and Stone Polishers, Rubbers and Sawyers, Tile and Marble Setters Helpers, Marble Mosaic and Terrazzo Helpers.

12. International Brotherhood of Painters and Allied Trades.

13. Operative Plasterers' and Cement Masons International Association.

14. United Association of Pipe Trades.

15. Slate, Tile, and Composition Roofers, Damp and Waterproof Workers' International.

16. Sheet Metal Workers International.

17. International Brotherhood of Teamsters, Chauffeurs, Warehousemen and Helpers of America.

All of the preceding except the teamsters (17) are affiliated with the Building Trades Council of the AFL-CIO.

Trailer See *Truck Types*.

Transformer An electrical device for changing the voltage of *Alternating Current*. It consists of an electrically separated primary and secondary winding around an *Iron* core, in the ratio to which the voltages are to be changed. An AU-

TOTRANSFORMER is one having a single coil with intermediate "taps" to effect the changed outgoing voltages. Transformers are rated in *Kilovolt Amps* (KVA) meaning volts x amps x 1,000.

Transit A *Surveying* instrument consisting of a telescope mounted on a tripod which can be rotated in a vertical and horizontal plane to measure angles by means of angular scales mounted on the transit. It can be used as a *Level*. A THEODOLITE is a very precise type of transit using micrometers to assist in precise angle determinations. Transits of European design are often called Theodolites.

Transite Trade name (Johns Manville) for a group of products composed chiefly of *Asbestos Cement*, such as pipe and wall panel sheets.

Transom An openable window over a doorway.

Transparent Mirrors See *Glass*.

Transverse Crosswise; at right angles to a long axis.

Trap In plumbing, a "U" shaped section of pipe so arranged that it is always full of water and will thereby prevent offensive odors from passing from the sewer up through a fixture.

Trapezoid A four-sided geometric figure of which two opposite sides are parallel.

Travertine A form of *Marble* characterized by a randomly pitted, or "wormy," surface appearance.

Tread The horizontal part of a stair step.

Treated Wood Wood that has been pressure impregnated with a chemical to give resistance to *Decay*, insects, or fire. See also: *Fire Treated Wood*.

Tremie Concrete *Concrete* deposited under water through a closed chute or spout.

Trencher See *Ditcher*.

Triaxial Compression Test A test used primarily for determining the *Bearing Capacity (of Soil)*. Carefully controlled loads are applied vertically and laterally to an undisturbed soil sample until failure occurs.

Trimmer A *Joist* or *Rafter* alongside an opening and into which the *Header* is framed.

Triple-Net Lease See *Lease*. (*Net-Net-Net.*)

Triplex See Apartment Nomenclature.

Trowel

1. A mason's tool for spreading *Mortar* onto *Masonry* units.

2. In *Concrete* construction, a tool used to give a final smooth finish to a concrete surface.

Trowelled Finish See *Plaster Finishes*.

Truck Crane A crane specifically designed and manufactured as a component part of a truck. It can be driven on the highway and when operated, can rotate approximately 270 degrees. Capacities range up to approximately 200 tons.

Truck Types Types of trucks include:

BOBTAIL TRUCK—Slang for a truck with a single rear axle as opposed to a double rear axle, as for a *Ten-Wheeler* described below.

BOTTOM-DUMP TRUCK-TRAILER—A type of truck-trailer used for hauling loose material such as earth, sand, or gravel and which empties the material out of the bottom as the truck is moving, through air operated gates on a bin-shaped bottom. (See Figure T-4.)

DUMP TRUCK—Either a *Bottom-Dump Truck-Trailer* or a *Rear Dump Truck*; the latter dumps by elevating the

Courtesy Hagel Construction Co.

FIGURE T-4   Bottom Dump Truck

front, causing material to slide out the back.

FLAT-BED TRUCK—A truck having a flat bed without sides.

LOW-BOY TRAILER—The bed of the truck is set lower than usual to facilitate loading and allow for an increase in the height of the equipment being transported.

REAR DUMP TRUCK—See *Dump Truck*.

SEMI-TRACTOR-TRAILER—A type of trailer supported at its front and pulled by a truck-like TRACTOR. Their overall lengths vary from state to state, but in general are limited to 60 feet. Their overall clearance height is 13' - 6" from the ground to the top of the trailer. When disconnected, they rest on a front "fifth wheel."

STAKE TRUCK—Similar to a *Flat Bed Truck*, but having slots around the perimeter of the bed for inserting removable sides.

TEN-WHEELER—Slang for a truck with a double rear axle and ten wheels.

TRACTOR-TRAILER TRUCK—See *Semi-Tractor-Trailer* above.

Trussed Joist See *Open Web Beam* (or *Joist*).

Trusses A truss is a single plane framework of individual structural members connected at their ends to form a series of triangles to span a large distance. The members forming the tops and bottoms of the truss are called CHORDS. The vertical or inclined members connecting the top and bottom chords are called the WEB MEMBERS. The most common types of trusses are shown in Figure T-5.

Trust Deed See *Deed*.

"T" Square A drawing instrument for drawing parallel, horizontal lines. It is now largely replaced by *Parallel Rules* or a *Drafting Machine*.

Tubing Generally, the term tubing applies to thinner walled sections than does *Pipe*. Tubing is available in *Aluminum*, *Brass*, *Copper*, *Lead*, *Plastic*, and *Steel* and its diameter size indicates the outside diameter rather than the inside diameter as with pipe. In addition to

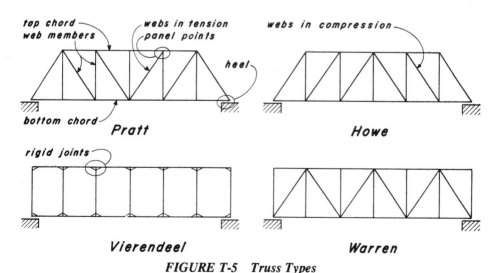

FIGURE T-5 *Truss Types*

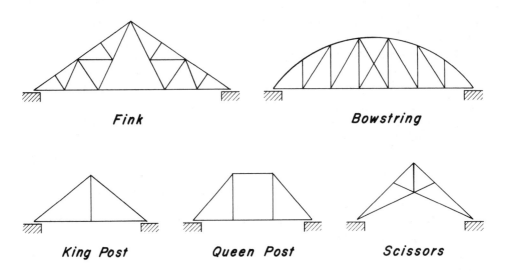

FIGURE T-5 *Truss Types (cont.)*

round sections, the term can also apply to square and rectangular shapes.

Tuck Pointing In *Masonry* construction, the filling in, with fresh *Mortar*, of cut-out or defective mortar joints.

Tungsten This metal has the highest melting point of any metal and is used as an alloying element in steel to increase strength and hardness at every high temperatures such as for high speed tools.

Tunnel Test A standard fire test used to determine *Flame Spread* along the surface of a material.

Turnbuckle A fitting for end-to-end connecting and tightening of two round rods. It consists of threaded holes at either end and a slotted hole in the center through which a leverage bar can be inserted for turning and tightening.

Turn-Key Job See *Package Deal*.

Turn-of-the-Nut Method A method of tightening *High Strength Bolts* whereby the nut is turned a predetermined number of revolutions, after being "finger-tight," to provide the proper *Tension*.

Turpentine See *Solvents*.

Two Phase Power See *Electric Service*.

Two-Way Slab See *Concrete Slabs, Waffle Slab*.

Type of Construction As used in *Building Codes*, a classification of all types of building construction into five categories ranging from an all concrete building (Type I), or equivalently fire resistive, to wood frame buildings (Type V).

U

"U" Block See *Concrete Blocks*.

"U" Bolt See *Bolts*.

U.L. Abbreviation for *Underwriters Laboratories*.

Ultimate Stress See *Stress*.

Ultrasonic Testing See *Weld Testing*.

Unconfined Compression Test In *Soil Mechanics*, a test for determining the strength of soil. A carefully controlled load is applied to the top of a laterally unconfined soil cylinder and progressively increased until failure occurs.

Underfloor Electrical Duct A *Wireway* consisting of a sheet metal, *Duct*-like enclosure within a floor——usually a concrete slab-on-grade—with spaced provisions for access through the finished floor. (See Figure U-1.)

Underpinning Improving an existing foundation, such as to provide for an added load, by constructing a new and larger foundation underneath.

Underwriters Laboratories (UL) An independent testing laboratory to evaluate products, materials, methods, and systems with respect to possible hazards, and to publish standards, specifications, and information to reduce such hazards. Approval is indicated by application of a "UL" label to the product or material which has been tested and then inspected in the manufacturer's plant. UL maintains various departments such as burglary, heating, air conditioning, fire protection, electrical, chemical, etc. Address: 207 E. Ohio St., Illinois 60611.

Undisturbed Soil Sample A sample of *Soil* withdrawn from the ground as a core, usually by a *Sampling Spoon*, such that it can be examined or tested as in its original state.

Unfinished Bolt See *Bolts*.

Uniform Building Code This model code is published and maintained by the International Conference of Building Officials. It is the largest and probably most influential of the *Building Code* groups. The Conference has a large technical staff and maintains various committees for research and educational purposes. They publish several texts and sponsor courses and workshops, as well as providing numerous special services for their members including plan checking. The Uniform Building Code has been adopted by over 1,000 municipalities. Code changes are processed yearly and are voted upon by the membership at their annual business meeting. A new edition of the code is published every three years.

FIGURE U-1　Underfloor Electrical Duct

Uniform Construction Index A system to standardize classification of products and construction related information. It has been prepared by the American Institute of Architects, Associated General Contractors, and the Construction Specification Institute. It comprises sixteen general divisions with a variety of subdivisions. (See Figure U-2.)

Uniform Plumbing Code A model plumbing code sponsored by the International Association of Plumbing and Mechanical Officials in Los Angles; it issues a revised code every three years and is now achieving increased usage.

Uniformly Distributed Load See *Load.*

Union See *Pipe Fittings.*

Unions See *Trade Unions.*

Unitary Air Conditioning See *Central Air Conditioning.*

Unit Cost A pricing system based upon the cost of individual items, as opposed to a *Lump Sum* (for an aggregate of items).

Unit Heater See *Heating Systems (Space Heater).*

Upset Threads The enlargement of the threaded portion of a *Bolt*. It is usually done to increase the net cross sectional area at the threads to at least that of the unthreaded portion.

Urban Redevelopment A community initiated process whereby cities attempt to eliminate and prevent deterioration of the physical and social environment. In addition to the removal or *Renovation* of structurally substandard buildings, the program alleviates inadequate educational, cultural and recreational facilities, diminishing tax revenue, traffic congestion, and air pollution. Financial gain is achieved through increased tax revenue. Such programs are usually administered by establishing a local Redevelopment Agency. The cost of such projects is shared by the participating city and the Federal government (under the U.S. Department of Housing and Urban Development); the latter contributing the majority of funds.

Urea Resin See *Glue*.

Urethane Foam The preferred name for *Polyurethane* used for foam. It is available as flexible sheets or slabs. Rigid foams are sprayed as liquids or poured into place. The foaming occurs due to chemical reaction to a blowing agent introduced prior to heating.

Urinal A bathroom fixture designed solely for male use.

U-Value See *Coefficient of Heat Transmission*.

DIVISION 1—GENERAL REQUIREMENTS
01010 SUMMARY OF WORK
01100 ALTERNATIVES
01200 PROJECT MEETINGS
01300 SUBMITTALS
01400 QUALITY CONTROL
01500 TEMPORARY FACILITIES & CONTROLS
01600 PRODUCTS
01700 PROJECT CLOSEOUT

DIVISION 2—SITE WORK
02010 SUBSURFACE EXPLORATION
02100 CLEARING
02110 DEMOLITION
02200 EARTHWORK
02250 SOIL TREATMENT
02300 PILE FOUNDATIONS
02350 CAISSONS
02400 SHORING
02500 SITE DRAINAGE
02550 SITE UTILITIES
02600 PAVING & SURFACING
02700 SITE IMPROVEMENTS
02800 LANDSCAPING
02850 RAILROAD WORK
02900 MARINE WORK
02950 TUNNELING

DIVISION 3—CONCRETE
03100 CONCRETE FORMWORK
03150 EXPANSION & CONTRACTION JOINTS
03200 CONCRETE REINFORCEMENT
033U0 CAST IN PLACE CONCRETE
03350 SPECIALLY FINISHED CONCRETE
03360 SPECIALLY PLACED CONCRETE
03400 PRECAST CONCRETE
03550 CEMENTITIOUS DECKS

DIVISION 4—MASONRY
04100 MORTAR
04150 MASONRY ACCESSORIES
04200 UNIT MASONRY
04400 STONE
04500 MASONRY RESTORATION & CLEANING
04550 REFRACTORIES

DIVISION 5—METALS
05100 STRUCTURAL METAL FRAMING
05200 METAL JOISTS
05300 METAL DECKING
05400 LIGHTGAGE METAL FRAMING
05500 METAL FABRICATIONS
05700 ORNAMENTAL METAL
05800 EXPANSION CONTROL

DIVISION 6—WOOD & PLASTICS
06100 ROUGH CARPENTRY
06130 HEAVY TIMBER CONSTRUCTION
06150 TRESTLES
06170 PREFABRICATED STRUCTURAL WOOD
06200 FINISH CARPENTRY
06300 WOOD TREATMENT
06400 ARCHITECTURAL WOODWORK
06500 PREFABRICATED STRUCTURAL PLASTICS
06600 PLASTIC FABRICATIONS

DIVISION 7—THERMAL & MOISTURE PROTECTION
07100 WATERPROOFING
07150 DAMPROOFING
07200 INSULATION
07300 SHINGLES & ROOFING TILES
07400 PREFORMED ROOFING & SIDING
07500 MEMBRANE ROOFING

07570 TRAFFIC TOPPING
07600 FLASHING & SHEET METAL
07800 ROOF ACCESSORIES
07900 SEALANTS

DIVISION 8—DOORS & WINDOWS
08100 METAL DOORS & FRAMES
08200 WOOD & PLASTIC DOORS
08300 SPECIAL DOORS
08400 ENTRANCES & STOREFRONTS
08500 METAL WINDOWS
08600 WOOD & PLASTIC WINDOWS
08650 SPECIAL WINDOWS
08700 HARDWARE & SPECIALTIES
08800 GLAZING
08900 WINDOW WALLS CURTAIN WALLS

DIVISION 9—FINISHES
09100 LATH & PLASTER
09250 GYPSUM WALLBOARD
09300 TILE
09400 TERRAZZO
09500 ACOUSTICAL TREATMENT
09540 CEILING SUSPENSION SYSTEMS
09550 WOOD FLOORING
09650 RESILIENT FLOORING
09680 CARPETING
09700 SPECIAL FLOORING
09760 FLOOR TREATMENT
09800 SPECIAL COATINGS
09900 PAINTING
09950 WALL COVERING

DIVISION 10—SPECIALTIES
10100 CHALKBOARDS & TACKBOARDS
10150 COMPARTMENTS & CUBICLES
10200 LOUVERS & VENTS
10240 GRILLES & SCREENS
10260 WALL & CORNER GUARDS
10270 ACCESS FLOORING
10280 SPECIALTY MODULES
10290 PEST CONTROL
10300 FIREPLACES
10350 FLAGPOLES
10400 IDENTIFYING DEVICES
10450 PEDESTRIAN CONTROL DEVICES
10500 LOCKERS
10530 PROTECTIVE COVERS
10550 POSTAL SPECIALTIES
10600 PARTITIONS
10650 SCALES
10670 STORAGE SHELVING
10700 SUN CONTROL DEVICES (EXTERIOR)
10750 TELEPHONE ENCLOSURES
10800 TOILET & BATH ACCESSORIES
10900 WARDROBE SPECIALTIES

DIVISION 11—EQUIPMENT
11050 BUILT IN MAINTENANCE EQUIPMENT
11100 BANK & VAULT EQUIPMENT
11150 COMMERCIAL EQUIPMENT
11170 CHECKROOM EQUIPMENT
11180 DARKROOM EQUIPMENT
11200 ECCLESIASTICAL EQUIPMENT
11300 EDUCATIONAL EQUIPMENT
11400 FOOD SERVICE EQUIPMENT
11480 VENDING EQUIPMENT
11500 ATHLETIC EQUIPMENT
11550 INDUSTRIAL EQUIPMENT
11600 LABORATORY EQUIPMENT
11630 LAUNDRY EQUIPMENT
11650 LIBRARY EQUIPMENT
11700 MEDICAL EQUIPMENT

11800 MORTUARY EQUIPMENT
11830 MUSICAL EQUIPMENT
11850 PARKING EQUIPMENT
11860 WASTE HANDLING EQUIPMENT
11870 LOADING DOCK EQUIPMENT
11880 DETENTION EQUIPMENT
11900 RESIDENTIAL EQUIPMENT
11970 THEATER & STAGE EQUIPMENT
11990 REGISTRATION EQUIPMENT

DIVISION 12—FURNISHINGS
12100 ARTWORK
12300 CABINETS & STORAGE
12500 WINDOW TREATMENT
12550 FABRICS
12600 FURNITURE
12670 RUGS & MATS
12700 SEATING
12800 FURNISHING ACCESSORIES

DIVISION 13—SPECIAL CONSTRUCTION
13010 AIR SUPPORTED STRUCTURES
13050 INTEGRATED ASSEMBLIES
13100 AUDIOMETRIC ROOM
13250 CLEAN ROOM
13350 HYPERBARIC ROOM
13400 INCINERATORS
13440 INSTRUMENTATION
13450 INSULATED ROOM
13500 INTEGRATED CEILING
13540 NUCLEAR REACTORS
13550 OBSERVATORY
13600 PREFABRICATED BUILDINGS
13700 SPECIAL PURPOSE ROOMS & BUILDINGS
13750 RADIATION PROTECTION
13770 SOUND & FIBRATION CONTROL
13800 VAULTS
13850 SWIMMING POOLS

DIVISION 14—CONVEYING SYSTEMS
14100 DUMBWAITERS
14200 ELEVATORS
14300 HOISTS & CRANES
14400 LIFTS
14500 MATERIAL HANDLING SYSTEMS
14570 TURNTABLES
14600 MOVING STAIRS & WALKS
14700 PNEUMATIC TUBE SYSTEMS
14800 POWERED SCAFFOLDING

DIVISION 15—MECHANICAL
15010 GENERAL PROVISIONS
15050 BASIC MATERIALS & METHODS
15180 INSULATION
15200 WATER SUPPLY & TREATMENT
15300 WASTE WATER DISPOSAL & TREATMENT
15400 PLUMBING
15500 FIRE PROTECTION
15600 POWER OR HEAT GENERATION
15650 REFRIGERATION
15700 LIQUID HEAT TRANSFER
15800 AIR DISTRIBUTION
15900 CONTROLS & INSTRUMENTATION

DIVISION 16—ELECTRICAL
16010 GENERAL PROVISIONS
16100 BASIC MATERIALS & METHODS
16200 POWER GENERATION
16300 POWER TRANSMISSION
16400 SERVICE & DISTRIBUTION
16500 LIGHTING
16600 SPECIAL SYSTEMS
16700 COMMUNICATIONS
16850 HEATING & COOLING
16900 CONTROLS & INSTRUMENTATION

FIGURE U-2 Uniform Construction Index—CSI Format

V

Vacuum Lifting Lifting sheet material—such as glass—by suction cup principle.

Valley The intersection at the bottom of two roof planes, as opposed to a *hip*.

Value Engineering A concept whereby building plans are reviewed solely for the purpose of finding cost saving revisions or changes which could be made without affecting the end result.

Valves In plumbing, a device to control the flow of liquid through a pipe. Types of valves include:

ANGLE VALVE—A valve, usually of the *Globe Valve* type, in which the inlet and outlet are at right angles to each other.

BACKFLOW VALVE—A device to prevent contaminated water from back-flowing into a *Potable* water supply system.

BALL COCK—A valve which is opened or closed by the fall or rise of a ball floating on the surface of water and connected by a lever to the valve. The valve opens or closes as the water level changes. A common example is the ball-cock arrangement used in the tank of most *Water Closets* (toilets).

BUTTERFLY VALVE—A valve consisting of a rotatable damper which, in the open position, is parallel with the flow and when rotated 90° closes the opening to stop the flow.

CHECK VALVE—A valve which prevents a liquid from flowing in more than one direction.

COCK—A valve type device with an inside rotating plug to shut off the flow of a liquid or gas.

FAUCET—A valve on the end of a water pipe by means of which water can be released from, or held within, the pipe.

GATE VALVE—A valve in which the flow of water is controlled by means of a circular disk fitting against, and sliding on, machine-smoothed faces, the motion of the disk being at right angles to the direction of flow. The disk is raised or lowered by turning a threaded stem connected to the handle of the valve. The opening of the valve is usually as large as the full bore of the pipe. Gate valves enable shut-off of certain pipe sections without the necessity of draining the entire system. Their use is preferred to *Globe Valves* because they offer less resistance to water flow.

GLOBE VALVE—A piping valve similar to a *Gate Valve* but having a stopper-like disk which screws down to seat over an opening which is at right angles to the direction of the flow.

PRESSURE REDUCING VALVE—A

valve which maintains a uniform pressure and keeps the system full. It opens when the pressure drops under 12 p.s.i. and closes against higher pressures.

PRESSURE RELIEF VALVE—An emergency type valve which bleeds water from pipes when the system pressure exceeds a set amount. It is spring operated and should be positioned where water discharge will not cause damage.

REDUCING VALVE—See *Pressure Reducing Valve*.

SHUT-OFF VALVE—Any device used in piping to shut off the flow of a fluid (liquid or gas).

Vanishing points In the making of *Perspective* drawings, points toward which horizontal lines of a plane converge giving the illusion of depth.

Vapor Barrier Any thin, water proof membrane used to prevent passage of moisture, such as under a concrete slab placed on the ground.

Variance A deviation from a *Zoning* law, usually granted when an extreme hardship would result, were the zoning law adhered to. Obtaining a zoning board hearing and approval requires approval of adjacent property owners. Zoning laws usually provide for procedures required to obtain variances.

Varnish See *Paint*.

Vehicle The liquid portion of a *Paint* which is mixed with the coloring *Pigment* and which, when exposed to air, dries to leave the pigment particles bonded into a solid, hard coating.

Vellum A special type of *Tracing Paper* which has a smooth, white finish and comes in various sizes ranging from 8-½" x 11" to 30" x 42". Although more expensive than some other types of paper, it is preferred because of its excellent reproducing quality and durability.

Veneer

1. In *Masonry* terminology, a facing of stone, masonry, or similar material applied over and against a structural wall for decorative effect.

2. In *Hardwood Plywood*, a thin layer of wood, such as applied as a decorative facing over a lesser quality plywood base. It is cut on a veneer machine and types of such veneer, based upon the manner in which they are cut from a log are described under *Hardwood Plywood*.

Veneer Plaster See *Thincoat High-Strength Plaster*.

Veneer Ties Metal anchor-like devices to secure masonry *Veneer* onto its backing.

Vent An opening for the release of air or gas into an open area to prevent pressure build-up. See also: *Drainage and Vent Piping*.

Ventilation The process of supplying or removing air, by natural or mechanical means, to or from any space. Such air may or may not have been conditioned.

Vent Piping See *Drainage and Vent Piping*.

Verge Rafter The last rafter at the end of a *Gable Roof*.

Vermiculite A very lightweight, mica-like mineral which expands upon heating and is used as an *Aggregate* in *Lightweight Concrete Fill* and insulating plaster.

Vertex The point of intersection of two angles, usually the highest point.

Vertical Grain (Wood) See *Grain (of Wood)*.

Vertical Pivoted Window See *Window Types*.

Vibration Isolator A mounting assembly, or device, used under equipment which could transfer objectional vibration or noise. It consists of a vibration absorbing resilient material or a similarly functioning spring-like mechanism.

Vibrator A device which, by means of vigorous shaking, aids in the placement of concrete into congested spaces. The most common type consists of a vibrating mechanism in an enclosed short tube, powered through a flexible hose, which is manually probed into the concrete as it is being placed. Another type can be applied against the outside of the *Formwork*.

Vibrating Driving See *Pile Driver*.

Verendeel Truss See *Trusses*.

Vinyl A general term used to classify a large and varied group of *Thermoplastics*. See also: *Polyvinyl Acetate, Polyvinyl Chloride, Vinyl Wall Covering, Plastic(s)*.

Vinyl Asbestos Tile See *Resilient Flooring*.

Vinyl Paint See *Paint*.

Vinyl Wall Covering (also called "vinyl wall fabric") A wallpaper-like wall surfacing material consisting of a *Vinyl* coating applied to a cloth backing. It is available in a wide variety of colors, patterns, and textures. Its chief advantage is its washability.

Viscosity In painting, the resistance of fluids to flow. It is the measure of the combined effects of *Cohesion* and *Adhesion*. *Shearing Stress* in solids is analogous to "Viscosity" in liquids.

Visual Grading (of lumber) See *Lumber Grading*.

Vitreous Descriptive of a glass-like surface appearance.

Vitrified For *Clay* products, the condition resulting when a material is heated in a kiln to a temperature sufficient to fuse grains and close pores, making the material impervious to fluids.

Vitrified Clay Pipe (V.C.P.) See *Pipe (Clay)*.

"V" Joint A longitudinal joint between edges of boards, having a "V" shaped groove, half of which is on each side of the joint. See also: *Woodworking Joints*.

Void An unfilled space.

Void Ratio A term usually used when referring to *Aggregate* or *Concrete* mixes. It is the volume ratio of space between particles to the total volume.

Volt The unit of measurement of *Voltage*.

Voltage The difference in electrical "pressure" in an electric *Circuit*; the force causing electrons to flow through a *Conductor*. The unit of measurement is the *Volt*, and is measured by a *Voltmeter*. See also: *Electric Service*.

Voltage Drop The difference in *Voltage* between the beginning and the end of a *Circuit*, due to resistance of the *Conductors*.

Voltmeter An instrument to measure *Voltage* (in *Volts*) between specific points in an electric *Circuit*.

Volumes (of Common Solids) For formulas to compute volumes of common solid shapes, see Figure V-1.

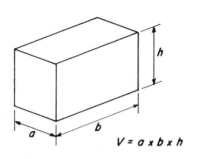

$$V = a \times b \times h$$

Cube or Rectangular Parallelopiped

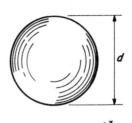

$$V = \frac{d^3}{1.91}$$

Sphere

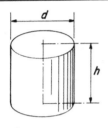

$$V = .785 \times d^2 \times h$$

Cylinder

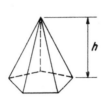

$$V = .33\, A_2 \times h$$

Pyramid

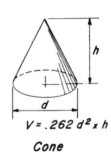

$$V = .262\, d^2 \times h$$

Cone

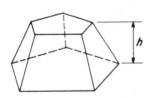

$$V = .33\,(A_1 + A_2 + \sqrt{A_1 \times A_2})\times h$$

Frustum of Pyramid

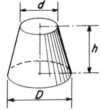

$$V = (.262\, d^2 + .262\, D^2 + .262\, d \times D)\times h$$

Frustum of Cone

note: A_1 = area at top A_2 = area at bottom

FIGURE V-1   *Volumes (of Common Solids)*

W

Waffle Slab A type of *Two-way Concrete Slab* where the *Formwork* is a grid of square pans, with the finished underside of the slab resembling a waffle.

Wainscot The lower portion of an interior wall when of a different material from the rest of the wall.

Walers In concrete *Formwork*, the horizontal restraining beams. Also, in *Shoring* of earth, a horizontal structural member to hold *Sheeting* in place and which is held in postion by *Struts*.

Walk-Up See *Apartment Nomenclature*.

Wallboard Any of a number of thin, manufactured sheets of material used to cover a wall surface. For example, *Gypsum Board, Particleboard*.

Walls Wall classifications include the following:

BEARING WALL—A wall that supports more than its own weight, as when it supports a roof or floor.

COMMON WALL—See *Party Wall*.

DIVISION WALL—A fire protection separation wall used to divide a building. It is used to divide a building whose floor area would otherwise exceed the maximum area allowed by the applicable *Building Code* or to comply with insur-ance requirements. Depending upon the Code, *Fire Doors* through a division wall may or may not be permitted.

FIRE WALL—A wall of fire resistant materials intended to slow the spread of fire from one area to another. The various types of wall constructions are rated by *Building Codes* in four classifications: one-, two-, three-, and four- hour fire walls and the fire protection of openings through such walls are also regulated. See also: *Fire Resistive Time Periods*.

NON-BEARING WALL—A wall which supports no load other than its own weight.

PARTY WALL—Same as COMMON WALL, a single wall common to two separate pieces of property.

SHEAR WALL—A wall which acts to brace a portion of a building against wind or earthquake forces. It resists, by its stiffness, forces applied parallel to its length.

Walnut See *Hardwood*.

Wane In *Lumber* terminology, occurrence of *Bark*, or lack of wood, from any cause on the edge or corner of a piece.

Warm Air Heating See *Heating Systems*.

Warping (of Lumber) Any distor-

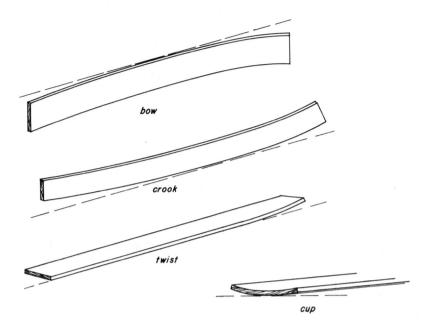

FIGURE W-1   **Kinds of Warp**

tion of a piece of *Lumber* such that it is not straight, or does not have a true plane surface (see Figure W-1). The four types of warp, which are determined by the way in which the piece is distorted, are:

CROOK—A distortion perpendicular to the smaller side of a board.

BOW—A distortion perpendicular to the long side of a board.

CUP—A cross section curvature.

TWIST—A rotational effect between ends.

Warranty An assurance that, at the risk of voiding the original contract, all statements and claims made in writing by the seller to the buyer are, in fact, as stated. See also: *Guaranty*.

Warren Truss See *Trusses*.

Wash Basin See *Lavatory, Sink*.

Washer A small metal device under the head or nut of a *Bolt* or *Screw* to spread the crushing force over a larger contact area. Washers are of two basic types: A CUT WASHER is round with a hole in the center and is manufactured by stamping out of metal strip, and a PLATE WASHER is a square metal plate with a hole. A BEVELLED WASHER is a steel washer which is tapered on one side so that a *Bolt* or rod can pass through it at an angle but still have full bearing of the nut against its retaining surface (See Figure W-2). A LOCK WASHER is one which has been cut through and the cut edges twisted in opposite directions to provide further "biting" power into the surface of the material as well as the bottom of the screw or bolt head. It is heat treated to exert spring tension between the fastener and the parent material.

Waste Pipe In plumbing, a pipe carrying the discharge from a *Sink*, shower,

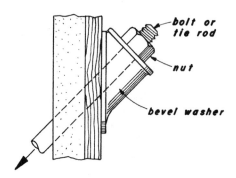

FIGURE W-2   *Bevel Washer*

Lavatory, and other source free from fecal matter, as opposed to *Soil Pipe* which carries discharge from a *Water Closet* or *Urinal*. See also: *Drainage and Vent Piping*.

Waste Stack In plumbing, a vertical *Waste Pipe*.

Water Base Paint See *Paint*.

Waterblasting Similar to *Sandblasting* but using a high velocity stream of water (5,000 to 10,000 p.s.i. depending on use) to remove paint or other coating materials, rust, and for general cleaning purposes. For special purposes such as smoothing concrete surfaces, sand is added to the water.

Water-Cement Ratio In *Concrete* mix design, the ratio of water to *Cement*. It is expressed in gallons of water per sack of cement, or on a weight-to-weight basis. This is the most important factor affecting the strength of concrete (the lower the water-cement ratio, the higher the strength of the concrete).

Water Closet A toilet.

Water Hammer A term applied to the noise made by a fast moving liquid inside of a *Pipe* when its flow is abruptly shut off.

Water Meter Any type of meter for measuring water flow through a *Pipe*, but usually referring to a water company meter set into a small concrete vault below ground near the place where water from the street *Main* enters the property. Water consumption, measured in cubic feet, is read periodically by a company employee for billing.

Waterproofing The making of a material or surface impervious to the passage of water. The two basic types of waterproofing are: *Membrane*, or surface *Waterproofing*, where a coating is put on to prevent water entering, such as a coating of *Asphalt*, a penetrating liquid, or *Polyethyline* sheeting; and INTEGRAL WATERPROOFING, where, in the case of concrete, an *Admixture* is incorporated in the mix to decrease porosity.

Water Reducing Agents See *Admixture*.

Water Seal See *Trap*.

Water Softener A device placed in a water supply line to "soften" incoming *Hard Water*, whereby ions of magnesium and calcium are exchanged for sodium ions in a bed of natural or synthetic silicates.

Water Stops Continuous strip-like sealing devices placed within a joint in a *Masonry* or *Concrete* wall to prevent water passing through the joint. They generally consist of a shaped strip of rubber or similar material embedded within and across the joint.

Water Table The upper surface of a body of *Ground Water*.

Watt In electricity, the unit of measurement of the amount of electrical power consumed. It is measured by a wattmeter. One KILOWATT (KW) is one thousand watts, and a KILOWATT HOUR (KWH) is one kilowatt acting for one hour.

Weather Stripping Sealing the spaces around doors and windows by attaching a strip of *Vinyl*, rubber, or felt to the edge of a door or window, or to its frame, to prevent or minimize passage of air or water.

Web That portion of a *Beam* or *Girder* between the *Flanges*. (For illustration see Figure F-4.)

Web Crippling A localized buckling type of failure of the *Web* of a *Beam* or *Girder* usually resulting from a *Concentrated Load* at that point.

Web Members See *Trusses*.

Weep Holes Holes through the bottom of a *Retaining Wall* to drain water from behind the walls and thereby prevent the buildup of *Hydrostatic Pressure*.

Welded Joints The two basic types of welds are the BUTT WELD, where the weld metal fills the space between parts to be joined, and the FILLET WELD, where a triangular (fillet-like) weld is placed in the corner between the parts (see Figure W-3). The value of such a weld is limited by the stress at the *Throat* (plane of minimum cross section). A PLUG WELD is a side-to-side joining of plates by filling a hole in one plate with weld metal that also fuses to the plate below; if such a hole is elongated it is called a SLOT WELD. A TACK WELD, or SPOT WELD, is a very short length of weld to temporarily join two pieces such as to hold them in position for further welding. See also: *Arc Welding, Resistance Welding, Welding,* and *Welding Symbols*.

Welded Wire Fabric Prefabricated reinforcement used in concrete work, consisting of parallel series of high strength, cold drawn wires welded together in square or rectangular grids (See Figure W-4). Welded wire fabric is specified by indicating the size of the wires in each direction and the spacing in each direction—the industry designated method is shown below. Note that the term "W" preceding the wire size indicates that it is smooth wire; if a "D" precedes the number it indicates deformed wire.

Industry Method of Designing Style

Longitudinal wire spacing   Transverse wire spacing

Example: 6 x 12 — W1.4xW1.4

Longitudinal wire size   Transverse wire size
(in hundredths of a square inch)

This system is replacing the older designation system where 6 x 12 — 10/10 meant 6″ x 12″ grid spacing with 10 gage wire in each direction.

Most commonly used wire spacings are 4″ x 4″ to 6″ x 6″. Rectangular and larger spacings are also available. Most commonly used wire gages are 6 gage and 10 gage (the latter being 1.4 hundredth of a square inch). Smaller gage wire fabric generally is manufactured in rolls 6′ wide;

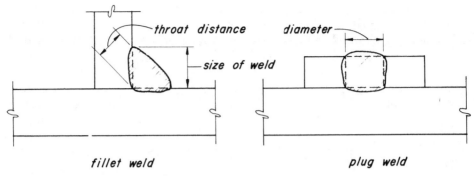

fillet weld   plug weld

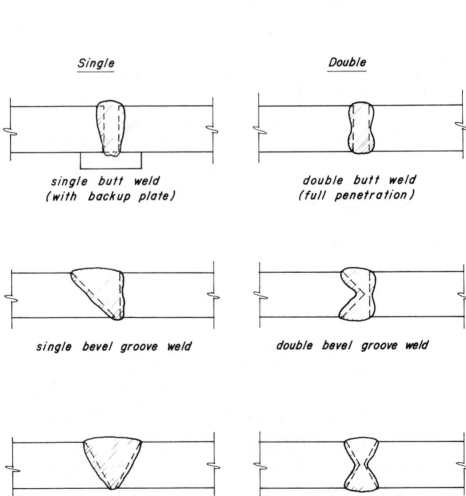

Single   Double

single butt weld   double butt weld
(with backup plate)   (full penetration)

single bevel groove weld   double bevel groove weld

single vee groove weld   double vee groove weld

FIGURE W-3   Welded Joints

FIGURE W-4   Welded Wire Fabric Mats Placed as Concrete Reinforcement

heavier gages are generally fabricated in mats 6' x 20'.

Welding The joining of metals by applying sufficient heat to melt and fuse two pieces together. In construction, the metals most often welded are *Steel* and *Aluminum* , and to a much lesser extent, *Magnesium* and *Titanium*. In *Welding* terminology, the metal parts being joined are called the *Base Metal* or *Parent Metal*. Additional metal added to the molten area, as from a welding rod or *Electrode*, is called the *Filler Metal*. Types of welding are classified according to the means of supplying heat and are, in order of their importance in construction: *Arc*
Welding, *Gas Welding and Cutting*, *Resistance Welding*.

Welding Electrodes See *Arc Welding*.

Welding Symbols Symbols used on drawings to indicate various types of *Welded Joints* are shown in Figure W-5.

Weld Testing The principal means of examining or testing for the quality of a weld. In addition to physical testing and examination, the following non-destructive testing methods are available:

DYE PENETRANT TESTING—A means whereby a dye is placed on the weld and subsequently rubbed off. Mi-

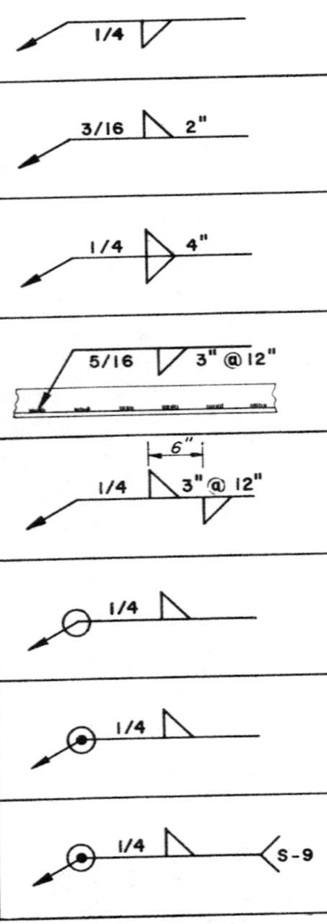

1/4	1/4" fillet weld, continuous on near side.
3/16 2"	3/16" fillet weld 2" long, on far side.
1/4 4"	1/4" fillet weld 4" long, on each side.
5/16 3"@12"	5/16" fillet welds 3" long, spaced 12" center to center on near side.
1/4 3"@12" 6"	1/4" fillet welds 3" long, spaced 12" center to center on both sides, with welds on opposite sides staggered.
1/4	1/4" fillet weld on far side, welded all around.
1/4	1/4" fillet weld on far side, field welded all around.
1/4 S-9	1/4" fillet weld on far side, field welded all around. See Specification Reference S-9.

Fillet Welds

FIGURE W-5 Basic Welding Symbols

Butt Welds		Square butt weld made on near side.
	1/8	Full penetration butt weld with 1/8" root opening, grind flush far side. Field welded.
		Beveled-grooved (one side) butt weld on near side.
	1/8	Beveled-grooved butt weld on both sides with 1/8" root opening, ground flush on far side.
		Beveled-grooved butt weld on far side with both abutting edges bevelled.
Plug (Slot) Welds	1/2	1/2" round plug weld on near side.
	1/2 2"	1/2" slot weld 2" long on far side.

nute cracks and voids which are penetrated by the dye can be seen.

MAGNETIC PARTICLE TESTING —A means of testing by applying finely ground iron particles, then brushing them away. By means of a magnetic detection system, remaining particles in minute cracks or other defects can be determined.

ULTRASONIC TESTING—A means whereby high frequency sound waves are directed at the joint and reflected back to a recorder by a transponder. The weld area can be scanned with the results displayed on an oscilloscope which will indicate voids or other defects in the weld.

X-RAY INSPECTION—A means whereby an X-ray film is made of welded joints such that voids or other defects in the weld show up on the film.

Well-Points A means of lowering a *Ground Water* level to prevent water from entering a deep excavation. Perforated pipes are driven or jetted into the ground around the perimeter of the area to be "de-watered" and pumps pump out the water seeping into the pipes.

Western Cedar See *Softwood*.

Western Hemlock See *Softwood*.

Western Redcedar See *Softwood*.

Wet Bulb Temperature The temperature registered by a thermometer whose bulb is covered by a moist wick exposed to moving air. Such measurements are used in determining *Relative Humidity*. By making a comparison between *Dry Bulb Temperature* and wet bulb temperature, the amount of moisture in the air can be determined and thereby the relative humidity.

Wet Mix Shotcrete *Pneumatic* placing of ready mixed concrete which is dis-

charged through a hose and mixed with air at the nozzle. It is sprayed onto a backing surface. See also: *Gunite*.

Wet Pipe System See *Automatic Fire Sprinkler System*.

Wet Standpipes See *Standpipe*.

White Cement See *Cement*.

White Fir See *Softwood*.

White Oak See *Hardwood*.

White Pine See *Softwood*.

White Spruce See *Softwood*.

Whiting An *Extender* used to make *Putty*, consisting of calcium carbonate in fine powder form.

Wicket Door Same as *Pilot Door*.

Wide Flange Beam A standard structural shape having wider *Flanges* than "I" beams. For illustration see *Structural Shapes*. See also: *Structural Steel Shape Designations*.

Winch A device for lifting or pulling by means of a *Rope* or *Cable*. The cable is wrapped around a grooved drum which is either hand or power driven.

Winders Stair *Treads* on a curving stairway where the outer end of the tread is wider than the inner end.

Wind Load See *Loads*.

Window Glass See *Glass*.

Window Types For illustration of most commonly used types of windows see Figure W-6.

Windrow A continuous ridge of loose *Soil* or rocks such as would be cast aside by a *Grader*.

Wiped Joint See *Pipe Joints*.

Wire See *Aluminum Wire, Copper Wire, and Steel Wire*.

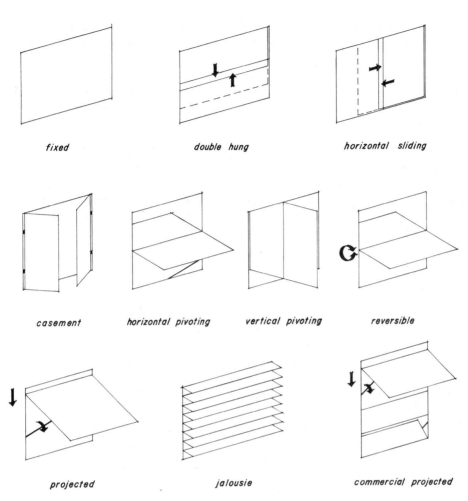

fixed   double hung   horizontal sliding

casement   horizontal pivoting   vertical pivoting   reversible

projected   jalousie   commercial projected

FIGURE W-6   Window Types

Wire Cloth Also called WIRE FABRIC or *Wire Screening*. It is a woven material consisting of metallic threads generally of steel, aluminum, or copper, and available in a wide variety of *Mesh* sizes—some so fine that smoke cannot be blown through them.

Wire Fabric See *Welded Wire Fabric*.

Wire Fencing See *Chain Link Fencing, Woven Wire Fencing*.

Wire Gage See *American Standard Wire Gage* and *Brown and Sharpe Wire Gage*.

Wire Glass See *Glass*.

Wire Mesh Reinforcement See *Welded Wire Fabric*.

Wire Rope See *Rope*.

Wire Screening See *Wire Cloth*.

Wireway In electrical work, a metal wiring enclosure, usually rectangular in cross section, with a fully removable front or top cover used for protection of the *Conductors* within it. Wireways are used for *Feeders*, *Branch Circuits*, and for interconnecting various pieces of electrical equipment.

Wiring The connecting of *Conductors* to form electrical *Circuits*. See also: *Aluminum Wire, Copper Wire, Conduit, Electric Service*.

Wood The term "wood" refers to the material whose strengthening and water conducting tissue is derived from tree stems; whereas "lumber" refers to the manufactured product of the sawmill. There are approximately 1000 identifiable species of wood in the U.S. (excluding Hawaii), of which a limited number—about 80—are used commercially to a significant extent for construction, furniture, pallets, paper, and related products.

Wood consists of about 70% *Cellulose*, 29% *Lignin*, and 1% various minerals. Wood cell walls are made up of cellulose encrusted with lignin and the lignin also acts as a cementing agent to bind the cell walls of the wood fibers together. Chemically, wood is chiefly carbon, hydrogen, and oxygen—all of which dissipate when the wood is burned; the remaining ash consists of the minerals present. The small core, consisting of soft tissue, of the cross section of a tree around which the annual growth rings occur is called PITH. ANNUAL RINGS are made of of alternating bands of light EARLYWOOD (also: "springwood") and dark LATEWOOD (also: "summerwood"). (Note: each annual ring consists of one earlywood and one latewood band.) Movement of water and food bearing sap from the roots to the leaves takes place in the outer rings called SAPWOOD. Growth (cell division) takes place in the cambium,

a layer of cells between the wood and the bark. The inner rings, which are inactive and often darker in color, are called HEARTWOOD.

The strength of wood is greatest along the direction of the *Grain* and considerably less in a direction across the grain. Its weight is generally between 20 and 50 lbs. per cubic foot.

The two botanical catagories of wood are *Softwood* and *Hardwood*. The most important use of softwoods is for structural products. Primary uses of *hardwood* are furniture, *Cabinetwork*, pallets, and *Veneer*.

See also: *Plywood*.

Wood Dowel See *Dowel*.

Wood Fiber Plaster A *Gypsum Plaster* containing fine paricles of wood fiber resulting in a lighter weight plaster with increased fire resistance.

Wood Framing For the nomenclature and illustration of various components of typical wood framed structures see Figures W-7a through W-7d.

Wood Framing Anchors See *Framing Anchors, Timber Connectors*.

Wood Grain See *Grain (of Wood)*.

Wood Grain Patterns For illustration of some common wood grain patterns see Figure W-8.

Wood Lath An older method for providing backing for the application of *Plaster* whereby small wood strips (¼" x 1-½" x 32") were nailed to wood *Studs* and spaced slightly apart to form a bonding groove for the plaster.

Wood Pile See *Piles*.

Wood Preservative Any substance that, for a reasonable length of time, is effective in preventing deterioration of *Wood* from *Decay* or harmful insects.

Wood Shingles See *Shingles*.

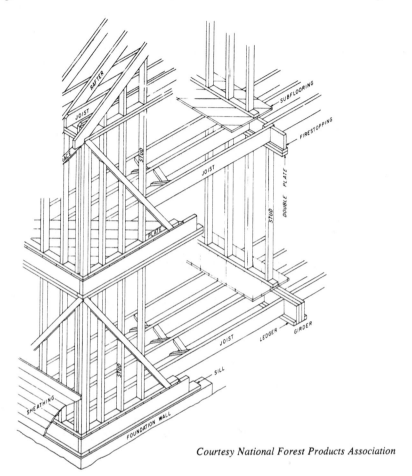

Courtesy National Forest Products Association

FIGURE W-7a Platform Frame Construction

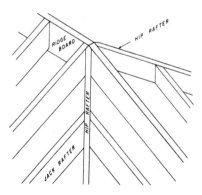

Courtesy National Forest Products Association

FIGURE W-7b Roof Framing at Hip Rafter

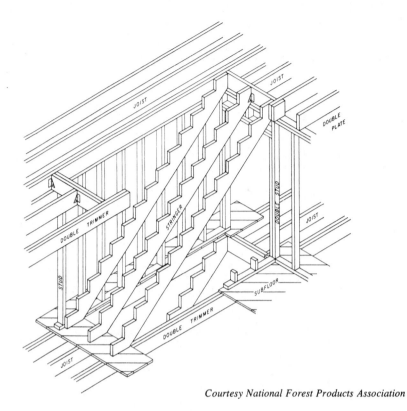

FIGURE W-7c   Interior Stairway Framing

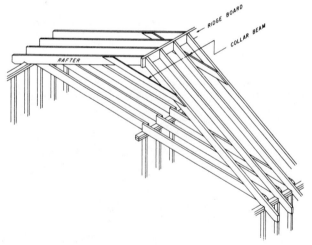

FIGURE W-7d   Roof Framing wtih Ceiling Joists Parallel to Rafters

a.   *Species: African Mahogany*

b.   *Species: Ash*

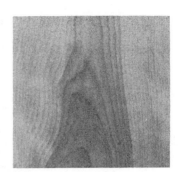

c.   *Species: Birch*

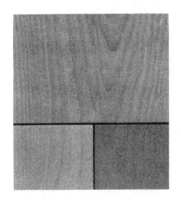

d.   *Species: Birch (heartwood)*

e.   *Species: Douglas Fir*

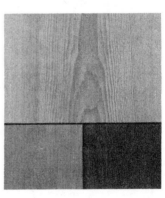

f.   *Species: Ponderosa Pine*

g.   *Species: Red Oak*

h.   *Species: Redwood*

i.   *Species: Walnut*

Courtesy Architectural Woodwork Institute

FIGURE W-8   Wood Grain Patterns

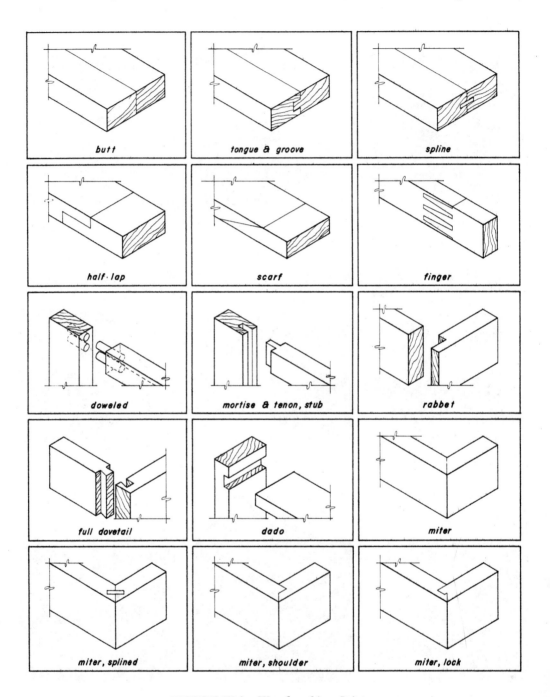

FIGURE W-9   Woodworking Joints

Woodworking Joints In carpentry, any of a number of ways of joining members. Commonly used joints are described below and illustrated in Figure W-9.

DADO JOINT—A joint having a rectangular groove cut into the surface of one board and into which the end of a joining board is fitted.

DAP JOINT—Made by notching one or both joining members.

DOVETAIL JOINT—Wedge-shaped protrusions on one board fitting into matching notches on a joining board.

FINGER JOINT—Similar to a *Scarf Joint* but using a series of "finger-like" projections on the end of a board which fit into matching slots in the abutting board.

LAP JOINT—A joint made by notching and lapping the boards being joined.

MITERED JOINT—A joint made by bevelling each of the meeting surfaces such that they come together at the outer and inner corners.

MORTISE AND TENON JOINT—Where a *Mortise* is cut into the side of one member to receive a protruding, matching piece from the joining member.

RABBET JOINT—A joint made by cutting a groove in an edge of one member into which a joining member fits.

TONGUE AND GROOVE—Matching protrusions on longitudinal edges of abutting members fitting together to assure alignment and prevent movement between adjacent boards.

SCARF JOINTS—An end-to-end splice between boards, usually in a *Glued Laminated Beam*, whereby the ends of the joining pieces are taper-cut so they can overlap but remain in the same place, thus increasing the contact surface between them.

Work Hardening The hardening effect given *Steel* by *Cold Working*.

Working Drawings A complete set of *Drawings* for a construction project which provide all information necessary for construction. The working drawings are usually a part of the *Contract Documents*. See also: *Shop Drawings*.

Working Stress See *Stress*.

Workmen's Compensation Insurance See *Insurance*.

Woven Wire Fabric Similar to *Welded Wire Fabric* except the wires are joined by weaving rathern than welding. It is primarily used as plaster lath reinforcing, usually over a paper backing.

Woven Wire Fencing Similar to *Chain Link Fencing* except the wires are "woven" such that a square or rectangular pattern is obtained.

Wrought Iron See *Iron*.

Wye Level See *Level*.

Wythe In *Masonry* wall construction, one vertical plane of masonry units.

X

X Bracing Crossed structural bracing members between *Columns*, resembling an ''X'', and usually of round rods with *Turnbuckles* for tightening.

X-Ray Inspection See *Weld Testing*.

Y

"Y" Fitting See *Pipe Fitting*.

Yellow Fir See *Softwood*.

Yellow Poplar See *Hardwood*.

Yield Point That point on a *Stress-Strain Curve* where, for a given material, an increase in *Stress* causes a permanent deformation. Also called the ELASTIC LIMIT.

Yield Stress See *Stress*.

Z

Zee Bar A rolled, extruded, or otherwise formed, metal structural shape in the form of a "Z". For illustration see *Structural Shapes*.

Zenith In surveying, a point directly overhead; the upward extension of a *Plumb* line.

Zinc A hard, silvery metal which is used primarily as a corrosion resistent coating on *Steel* due to its ease in melting and high resistance to *Rust*. See Also: *Galvanizing*.

Zoning

1. The division of a city into areas (zones) restricting by law the use of the land within the zone, such as for single family residences only, apartments, retail stores, manufacturing, etc.

2. In *Air Conditioning*, the control of the temperature in one room or a group of rooms, independently from other rooms.